Findlay's practical physical chemistry

Findlay's practical physical chemistry

Ninth edition

Revised and edited by

B. P. Levitt

Lecturer in Physical Chemistry,
Imperial College, London

Previously revised (1954) by

J. A. Kitchener

Imperial College, London

Longman

LONGMAN GROUP LIMITED
London

Associated companies, branches and representatives throughout the world

First edition 1906

Eighth Edition (revised by J. A. Kitchener) 1954

Ninth Edition (revised by B. P. Levitt) 1973

ISBN 0 582 44222 2

Printed in Great Britain by
William Clowes & Sons, Limited,
London, Beccles and Colchester

From the reviser's preface to the eighth edition

For nearly half a century Professor Findlay's 'Practical Physical Chemistry' has been a familiar guide to University students of Chemistry, and its seven previous editions testify to the value of its presentation. Consequently, it is with a sense of privilege that I have undertaken the work of preparing the eighth edition, and my aim has been to bring the technical details up to date without changing the general character of the book. However, the development of the subject has already reached the stage where there are more techniques available than can be illustrated in the time usually assigned to practical chemistry; consequently, experiments which merely serve to illustrate a point of theory and employ only common analytical apparatus must tend to give place to those which offer the student both experience with an unfamiliar piece of physico-chemical apparatus and also a useful theoretical exercise.

The experiments have been designed to illustrate the common techniques of physical chemistry *in their usual, basic form*, avoiding, on the one hand, crude improvised apparatus and, on the other hand, elaborate instruments or complicated combinations of operations. Advanced research techniques such as those used in X-ray diffraction, infra-red spectrometry and radio-chemistry, are outside the scope of this book, but a few 'key' references to specialized monographs have been included for the benefit of advanced students.

J. A. KITCHENER

Imperial College of Science and Technology, London
August, 1953

Reviser's preface to the ninth edition

My thoughts on revising the 9th edition closely parallel Dr Kitchener's on revising the 8th edition. In particular it is a tribute to his foresightedness that this book has continued in use without revision for almost twenty years. The 9th edition includes material on new techniques including a chapter on electrical measurements and transducers. There are new experiments: the thermistor, solubility and ionic strength, molecular spectroscopy, a photometric investigation of the NO_2–N_2O_4 equilibrium, absorption spectra in solution, heats of mixing, the rotating disc electrode, the thermodynamic properties of alloys using the amalgam cell, homogeneous catalysis, adsorption of gases on solids, gas chromatography and the rotational viscometer. The remaining experiments have been revised, particularly where modern instruments such as the potentiometric recorder make them less tedious or enable the average student to complete experiments previously accessible only to the skilled experimentalist. The introductory material has been modernized, particularly the chapters on electrochemistry and kinetics. The SI system of units has been used throughout.

I should like to thank Dr Kitchener for his advice and encouragement. The new and revised experiments are nearly all in use in the physical chemistry laboratory course at Imperial College; I should like to thank my colleagues who originated or developed these experiments, and who have kindly allowed me to incorporate material from their laboratory notes in this edition. In particular, Dr Spiro has patiently guided me past innumerable electrochemical pitfalls; if I have succumbed to a few still, it is not his fault.

Finally, I should be grateful to receive corrections or suggestions for the improvement of this book. Over the years it has become very much a co-operative effort of the revisers and the users, and I hope that this will continue in the future.

B. P. LEVITT

Department of Chemistry,
Imperial College of Science and Technology,
London, 1972

Contents

List of experiments

Physical properties

Thermodynamics of dilute solutions

Surface chemistry

Introductory note on arrangement, bibliographies and symbols

The arrangement of this book

Chapters 1–4 deal with topics of general importance throughout practical physical chemistry, namely, the treatment of experimental data, the determination of weight, volume and length, the measurement and control of temperature, and electrical measurements.

In the remaining eleven chapters, physical chemistry is treated systematically in its main, logical divisions, starting with the properties of pure substances in a single phase, and progressing through phase equilibria, etc., to the more complex systems such as those involving electrical energy, surface effects, etc.

Each chapter is sub-divided into self-contained *sections*, and these are labelled alphabetically, and indicated at the head of each page. In general, each section begins with a brief survey of the theory, and then describes the chief experimental methods employed in the field. Instructions for carrying out experiments are given in considerable detail, and the method of working out the results is indicated.

Bibliographies

At the end of each section there is a short list of relevant standard works, monographs, and original papers (indicated in the text by superscript reference numbers), in that order.

The *standard works*, which are quoted by author only, are as follows:

Partington, *An Advanced Treatise of Physical Chemistry*, Vols 1–5, 1949–54 (Longmans Green, London).

Weissberger (ed.), *Technique of Organic Chemistry*, Vol. I, 4 Parts, Physical Methods, 3rd edn, 1960 (Interscience Publishers, New York). Volumes of the 4th edition are now appearing under the title *Physical Methods of Chemistry*, Vol. I. Parts 1A and 2A contain a general account of electronics and instrumentation.

Tables of physico-chemical properties can be found in *Handbook of Chemistry and Physics* (Chemical Rubber Publishing Co., Cleveland, Ohio), published annually; Kaye and Laby, *Table of Physical and Chemical Constants*, 13th edn, 1966 (Longmans, London); Timmerman, *Physico-Chemical Constants of Pure Organic Compounds*, Vol. 1, 1950; Vol. 2, 1965 (Elsevier, Amsterdam).

Appendix I contains a bibliography of physico-chemical techniques not dealt with in this book.

Symbols and abbreviations

The symbols and abbreviations used in this book conform (with a few exceptions) to the recommendations of the Symbols Committee of the Royal Society, the Chemical Society, the Faraday Society, and the Physical Society. (*Quantities, Units and Symbols,* 1971 (Royal Society, London).)

In addition to recommending specific letters to represent physical quantities (e.g. U for internal energy, E for electromotive force, etc.), the report advised use of the following typographic conventions for the different kinds of symbol:

Physical variables, such as pressure, temperature, concentration, in *italic* (sloping) type; thus, p, T, c.

General constants, such as Avogadro's number, the gas constant, Planck's constant, in ***bold italics***; thus, $\boldsymbol{N}$, $\boldsymbol{R}$, $\boldsymbol{h}$.

Internationally accepted units, such as metre, kilogram, ohm, hydrogen-ion exponent, in roman (upright) letters without a stop; thus, m, kg, Ω, pH. Units named after a person are given a capital; thus, V for volt.

Chemical elements in roman type; thus Li.

Mathematical operators in upright type; $\log_{10}p$, $\mathrm{d}y/\mathrm{d}t$, ΔU.

Abbreviations, other than units, in lower case letters with a full stop. A list of common abbreviations was given, and the following are a few examples frequently met in physical chemistry.

alternating current	a.c.	melting point	m.p.
atmosphere	atm.	molecular weight	mol. wt.
boiling point	b.p	per cent	%
concentration	concn.	potential difference	p.d.
dilute	dil.	relative humidity	r.h.
direct current	d.c.	solution	soln.
electromotive force	e.m.f.	specific gravity	sp. gr.
freezing point	f.p.	ultra-violet	u.v.
mole	mol	vapour pressure	v.p.

Units

The SI system of units is based on the fundamental units of mass (the kilogram), length (the metre), and time (the second). The most important changes from earlier systems which occur in this book are:

(a) the unit of FORCE is the newton ($N = kg\,m\,s^{-2}$); PRESSURE is measured in newtons per square metre ($N\,m^{-2}$): $1\,atm \equiv 760\,mm\,Hg = 1{\cdot}013 \times 10^5\,N\,m^{-2}$.
(b) ENERGY is always measured in joules (J), never in calories. The gas constant $\boldsymbol{R}$ is $8{\cdot}314\,J\,mol^{-1}\,K^{-1}$; the units of $PV = n\boldsymbol{R}T$ with P in $N\,m^{-2}$, V in m^3 and T in K, are joules (J).
(c) Electrical units are all derived from the practical unit of CURRENT, the ampere (A); the unit of CHARGE is the coulomb (C = A s), not the e.s.u. The volt (V) and ohm (Ω) are unchanged. Magnetic units are also derived from the ampere; e.m.u. are no longer used.
(d) The mole is defined as an Avogadro's number ($\boldsymbol{N}$) of elementary units (atoms, molecules, electrons, etc.), where $\boldsymbol{N}$ is the number of ^{12}C atoms in 0·012 kg of carbon-12. $\boldsymbol{N} = 6{\cdot}022 \times 10^{23}\,mol^{-1}$.
(e) The units of concentration used in this book are now $mol\,m^{-3}$ (replacing $M = mol\,l^{-1}$) and molality = mol (kg solvent)$^{-1}$ (which is unchanged). The use of $mol\,m^{-3}$ instead of $mol\,dm^{-3}$ ($\approx M$) avoids factors of 1 000 in electrochemical equations, and gives whole numbers for commonly used solutions: e.g. 0·1 M is replaced by 100 $mol\,m^{-3}$. Unfortunately no symbol has yet been adopted for this unit.

The equivalent and derived quantities (e.g. N = normality; equivalent conductance) are no longer favoured: a stoichiometric coefficient is used where necessary.

BIBLIOGRAPHY

McGlashan, *Physico-Chemical Quantities and Units*, 1968, Royal Institute of Chemistry Monograph for Teachers No. 15.

1
The accuracy of measurements and treatment of results

1A Accuracy of measurements

1. INTRODUCTION: THE ROLE OF MEASUREMENTS

When the scientific method is brought to bear on a new problem, the first stage consists of *experiment*, the second *correlation* of observations, and the third the construction of a *theory* to explain the phenomena. Further experiments may then be needed to test the validity of the theory and extend the scope of the investigation. Innumerable examples could be found of the power of this approach in the various sciences.

Physics can claim to be the most advanced of the sciences, since it has reached the stage where the majority of its phenomena can be fitted into a scheme of *quantitative* theory. Consequently, experimental physical chemistry—that is, the investigation of the physical properties of chemical substances and the relation of these properties to chemical reactions—is concerned largely with *quantitative measurements*. *Qualitative* observations have their place in physical chemistry, but quantitative studies provide concrete data by which a hypothesis can be stringently tested.

It is true that a quantitative framework of theory is not always available for interpreting a new set of measurements. For example, at the present time the mechanical behaviour of suspensions (e.g. clay pastes) is being much studied (see chapter 15) because of the great technical importance of such materials, but a fundamental, quantitative 'treatment' of phenomena such as thixotropy (section 15C) has not yet been developed. Nevertheless, quantitative measurements are valuable in this field, although at present they must be classified as purely *empirical*—that is, the result of experiment without direct theoretical interpretation. The data provide a foundation upon which a theory can eventually be built.

It is therefore appropriate to begin a book on practical physical chemistry with a consideration of *measurements*, their trustworthiness, and the methods employed for using them to best advantage.

2. ATTAINING A SUITABLE ACCURACY

No measurement is *exact*, excepting, perhaps, the counting of integral units such as the number of α-particles entering a Geiger counter. Measurements of mass, length, time, and all the derived properties such as volume, density, viscosity, etc., are inevitably of limited accuracy, and every such measurement is valid to a certain degree only. It is not always realized that one should attach as much importance to ascertaining the degree of precision of a measurement as to recording the result. Consider, for example, an imaginary investigation of the effect of, say, glycerol on the surface tension of water. Suppose the surface tension of pure water had been measured once and found to be 0·073 $N\ m^{-1}$, and then glycerol was added and a second measurement of surface tension gave 0·072 $N\ m^{-1}$; would it be correct to conclude that glycerol lowers the surface tension of water? The answer obviously depends on how precise the two measurements really were. Perhaps they were quite rough, and a few repetitions would give 0·073, 0·074, 0·072, 0·072 for the first, and 0·072, 0·073, 0·072, 0·074 for the second. In this case the original difference between 0·073 and 0·072 is said to be '*not significant*'. It is clearly *essential* to be able to attach to the figures 0·073 and 0·072 some index of their reliability, because without such an index these measurements are quite worthless. It is the principal object of this chapter to consider how uncertainties attaching to physical measurements arise and how they can be dealt with.

The precision of an experiment can always be improved by using better instruments, by taking more care in the measurements, and by giving attention to the method of determination employed. The question is therefore implicit in the planning of every experiment—how accurately need the measurements be made? The answer will depend entirely on the purpose of the experiment. Thus, in the determination of important physical constants such as the velocity of light, Faraday's constant, and atomic weights, extreme care and attention to every detail are called for, but many of the measurements needed in physical chemistry can be made sufficiently accurately for the purpose in hand and with greater speed and convenience without pressing all the determinations to the limit. For example, in finding the 'order' of a chemical reaction (chapter 14) it is unnecessary to determine the concentration of the reactants with the greatest accuracy possible to analytical chemistry: this would be a waste of effort, if the object of the work is simply to decide whether the reaction is first order or second order.

Clearly the proper approach to an experiment is first to decide the *minimum* overall accuracy required for the purpose, and then to design the experiment so that each link in the experimental chain is sufficiently accurately measured. Throughout this book, however, it will be assumed that it is desired to achieve the best accuracy which can be obtained conveniently with the instruments normally employed in physical chemistry laboratories. Naturally, for special researches it is possible to obtain or design instruments capable of attaining a still higher order of precision. For example, ordinary potentiometers (chapter 4) serve to determine 1 volt to 0·1 millivolt, i.e. 1 in 10^4, but precision 'vernier' potentiometers are obtainable with which an accuracy of 1 in 10^6 can be achieved.

3. SOURCES AND TYPES OF 'ERROR'

Having seen that a knowledge of the precision of measurement is as important as the measurement itself, it is next necessary to consider where the uncertainties arise.

Even simple quantities such as *length* have to be measured against some kind of scale which is itself subject to some uncertainty in manufacture. Further, the operation of comparing the object with the scale involves an element of judgment, and two observers may differ slightly in their decision as to the reading, particularly as it is generally necessary to *estimate* fractions of a division engraved on the scale. The final measurement of length is therefore seen to be subject to uncertainties due to the scale and to the using of the scale. These are typical examples of what may be called 'instrument errors' and 'random errors of observation'. The technical term 'error' should be understood to mean a 'departure from the true value', not a mistake on the part of the observer.

If the quantity to be determined is more complex (e.g. molecular refractivity—see chapter 9), a considerable number of measurements and operations may be involved before the final value is obtained. Errors in the individual measurements may or may not (in different cases) have an important effect on the final result (see below), but, in addition, another kind of error may be incurred, namely an 'error of method'. There is generally a theoretical relationship connecting the measured quantities with the final quantity; often, approximations have to be made *in the theory*, and it may happen, therefore, that the final result is of limited accuracy through no fault of the measurements. For example, consider the measurement of angular rotation of a suspended system by means of a galvanometer mirror with lamp and scale. The displacement of the spot of light on the scale is often considered to be proportional to the *angular deflection of the suspension*, but this is really only an approximation which assumes that $\tan \theta$ is proportional to θ. It is fairly accurate for very small angles, but less accurate as the angle increases. For example, $\tan 6° = 0{\cdot}1051$, but $\tan 12° = 0{\cdot}2126$; therefore if the method were used without correction an error of about 1% would inevitably be incurred with deflections of this order, although the actual experimental quantity—the scale reading of the spot of light—might be measured much more accurately.

This example of an error of method is a very simple one, and it would be easy in this case to refine the theory in order to eliminate the approximation altogether. In many cases, however, the theory behind an experiment may be already difficult, and certain assumptions and approximations may be unavoidable. With difficult research problems of this kind it is advisable to determine the required quantity, if possible, by two quite independent methods, as a check.

For the present purposes it will be sufficient simply to note the existence of *errors of method*. The more practical problems presented by *instrument errors* and *random errors* can, however, be treated further.

4. INSTRUMENT ERRORS

No instrument is perfect. However skilled the mechanic who makes it, there is a limit to its accuracy, and it is important for the observer using the instrument to know where this limit lies. All reputable instrument makers provide *calibration data*, stating the reliability of the particular instrument as determined in a physical standards laboratory, or *specifications* which give the accuracy which the instrument can attain. These should be read with care. For example, meters are often said to be calibrated to '1% f.s.d.'—i.e. to 1% of full scale deflection. If the actual reading is about 10% of full scale deflection, this specification only guarantees an accuracy of 10% of the reading!

After a period of use, however, or if no calibrations are available, it may be necessary to devise a method of checking and, if necessary, calibrating the instrument. The checking or calibration of an instrument usually requires the use of reliable *standards*—for example, standard weights, thermometers, resistances, voltaic cells (chapter 4), etc.—which have themselves been checked and preferably supplied with a certificate from a standards laboratory such as the National Physical Laboratory at Teddington, Middlesex. In research work where high precision is important, *all* instruments should be carefully calibrated. The calibration of instruments such as electrical resistances and potentiometers is best left to the makers, although a physical chemist should be prepared to make critical tests of such instruments.

Even if an instrument is considered to be very reliable, e.g. a refractometer by a reputable maker, it should still be checked at one or two points—for example, by determinations with substances of known physical constants. At this stage it will be realized that two distinct considerations are always involved in making a measurement, namely, the *reproducibility* of the reading, and the *absolute accuracy* of the value obtained.

If the operations of making a measurement are repeated a number of times some idea is obtained of the *reproducibility* of the result. Probably slightly different results will be obtained each time (especially if a number of operations are involved), and the reproducibility of the reading will depend on the care and skill of the operator and on the imperfections of the instrument being used. On the other hand, the absolute accuracy might be considerably less than the reproducibility would lead one to expect, as, for example, if the instrument were out of adjustment. Measurements with pure substances can frequently help to decide how reliable an instrument really is.

This raises the question of instrument *corrections*, because unless all the necessary corrections are applied the full accuracy of the instrument is not obtained. Well-known examples of determinations where corrections must obviously be made are (a) weight, for buoyancy of the air (chapter 2), (b) temperature, for emergent stem correction (chapter 3), (c) barometric pressure, for expansion of the scale and gravitational field (chapter 5), (d) thermocouple reading, for cold-junction temperature (chapter 3). Wherever such corrections would make a significant change in the reading they should certainly be applied, and in physical chemistry any reported value is assumed to have been corrected.

There still remains the question of instrument defects, revealed by tests with substances of accurately known constants, even after all obvious corrections have been applied. The most common imperfections are (a) 'zero' error due to displacement of a scale, (b) constant or proportional error arising from error in an electrical resistance, (c) 'backlash' in an adjustable part of an instrument, due to poor mechanical construction or to wear, (d) eccentricity of a circular scale. In many cases it is possible to apply corrections to minimize errors from such sources; for example, the zero error may be determined, the effect of backlash can be largely eliminated by moving the adjustable part in the same direction every time, and eccentricity of circular scales can be overcome by taking readings on both sides of the scale (chapter 9).

With some instruments, however, a check at one point of the scale provides no guarantee that other parts of the scale are equally accurate. Thermometers and burettes are examples of instruments which ought to be checked at a number of points.

5.1. RANDOM ERRORS OF MEASUREMENT

However good an instrument may be, there is a limit to the *reproducibility* of measurements. This may arise from slight difficulty in making the final adjustment (as in matching the fields in a polarimeter—chapter 9), or in estimating fractions of a division on the scale (as on a thermometer), or from small irreproducible factors such as the sticking of a mercury meniscus in a manometer. In other cases the limit of reproducibility may be outside the instrumental measurements; it may be associated with the operations of preparing the material for measurement. For example, the chief error in gravimetric analysis or determination of gas density (chapter 5) is rarely the final weighing, but is generally bound up with the preparation and purity of the substance to be weighed, and the handling, drying, and temperature control of the crucible or bulb in which the substance is to be weighed.

When all obvious sources of error have been removed, it is still found that a number of complete repetitions of the work lead to slightly different values. These 'random errors' are usually only in the last place of figures which can be recorded, although sometimes, when the irreproducibility lies with the operations, the random error may be found to be much greater than the random errors of the instruments.

Random errors can be reduced by practical skill, care, and attention to detail. A good example is the measurement of the optical rotation of a solution (chapter 9); the determination involves a subjective judgment of the setting of the polarimeter to a position where the two parts of the optical field of view are of equal intensity. The reproducibility will be found to increase as the observer becomes more practised, and it is also improved by darkening the room and arranging a comfortable position for the observer in order to reduce eye strain. However, sometimes improvement which is apparently due to factors of this sort in fact arises from the subconscious tendency of a human observer to make successive readings agree!

It is still necessary to decide the magnitude of the uncertainties introduced by random errors, and to find some method of reporting and dealing with them.

Suppose a number of repetitions of a measurement are made by the same technique, with the same apparatus and care, and by the same observer (or observers). The final results show small random differences, the values being sometimes slightly higher and sometimes lower. How can the utmost information be extracted from the results? Obviously the set of repeated measurements conveys more information than a single observation. Firstly, the *arithmetic mean* of the results is almost certainly a more reliable value than a single observation; secondly, the observed deviations of the observations from the mean give an indication of the degree of reliability to be attached to a single observation (or to a mean of a number of observations). For an adequate understanding of the proper treatment of random errors it is necessary to introduce some elementary theory of statistics (see below).

5.2. The problem can be most readily understood by considering an imaginary set of measurements of a length. Suppose, for example, the surface tension of a liquid is to be measured by the well-known capillary rise method. One quantity required is the radius of the tube, and the uniformity of the bore must also be examined. This is usually done by measuring the length of a weighed thread of mercury at different positions along the tube by means of a vernier travelling microscope. A good vernier microscope can be *read* to 0·001 cm, but doubtless if repeated measurements of the length of the thread of mercury were made there would be appreciable differences between them, arising from difficulties in setting the microscope, moving the mercury (with attendant 'sticking' due to hysteresis of the contact angle, etc.). A set of consecutive readings might well give figures such as the following: 5·216, 5·213, 5·214, 5·218, 5·215, 5·213, 5·215, 5·211, 5·217, 5·216, 5·212, 5·213, 5·214, 5·214, 5·212.

The problem is to find a method of extracting a result of maximum accuracy from the data, together with an index of reliability.

Figure 1A.1(a) represents the sequence of results diagrammatically. It becomes evident as the repetitions proceed, that the 'true' length probably lies somewhere between 5·215 and 5·213, and the figures obtained are merely random deviations on one side or other of the true result. This is seen clearly from Fig. 1A.1(b), in which, at

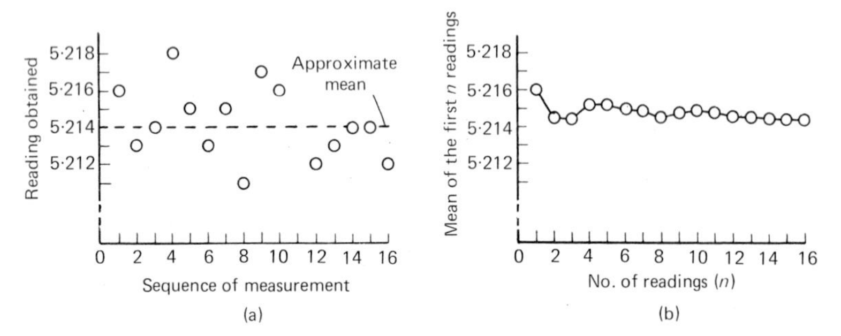

FIG. 1A.1 Charts representing random deviations in the apparent length of a mercury thread, measured 16 times with a vernier microscope. (a) The sequence of the readings obtained. (b) The mean of the readings, calculated after each additional reading.

each stage, the mean of all the points so far obtained is plotted. It is evident that as the number of readings increases, the mean settles down gradually towards a constant value, which is slightly greater than 5·2140 but less than 5·2145. After sixteen readings there is little doubt that, to the nearest thousandth of a centimetre, the best result is 5·214, but the fourth decimal place is still in doubt.

The original readings were evidently deviating at random on either side of 5·214 by amounts up to 3 or 4 units on the last place. However, small deviations are more frequent than large, as can be seen by plotting a *frequency histogram* which shows the number of observations falling in each interval of length (Fig. 1A.2). The observations are evidently distributed more or less systematically about 5·214, positive and negative deviations of a given magnitude being approximately equally probable. A symmetrical histogram with a well-defined maximum is not obtained, however, unless a considerable number of observations are available; for example, the first six observations in the present series give a very peculiar histogram (see the numbering of observations in Fig. 1A.2), and the 'best' value could not be established to the third decimal place with any confidence.

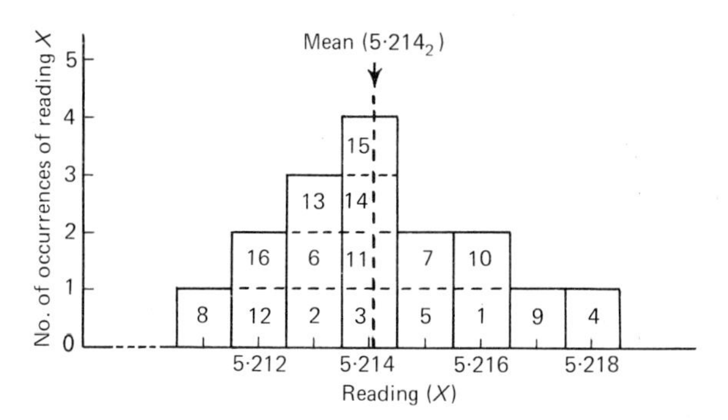

FIG. 1A.2 Random deviations plotted as a frequency histogram (data from Fig. 1A.1).

On the other hand, if a very large number of observations were made, the frequency histogram would eventually become a practically smooth '*distribution curve*', the maximum of which would give the 'most probable value' with considerable precision. When only a small number of observations are available, it is useful to *assume* that they are a representative sample of an infinite number of observations, to which the well-established statistics of probability could be applied. If the deviations of the readings were entirely random (i.e. arose from a large number of small, independent positive or negative deflections), the frequency distribution would follow a theoretical law known as the 'normal frequency distribution', or Gauss's 'normal error curve'.[1] This branch of statistics is of great importance in some of the sciences into which physical chemistry sometimes trespasses, notably biology and agricultural chemistry. A brief account of the normal distribution curve is therefore given below. Some of its properties are used in the treatment of errors of observation in physical determinations.

5.3. THE NORMAL DISTRIBUTION (OR 'ERRORS') CURVE

Consider some property (X) of a large number (n) of samples drawn from a given batch of similar items which differ from one another only by small random deviations—for example, repetitions of a physical measurement, weights of individual ball-bearings from the same machine, size of wheat grains from the same sack, survival time of insect pests on a crop sprayed with insecticide, etc. The X values measured on individual samples will be X_1, X_2, X_3, etc., and will scatter about the arithmetic mean value, $\overline{X}$, which is defined by

$$\text{Arithmetic mean, } \overline{X} = \frac{X_1+X_2+X_3+\cdots}{n} = \frac{1}{n}\sum X$$

It can be shown that if n is very large the frequency distribution curve (corresponding to the frequency histogram in Fig. 1A.2) is given by the equation

$$y = \frac{1}{\sigma\sqrt{2\pi}} \exp\left[\frac{-(X-\overline{X})^2}{2\sigma^2}\right]$$

This distribution is shown graphically in Fig. 1A.3 (curve *A*). In the equation, y has the meaning of a probability coefficient; any element of area under the distribution curve corresponding to an increment from X to $(X+\mathrm{d}X)$, gives the *probability* that a single sample drawn from the batch will show an X value lying in the

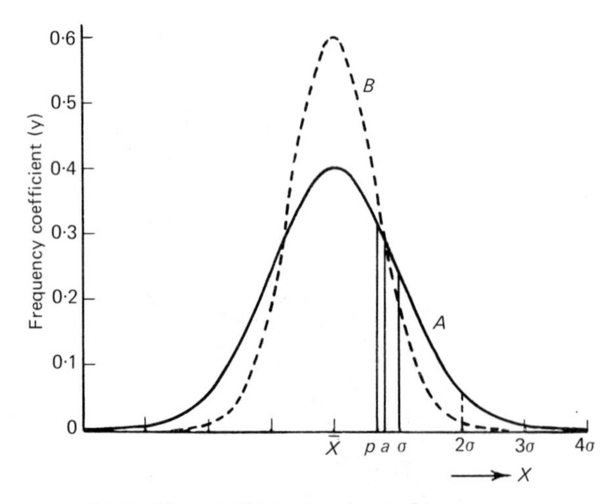

FIG. 1A.3 Normal distribution ('error') curves.

above range, X to $(X+\mathrm{d}X)$. The equation in the form given above has been arranged so that the total area under the curve (i.e. $\int_0^\infty y\,\mathrm{d}X$) is equal to unity, and therefore the area under the curve between X_1 and X_2 gives directly the fraction of the observations which lie between X_1 and X_2. Clearly, the most probable value to occur is that at the maximum of the curve (y a maximum), and this corresponds to the arithmetic mean of all the samples, namely $\overline{X}$.

On the other hand, a *single* observation is more likely to lie at some distance from $\overline{X}$, as, indeed, do almost all the observations.

The quantity σ in the normal frequency equation is a constant for a particular batch of samples, and it measures the *spread* of the curve—that is, the extent to which the X values deviate from $\overline{X}$. Curves A and B in Fig. 1A.3 are typical normal distribution curves for two different batches which show the same mean value but different degrees of spread; on the average batch A shows greater deviations from $\overline{X}$ than does batch B. Such differing spreads of results might be obtained if a physical quantity were measured many times, first with a poorly made instrument and then with a very good one, or by a careless and a skilled observer using the same instrument.

The normal distribution curve is symmetrical about $\overline{X}$, since it is of the form $y = a \exp[b(X-\overline{X})^2]$, or

$$X = \overline{X} \pm \sqrt{(1/b)\log_e(y/a)}$$

The degree of spread of the data from $\overline{X}$ can be expressed in a number of ways. In statistics the quantity σ is most commonly used. This is called the *standard deviation* (σ), and is actually equal to the *root mean square deviation*, i.e. the standard deviation is defined (provided n is large), by

$$\sigma = \sqrt{(1/n)\sum(X-\overline{X})^2}$$

It can be shown,[1] however, that if the number (n) of samples examined is small the estimate of σ obtained from them is, on the average, low by a factor of $\sqrt{(n-1)/n}$, and, consequently a better ('unbiased') estimate of σ is obtained by computing the standard deviation by the formula

$$\text{standard deviation, } \sigma = \sqrt{\frac{\sum(X-\overline{X})^2}{n-1}}$$

σ corresponds to the point of inflexion on the side of the probability curve.

Another possibility for measuring the spread is to use the *average or mean deviation* (a), which is simply the mean difference between X and $\overline{X}$, without regard to sign, i.e.

$$\text{mean deviation, } a = \frac{1}{n}\sum|(X-\overline{X})|$$

Still another measure of the spread is the *probable error* (p); this is a deviation such that there is an equal probability of finding deviations larger than p and smaller than p. In Fig. 1A.3 the ordinate through $(\bar{X}+p)$ bisects the *area* under the right-hand branch of the curve, and that through $(\bar{X}-p)$ does the same for the left-hand branch.

It can be shown that, for a normal distribution curve, these three ways of measuring the spread are slightly different in magnitude, but are related to one another by constant factors, namely $p = 0{\cdot}6745\sigma$ and $a = 0{\cdot}7979\sigma$ or approximately $p:a:\sigma = 3\frac{1}{2}:4:5$. Their relative positions on the curve are shown in Fig. 1A.3; 68·2% of the observations deviate by less than σ from the mean, and 95·5% are within 2σ. Only 1 in 400 should deviate by more than 3σ, and any such points are highly suspect and likely to be due to gross mistakes. Numerical values for the ordinates of the normal probability curve and also the area under the curve (the cumulative percentage and derivatives) can be found in statistical tables.

Abnormal distribution curves (e.g. 'skewed') will not be considered here.[2]

BIBLIOGRAPHY 1A: Accuracy of measurements

Moroney, *Facts from Figures*, 1956 (Penguin, London).
MacDonald, *Mathematics and Statistics for Engineers and Scientists*, 1966 (Van Nostrand, London).
Tippet, *Technological Applications of Statistics*, 1952 (Williams and Norgate, London).
Gore, *Statistical Methods for Chemical Experimentation*, 1952 (Interscience, New York).

[1] See, for example, Yule and Kendall, *An Introduction to the Theory of Statistics*, 14th edn, 1958 (Griffin, London).

[2] See, for example, Peatman, *Descriptive and Sampling Statistics*, 1947 (Harper and Bros, New York).

1B The treatment of results

1. STATISTICAL TREATMENT OF ERRORS

It is clearly impracticable (and, indeed, unnecessary) to make a large number of repetitions of every physical measurement in order to study the distribution of the experimental errors. Instead, it may be assumed (in absence of evidence to the contrary) that if such laborious repetitions were carried out they would follow a normal errors curve. A few repetitions of one selected determination will therefore suffice to give an adequate indication of the reliability of the results.

Consider, for example, the first six readings of the set which has been discussed in section 1A(5.2). The arithmetic mean ($\bar{X}$) is 5·214(8); the average deviation (a) from the mean is $\pm 0{\cdot}0015$, and the standard deviation (σ) is $\pm 0{\cdot}0020$. The ratio a/σ for this set is fortuitously close to that expected for a complete normal distribution. The most probable error to expect (p) on a *single* measurement would therefore be about $0{\cdot}67 \times \sigma = \pm 0{\cdot}0013$. This information indicates the chance of getting near the correct value. If a *single* reading were taken, the probability is one-half that it would not differ from the true value by more than $\pm 0{\cdot}0013$. If the precision of the readings had already been known from previous work, the first result of the above series could have been reported as $5{\cdot}216 \pm 0{\cdot}001(3)$. The $\pm$ quantity written after a result is referred to as 'the limits of error'. This does *not* mean that the true value must inevitably lie between 5·214(7) and 5·217(3), but that there is a 50% chance that it does so. (In point of fact, the subsequent repetitions show that it probably lies just outside at about 5·214(2).)

Instead of quoting the 'probable error', the 'average deviation' or the 'standard deviation' would have conveyed the same information. The probable error has been most used in physical sciences in the past, but it offers no real advantage over σ, which is more directly derived from statistical theory. For purposes where the highest accuracy is required, it is probably best to report the 'limits of error' specifically as the standard deviation σ. For most purposes, however, the average deviation is quite good enough, and is more quickly calculated. In any case, unless a large number of repetitions are made, a will not differ significantly from p or σ. The result could therefore be reported as $5{\cdot}216 \pm 0{\cdot}001(5)$, meaning that determinations of this kind generally have an error of about $\pm 0{\cdot}001(5)$.

In general we are not so much interested in the reliability of a single reading as with a possible difference between the mean of our limited number of readings and the 'true' mean which would be found if we took a very large number. The mean is obviously likely to be closer to the true mean than a single determination would be, and its limits of error are therefore smaller. It can be shown by statistical theory that if n equally reliable measurements are made, and the probable, average and standard errors on a single measurement are known from previous measurements to be p, a and σ respectively, then the corresponding *errors on the mean value* are $p/\sqrt{n}$, $a/\sqrt{n}$ and $\sigma/\sqrt{n}$. Thus, four repetitions would halve the uncertainty, but it would require 100 repetitions to get a mean with only one-tenth of the uncertainty associated with a single measurement. This is only true if n is large: the reliability of the mean of a small number of readings is less than these limits. The difference is small for $n > 10$. More precise assessment of the accuracy of means can be made using Students t distribution.[1]

Application of this relationship to the original sixteen readings leads to a final statement that the 'best' value is 5·2148, and the standard deviation (s.d.) on this result is $\pm 0{\cdot}0020/\sqrt{16} = \pm 0{\cdot}0005$. The fourth place of decimals is now seen to be just significant, whereas on a single determination the third place of decimals was subject to undertainty of one or two units.

Note that the uncertainty (usually σ) attached to a result should always be quoted as a quantity (10 ± 1), never as a proportion $(10 \pm 10\%)$.

2. COMBINATION OF ERRORS

Most final results in physical chemistry are produced by calculations in which a number of measurements are combined. Thus, in arriving at the density of a gas (chapter 5), the results of several weighings and measurements of temperature and barometric pressure are employed. It is important to ascertain how the errors on the individual measurements affect the final result. This problem is sometimes called 'the propagation of precision indices'. Since the 'errors' quoted after a result are statistical indices rather than definite limits, statistical theory is employed to find valid methods of combining the uncertainties. The following rules emerge:[2] they can be applied to any index of precision (e.g. σ, p or a).

2.1. *Sums or differences.* Suppose the quantities A and B are to be added (or subtracted) to give the quantity C, and the indices of precision (uncertainty) on A and B are a and b respectively, and the index of precision on C is c, such that $A(\pm a) + B(\pm b) = C(\pm c)$, then

$$c = \sqrt{a^2 + b^2}$$

The error on a sum or difference is therefore greater than that on one of the quantities but less than the sum of the errors.

2.2. *Products or quotients.* If

$$A(\pm a) \times B(\pm b) = C(\pm c)$$

or

$$A(\pm a)/B(\pm b) = C(\pm c)$$

then

$$c = C\sqrt{\left(\frac{a}{A}\right)^2 + \left(\frac{b}{B}\right)^2}$$

i.e. the *percentage* (or *fractional*) error on a product or quotient is equal to the square root of the sum of the squares of the percentage (or fractional) errors on the terms multiplied (or divided).

2.3. *General case.* Functions more complicated than either **2.1** or **2.2** can be treated by the general rule:

If $U = f(X, Y, \ldots)$ and the errors on $U, X, Y, \ldots$ are $\pm\Delta U$, $\pm\Delta X$, $\pm\Delta Y, \ldots$, then

$$(\Delta U)^2 = \left(\frac{\partial U}{\partial X}\right)^2 (\Delta X)^2 + \left(\frac{\partial U}{\partial Y}\right)^2 (\Delta Y)^2$$

In most cases it is found that there are one or two principal sources of error, and that all the other measurements incur negligible errors.

Error in the coefficients a and b when a straight line $y = ax + b$ is fitted to a set of values of x and y is discussed in section 1B(7).

3. *EXAMPLES*

3.1. If a vernier microscope can be read with an uncertainty of $\pm 0{\cdot}002$ cm, what is the uncertainty on the reading of a length?

Since the length is the difference of two microscope readings of equal accuracy, e.g.

$$L_1(\pm 0{\cdot}002) - L_2(\pm 0{\cdot}002)$$

then the uncertainty on the length is

$$\sqrt{0{\cdot}002^2 + 0{\cdot}002^2} \approx \pm 0{\cdot}003 \text{ cm}$$

3.2. What is the probable error in a determination of the density of mercury ($\rho \approx 13{\cdot}6$ g cm^{-3}) in a 25 cm^3 pyknometer of weight 10 g, assuming that weighings, each reproducible to 1 part in 10 000, are the only significant sources of error?

Density is obtained by

$$\rho = \frac{\text{wt of mercury}}{\text{wt of water}}$$

$$= \frac{(\text{wt of pyknometer} + \text{mercury}) - (\text{wt of pyknometer})}{(\text{wt of pyknometer} + \text{water}) - (\text{wt of pyknometer})}$$

The weight of mercury will be about $25 \times 13{\cdot}6 = 340$ g, and the error in the numerator, by rule **2.1**, will be about

$$\sqrt{\left(\frac{350}{10\,000}\right)^2 + \left(\frac{10}{10\,000}\right)^2} \approx \pm 0{\cdot}035.$$

It is seen that the error arising from the weight of empty pyknometer is quite negligible compared with the error from the weight of (pyknometer + mercury).

Similarly, the uncertainty on the weight of water is

$$\sqrt{\left(\frac{25+10}{10\,000}\right)^2 + \left(\frac{10}{10\,000}\right)^2} \approx \pm 0{\cdot}004$$

If ρ is the density of mercury and the uncertainty is $\pm\Delta\rho$, then

$$\rho(\pm\Delta\rho) \approx \frac{340(\pm 0{\cdot}035)}{25(\pm 0{\cdot}004)}$$

By rule **2.2**

$$\frac{\Delta\rho}{\rho} = \sqrt{\left(\frac{0{\cdot}035}{340}\right)^2 + \left(\frac{0{\cdot}004}{25}\right)^2} = 1{\cdot}6 \times 10^{-4}$$

The error on the density is thus somewhat less than 2 parts in 10 000, in spite of the fact that *four* weighings, each uncertain to 1 part in 10 000, were involved.

3.3. The molecular refractivity of a substance (chapter 8) is defined by

$$[R] = \frac{n^2-1}{n^2+2} \cdot \frac{M}{d}$$

where n = refractive index, M = molecular weight, d = density. What is the percentage error on $[R]$ for benzene, if $n = 1{\cdot}498 \pm 0{\cdot}002$, $d = 879 \pm 1$ kg m^{-3} and $M = 0{\cdot}07808$ kg mol^{-1}?

The error on $[R]$ could be obtained by stages, using rules **2.1** and **2.2**, but is best arrived at by the general rule **2.3**. This gives

$$(\Delta[R])^2 = \left(\frac{\partial[R]}{\partial n}\right)^2 (\Delta n)^2 + \left(\frac{\partial[R]}{\partial d}\right)^2 (\Delta d)^2$$

On substituting values, $[R]$ is found to be $2{\cdot}604\times10^{-5}$, and $\Delta[R]=\pm9\times10^{-8}$, and thus the percentage error is $\pm0{\cdot}3\%$.

4. NUMBER OF FIGURES TO BE EMPLOYED

Every measurement should be recorded to as many figures as are significant—no more, as that would implicitly claim a' greater accuracy than justified, and no fewer, as that would incur loss of accuracy. It is understood that the last figure quoted may be slightly in doubt, but it is probably correct to the nearest integer. Thus a burette titration given as 25·42 cm^3 is presumed to be valid to within $\pm0{\cdot}005$ cm^3, and thus the end-point must have been sharp to within a fraction of a single drop of titrant. The last figure is obtained by estimating tenths of a division on the scale.

Where the last figure is very doubtful, but dropping it might lose a little accuracy, it is useful to give it in parentheses or as a subscript (e.g. 0·512(4) or $0{\cdot}512_4$); the implication here is that accuracy is better than $\pm0{\cdot}001$ but not as good as 0·0001.

When measured values are employed in a calculation, the same principles are followed—figures are retained to the limit of significance, but not beyond. For example, in adding 25·42 cm^3 to 0·512 cm^3, the sum should be quoted as 25·93, *not* 25·932 cm^3, because the uncertainty is at least $\pm0{\cdot}01$ cm^3.

In multiplication or division it is only the *relative* or *proportional* error that matters, not the apparent or absolute value, and a given relative error in one of the numbers will produce a corresponding relative error in the result.

Suppose, for example, that one has to multiply two quantities given as 2·3416 and 2·55, and suppose each of these numbers to have the maximum apparent error; then the relative error in the first number is about 5 in 230 000, and the error in the second number is about 5 in 2 600. Evidently, therefore, the result of the multiplication will also have an error of about 5 in 2 600, or 0·2%. Consequently it would be incorrect to perform the multiplication in the ordinary manner, and write the result as 5·971080, for this result has a derived error of 0·2%, or of about 1 unit in the second place of decimals. All the figures after this are therefore meaningless, and should be discarded, the result being written 5·97.

When a number has to be *rounded off* to reduce it to a significant number of figures, the rule is that, for a 5 or over in the last figure rejected, the last retained figure is increased by one. Thus, 5·13462 if rounded to five figures becomes 5·1346, to four figures 5·135, to three figures 5·13.

5. METHODS OF CALCULATION

If ordinary, long, laborious methods of arithmetic calculation are used they are liable to involve manipulation of a large number of useless figures which are afterwards discarded as meaningless. Time and mental energy can be saved by adopting abbreviated methods of multiplication and division, and by the use of logarithms and of the slide rule. These methods are also preferable because the chance of making an arithmetic error is reduced.

However, electronic calculating machines are now widely used, because they reduce both labour of calculation and the chance of error. In this case it is pointless to drop non-significant figures at intermediate stages in the calculation, unless this is forced by the operation of the machine. Machines which can store at least two intermediate results in memories as well as perform the usual arithmetical operations are most suitable for the physical chemistry laboratory. Machines which can memorize a series of operations are particularly useful, so are those which can take logarithms or extract square roots.

In making calculations with the aid of logarithms, the precautions adopted in the preceding methods for the avoidance of unnecessary figures are introduced automatically, if it be so arranged that the number of figures in the logarithm is greater by one than the number of figures in the least accurate of the numbers involved in the calculation. In this way one ensures that the error in the result shall not be greater than the error in the numbers from which the result is obtained. If one had to multiply $2{\cdot}54 \times 4{\cdot}3664 \times 0{\cdot}89676$, one should use 4-place logarithm tables, and the second and third numbers should be rounded off to 4·366 and 0·8968.

The error inherent in the logarithm itself decreases with the number of places in the logarithm, each additional figure in the logarithm being accompanied by about a tenfold decrease in the error. In the case of 4-place logarithms, the maximum possible error introduced into a calculation through their use may be taken as about 1 in 3 000. For work of moderate accuracy, 4-place logarithms will be sufficient; but in some cases, the error so introduced is greater than that due to experiment, e.g. in determinations of density. In the latter cases, therefore, where the accuracy of the calculation is desired to be equal to the accuracy of the experiment, logarithms with 5 or 6, and even, in more exceptional cases, 7 places should be used.

In few, or none, of the experiments described in the following pages will an accuracy in calculation be called for greater than can be obtained by the use of 4-place, or, at most, 5-place logarithms. Frequently the accuracy required in practical physical chemistry will be considerably less.

For all calculations where the required accuracy of experiment or calculation is not greater than about 1 in 500, the slide rule is of great assistance. It is also valuable for rapid, approximate checking of more accurate calculations. With this instrument various degrees of accuracy can be obtained according to the size of the rule, but with the ordinary size of slide rule (25 cm in length) the accuracy obtainable may be put at about 1 in 500 to 1 in 800. Considerably better accuracy can be obtained with special forms of slide rule such as the 'Otis King': this has cylindrical scales equivalent to a 30 in slide rule and can be read to about 1 in 2 000.

Some types of modern physical research work (for example, X-ray crystallography) call for a great deal of accurate computation. The digital computer is essential for this work, and every chemist should learn to program these machines. However, this subject is beyond the scope of this book.[1]

6. PRESENTATION OF RESULTS

Experimental work should be recorded at once in fullest possible detail in a bound laboratory notebook. (Loose sheets are to be deprecated.) It is to be expected that much that is recorded will never be used, because only final, successful measurements will be adopted. The final result of the work will usually be presented as a report or paper for publication. Such reports should contain all *essential* information about the work (according to the purpose of the report), but nothing superfluous. In modern scientific writing short, precise sentences are favoured, figures of speech are avoided, and an impersonal style of writing is generally adopted.

The most common lay-out for a scientific report is roughly as follows. Firstly, a brief introduction states the purpose of the work, and the principles on which the experimental work is based. Then follows a section giving all necessary details of how the practical work was carried out. Next comes a presentation of the experimental results together with evidence of their degree of reliability. Finally, a 'discussion' section considers the final conclusions resulting from the whole work, and perhaps provides a theory to account for the observations. A common fault is omission of clear statements of the significance of the work. When data or evidence from other sources is quoted it is very desirable to find and refer to the *original* source of the information, rather than to employ data on trust at second hand.

In undergraduate laboratory courses details of the theory and experimental procedure are often given in books or on sheets. In this case the material should *not* be copied word for word into the report. It should be briefly summarized with a reference to the original source. Special precautions found necessary to obtain good results should, however, be reported in full.

The numerical results of practical work are most conveniently presented in the form of tables, supplemented, where necessary, by a summary of other data necessary to the calculations. It should be possible for a competent person reading the report to check any but elementary calculations. The general form of the results is most readily seen if they can be given also as a graph on which the experimental points are clearly plotted. The size of circles, triangles, squares or other symbols used to characterize the points should be approximately equal to the limits of error on the results. The abscissae should give the independent variable and the ordinates the dependent quantity. The scales should be so arranged that the data make full use of the paper, and they should be clearly graduated and the units stated.

Wherever possible, it is an advantage to employ in a graph such mathematical functions as will result in a straight line or a smooth curve of gentle curvature. For example, the vapour pressure of a liquid increases very steeply with temperature, but a graph of log (vapour pressure) is practically a straight line when plotted against $(1/T)$ (chapter 5).

A straight line function is especially valuable for *interpolation* or *smoothing* of a set of results—a matter which deserves some consideration. A great many investigations in physical chemistry lead to a series of values of some function y for different values of an independent variable x. In most cases it is either known from theory or becomes apparent from the regular trend of the results that a definite functional relationship exists between y and x, i.e. $y=f(x)$.*

* There is also, of course, another important type of investigation—rarely encountered in physical chemistry, but very common in biology, agricultural chemistry, operational research, etc.—in which it is desired to find whether any *correlation* does exist between two properties; for example, whether the concen-

The ideal treatment of the results is, then, to discover the nature of the function and express the relationship as a mathematical equation. The final equation then summarizes the result of the work in a more convenient form than a table of experimental data, and it is probably more reliable because it makes use of all the points. Further, it can be used for interpolation between the experimental values, and, with less confidence, for extrapolation outside the experimental range.

The simplest method of fitting an equation to the data is to plot a graph of suitable functions to produce a straight line relationship and draw 'the best' straight line by judgment, the aim being to select a line to run as close as possible to the greatest possible number of points. A less arbitrary procedure is that known as the 'method of least squares'.

7. FITTING EQUATIONS BY THE METHOD OF LEAST SQUARES

The criterion for the 'best' fit is that the sum of the squares of the deviations of the experimental points from the chosen line should be as small as possible; this can be shown to correspond to the condition that the arithmetic mean should be the 'best' representation of a series of readings of the same quantity. It would hold if the deviations from the line followed the normal errors distribution. The method is applicable to many forms of equation, but is chiefly used for straight lines—the only case considered here.

Suppose the data consist of a set of n points, the coordinates of which ($x_1 y_1, x_2 y_2, \ldots, x_n y_n$) are tabulated. Suppose, further, the values of the independent variable x can be regarded as subject to negligible error, all the deviations being in the y values. By hypothesis, these results are scattered at random from the 'true' line, the equation of which is taken as $y = a + bx$. The y deviations can then be written down. Thus, for a typical point i, y_i should have been at $(a + bx_i)$, and the deviation is therefore $y_i - (a + bx_i)$. The sum of the squares of the deviations is $\sum (y_i - a - bx_i)^2$ summed over all n points and this quantity must be made a minimum by

tration of certain trace elements in the soil affects crop yields, or whether there is any connection between the silicon content and the sulphur content of pig-iron produced by a blast furnace, etc. This field of study requires special statistical methods by which 'coefficients of correlation' or significance tests can be applied. These methods are outside the scope of the present work; they may be found in most textbooks of statistics.[3]

suitable choice for the values of a and b. The condition for a minimum is obtained algebraically by differentiating the expression with respect to a and b, putting the derivatives equal to zero, and solving the resulting pair of simultaneous equations. *The final working formulae for obtaining the 'best' constants are*

$$a = \frac{\sum x \sum xy - \sum y \sum x^2}{(\sum x)^2 - n \sum x^2} \quad \text{and } b = \frac{\sum x \sum y - n \sum xy}{(\sum x)^2 - n \sum x^2}$$

It is therefore necessary to draw up columns of the x and y values of the experimental points, and then tabulate x^2 and xy and add each column for insertion of the sums in the above formulae. The limits of error attaching to the constants a and b of the fitted equation ($y = a + bx$) can be calculated if the limits on a single observation are known; if, for a given value of x, the y value is $(y \pm \Delta y)$ where Δy is the standard deviation, probable error or mean deviation, then the corresponding uncertainty on a is $\pm \Delta y / \sqrt{n}$ and on b is $\pm \Delta y / \sqrt{\sum (x - \bar{x})^2}$.

The standard deviation of y from the line is

$$\left\{ \frac{1}{n} \sum_n (y - a - bx)^2 \right\}^{\frac{1}{2}} = n^{-\frac{1}{2}} \{ \sum y^2 + b^2 \sum n^2 - 2b \sum xy + ab \sum x - 2a \sum y + na^2 \}^{\frac{1}{2}}$$

All these quantities except $\sum y^2$ have already been calculated. These lengthy calculations are best done on a calculating machine, or for very large numbers of points on the computer; standard statistical programs are available. Note that, as for a single variable, these formulae are exact only if n is not too small.

STATISTICAL EXPERIMENT

Investigate the reproducibility of a 5 cm³ pipette, and determine whether the addition of a wetting agent to the solution employed affects the volume which the pipette delivers and the reproducibility of the volume.

Procedure. The method consists simply in weighing successive volumes of water delivered by the pipette.

Clean a 50 cm³ stoppered standard flask, a 5 cm³ pipette and a stoppered bottle of about 250 cm³ capacity. Take the temperature of the balance case. Fill the bottle with distilled water and adjust the temperature to that of the balance. Then proceed to weigh the flask with successive additions of 5 cm³ portions of water from the pipette until ten portions have been added. The weighings should be made to the nearest milligram. A standard pipetting technique (drainage time, etc.) must be adopted.

Repeat the set of weighings using now water to which a small quantity of synthetic detergent has been added. Since the object of the detergent is to depress the surface tension, it is not necessary to add more than the minimum amount required to produce a stable foam when the solution is shaken. The change in density resulting from addition of the detergent will be negligible.

8. TREATMENT OF RESULTS

We will illustrate the treatment of the results by an example. The weight W g of water delivered by a 10 cm^3 pipette is conveniently expressed as $x = 1\,000\,(W-10)$ mg. For 30 consecutive weighings without detergent $\sum x = -631$, $\sum x^2 = 24\,892$; for 20 runs with detergent $\sum x = 56$, $\sum x^2 = 3\,838$. The pipette was calibrated and the experiments performed at 20°C where $\rho_{H_2O} = 998{\cdot}23$ kg m^{-3}. We want to determine:

8.1. whether the results reveal any error in the manufacturer's original calibration of the pipette,
8.2. whether there is any significant difference in the amounts delivered with and without the detergent,
8.3. whether the deliveries were significantly more reproducible in one case than the other, and
8.4. whether the use of detergent should be recommended as standard practice in future.

Preliminary calculations

	Without detergent	*With detergent*
Mean $\bar{x} = (\sum x)/n$	$-631/30 = -21{\cdot}0$	$56/20 = 2{\cdot}8$
Sum of squares of deviations $= \sum (x-\bar{x})^2$ $= \sum x^2 - 2\bar{x}\sum x - n\bar{x}^2$ $= \sum x^2 - ((\sum x)^2/n)$	$24\,892 - (631^2/30)$ $= 11\,610$	$3\,838 - (56^2/20)$ $= 3\,681$
Variance of population, $V = \sum (x-\bar{x})^2/(n-1)$	$387 \times 30/29 = 400$	$184 \times 20/19 = 194$
Standard deviation, $s = \sqrt{V}$	$\sqrt{400} = 20$	$\sqrt{194} = 14$
Variance of mean, $V_m = V/n$	$400/30 = 13{\cdot}3$	$194/20 = 9{\cdot}7$
Standard deviation of mean, $s_m = \sqrt{V_m}$	$\sqrt{13{\cdot}3} = 3{\cdot}65$	$\sqrt{9{\cdot}7} = 3{\cdot}1$

8.1. Weight of 10 cm^3 of water at 20°C = 9·9823 g;

$$x = 1\,000(9{\cdot}9823 - 10) = -17{\cdot}7 \text{ mg.}$$

Difference between this and the observed mean $= -21{\cdot}0-(-17{\cdot}7)=$ $-3{\cdot}3$ mg $= 3{\cdot}3/3{\cdot}65 = 0{\cdot}9$ standard deviations of mean.

Deviations of 1 s.d. or more occur by chance in about one-third of all determinations. We conclude that an error of 0·9 s.d. might well occur by chance, and that the results show no significant error in the manufacturer's calibration. Strictly we should use the Student's t distribution. Entering the tables with $t = 0{\cdot}9$ and 29 degrees of freedom for the distribution of the mean we obtain a similar result. The reader is referred to statistical text books for this distribution: see section 1B(1).

8.2. On a null hypothesis all the weighings form part of the same population, whose mean is

$$\bar{x} = (\sum x)/n = (-631 + 56)/50 = -11{\cdot}5,$$

and whose variance is

$$(1/(n-1))\,(\sum x^2 - (\sum x)^2/n) = (1/49)\,(24\,892 + 3\,838 - (-631 + 56)^2/50) = 451.$$

For samples of 30 and 20 weighings the variances of the means will be $451/30 = 15$ and $451/20 = 22{\cdot}5$ respectively; the corresponding standard deviations will be 3·9 and 4·7. For runs without detergent, the difference between the sample mean and the population mean is $-21-(-11{\cdot}5) = -9{\cdot}5$ mg or $9{\cdot}5/3{\cdot}9 = 2{\cdot}4$ standard deviations. The probability that a difference this large would occur by chance is about 0·016. For the runs with detergent we have $2{\cdot}8-(11{\cdot}5)=$ 14·3 mg or $14{\cdot}3/4{\cdot}7 = 3{\cdot}0$ standard deviations, probability 0·0026. The chance that both samples will be this far out is then only $0{\cdot}016 \times 0{\cdot}0026 = 4 \times 10^{-5}$. The difference in the means is certainly significant. Strictly this should be tested using the Student's t distribution.

It would be necessary to recalibrate pipettes to be used with detergent.

8.3. To test the relative reproducibility of the two methods we calculate Snedicor's F = ratio of variances of two populations[1] = 400/194 = 2·1. Entering statistical tables with 29 degrees of freedom for the larger variance, and 19 for the smaller, the ratio would be as large or larger than this by chance only about once in 50. The reduction in variance, i.e. the greater reproducibility with detergent, is therefore significant. However it could just as well be due to increasing skill in pipetting because the set of runs using detergent were made last. It would have been better to do sets of say ten weighings with and without detergent alternately.

8.4. The difference in s.d. for single weighings is 20 − 14 = 6 mg, and this would be the increase in accuracy using detergent. However this is only six parts in 10 000 and is quite negligible compared to other errors in titration. It is thus pointless to recommend the use of detergent in practice.

It will be noticed that the statistical treatment of the data does not lead to absolutely certain conclusions—only to conclusions of a definite degree of probability. If the degree of confidence (in a conclusion) indicated by a statistical test is not great enough to allow a satisfactory decision to be made, the data must be considered inconclusive, and the only thing to be done to get a more definite result is to make a much larger number of observations or more accurate ones.

BIBLIOGRAPHY 1B: Treatment of results
See bibliography for 1A

Worthing and Geffner, *Treatment of Experimental Data*, 1943 (John Wiley, New York).

Brownlee, *Industrial Experimentation*, 4th edn, 1949 (H.M. Stationery Office, London).

[1] Moroney, *Facts from Figures*, 1956 (Penguin, London).

[2] James, Whitehead and O'Brien, *A FORTRAN Programming Course*, 1970 (Prentice-Hall, London); Golden, *FORTRAN IV Programming and Computing*, 1965 (Prentice-Hall, New Jersey); Weissberger, *Physical Methods of Chemistry*, Vol. I, Pt 1B, chapter IX, 1971 (Wiley-Interscience, New York).

[3] E. G. Fisher, *Statistical Methods for Research Workers*, 13th edn, 1958 (Oliver and Boyd, Edinburgh); Ezekiel, *Methods of Correlation and Regression Analysis*, 3rd edn, 1959 (John Wiley, New York).

2

Determination of weight, volume and length

Determination of weight, volume and length

Almost all experimental work in physical chemistry involves determinations of *weight*, *volume* or *length*. A knowledge of the most appropriate methods of measuring these quantities in different circumstances is essential; the general methods are well known, but it is important to consider the degree of accuracy that can be achieved with the various instruments, and the technique and precautions which must be adopted to obtain the highest possible accuracy. Calibration of weights and of volumetric apparatus is rarely performed nowadays by the average chemist as apparatus can be brought with the required accuracy.

2A Determination of weight

1. THE BALANCE

The modern analytical balance is a highly refined and indispensable instrument, and the determinations of weight can be made very accurately without great difficulty; for example, an object of about 100 g can readily be weighed correct to the nearest milligram, i.e. 1 part in 10^5. No other quantity can be determined with this accuracy in an ordinary laboratory. In fact, the accuracy of the balance is generally greater than the accuracy with which a body can be defined or reproduced. Thus, in weighing glass vessels or other material used in an experiment, the difference in weight due to *handling*, manner of drying, etc., often amounts to several tenths of a milligram, while the balance itself might be capable of weighing to 0·1 mg. A *significant* accuracy of 1 in 10^5 is not achieved without (a) a balance of excellent workmanship and in proper adjustment, (b) rigorous attention to the technique of weighing, handling, etc., and (c) a set of carefully calibrated weights.

2. THE AIR-DAMPED, 'APERIODIC' BALANCE

Most of the time-consuming tedium of weighing has been eliminated by the air-damped balance with automatic fractional weights. Figure 2A.1 shows a modern balance with these features. The oscillations of the beam are heavily damped by the viscous resistance of

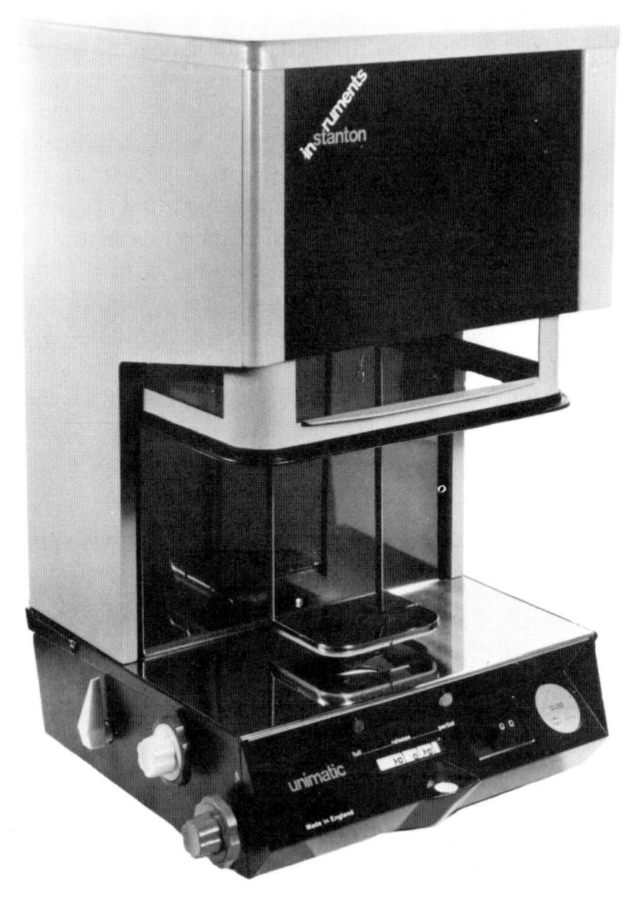

FIG. 2A.1 Aperiodic balance with weights applied by external dials. (Courtesy Stanton Instruments Ltd.)

air in the small gaps between pairs of concentric cups, the outer ones being fixed and the others attached to the balance arms. After a few seconds the balance settles down into a position of equilibrium, and the steady deflection is read on a small screen. The usual pointer is replaced by a minute glass scale, and a magnified image of this scale is projected by an optical system on to the screen. The sensitivity of the balance is carefully adjusted to make the image of the scale read directly in milligrams and tenths of milligrams. In using this balance, the balance case is closed and the tens, units, tenths and hundredths of a gram are added in the form of rings of wire which can be lowered into position on a cross strut on the right-hand arm by rotating graduated dials on the outside of the balance case. These balances have only a single pan. After balancing to the nearest 0·01 g, the next two places of decimals are read off as soon as the image of the graduated scale has come to rest on the screen.

Before using the balance, the position of the zero point should be checked with the pans unloaded: the scale is adjusted to read zero.

It should be unnecessary to emphasize here that accurate weighings can be expected only if the balance is kept clean and free from dust, and if the beam is released and arrested in such a manner as not to cause jarring of the knife-edges. The beam must never be released or arrested with a jerk. The balance should not be exposed to unequal heating, and should not, therefore, be placed in a window exposed to direct sunlight. A balance in general use in a large laboratory should be regularly inspected, and, if necessary, cleaned and adjusted by a competent instrument mechanic.

3. CORRECTION FOR THE BUOYANCY OF AIR

In accurate determinations of the mass of a body by weighing, the apparent weight, i.e. the sum of the face values of the weights used, must be corrected for the buoyancy of the air. A body will appear lighter by an amount equal to the weight of air displaced. The greater the difference between the density of the body and of the weights used to counterpoise it, the greater the correction.

If a body of density d kg m^{-3} is counterpoised by weights whose mass (or weight *in vacuo*) is G kg, the volume of the body will be G/d m^3. Its weight will therefore be diminished by $1{\cdot}2G/d$ kg, where 1·2 kg is the weight of 1 m^3 of air under normal conditions: room temperature, 1 atm pressure and average humidity. The mass of the body would therefore be $(G+1{\cdot}2G/d)$ kg if the counterpoise has negligible volume. A counterpoise of density b kg m^{-3} has volume G/b m^3 and its weight is diminished by $1{\cdot}2G/b$ kg. The mass of the body counterpoised by weights totalling G kg is thus $G(1+1{\cdot}2/d-1{\cdot}2/b)$ kg.

Values of b for various materials are given in the table below. Note that these values are nominal: the volumes of the weights are adjusted so they have the overall densities shown. The value of b should be given with the weights. Most modern weights are of stainless steel. Corrections for single pan balances are made in the same way.

TABLE 2A.1 Values of nominal density (kg m^{-3}) for balance weights

Brass	8 400 ('scientific brass')
Brass	8 140 ('B.O.T. brass')
Stainless steel	8 000 (world except U.S.A.)
	8 400 (U.S.A.)
Nichrome	8 400 (usually, but subject to some variation)

It should be pointed out that although the above correction is adequate for most work it is not quite accurate enough if a precision of 0·1 mg in 50 g (100 cm^3 volume) is required, since the density of the air varies with temperature, pressure and humidity.

Because of the uncertainties associated with correction for buoyancy it is usual, when weighing large objects accurately, to employ a counterpoise of approximately the same size as the object. Vacuum corrections are then needed only for the small differences between the volumes of the objects on the two scale-pans. An example of this procedure is found in the determination of the density of a gas by the globe method (chapter 4).

Other precautions in accurate weighing

Appreciable errors are incurred if objects are not allowed enough time to reach the same temperature as the balance. A platinum crucible needs at least 20 min in a desiccator after heating and then 10 min in the balance, and larger objects may need considerably longer. Clearly, a trial should be made to find the minimum that will suffice.

The surface condition of large pieces of glassware is not easily reproducible, but wiping with a clean chamois leather is usually recommended, and heating by the warmth of the hand must be minimized. Needless to say, chemicals should always be weighed in a closed vessel. The balance pans may be dusted with a camel-hair brush.

Other methods for determination of weight

Various modifications of the ordinary balance are available for large or small masses; thus, an 'assay' balance may be sensitive to

0·01 mg and take a load of 2 g, and a modern electromagnetic micro-chemical balance will weigh 20 g with a sensitivity of 1 microgram.

Torsion wire balances are useful for rough weighing and for repetitive work with similar objects (e.g. weighing electric light filaments); loads up to 0·5 kg can be weighed with an accuracy of 0·1%. Torsion balances are found of use in measuring surface tension (chapter 6).

In research work it is sometimes necessary to carry out a weighing inside a glass apparatus, e.g. for studying adsorption of gases by solids; the methods used include the commercially available electromagnetic balance and the tungsten wire micro-balance of Gulbranson[1] (load 0·7 g, sensitivity 3×10^{-7} g). With modern forms of the quartz micro-balance[2] it is possible to detect the increase of weight accompanying adsorption of a *monolayer* of gas on a piece of copper foil. Various automatically recording balances have been developed for following changes of weight.

BIBLIOGRAPHY 2A: Determination of weight

Weissberger, *Physical Methods of Chemistry*, Vol. I, Pt 1, 1971 (Wiley-Interscience, New York).

[1] Gulbranson, *Rev. Sci. Instrum.*, 1944, **15**, 201.

[2] R. S. Bradley, *J. Sci. Instrum.*, 1953, **30**, 84; J. Rhodin, *Amer. Chem. Soc.*, 1950, **72**, 4 343.

2B Calibration of volumetric apparatus

1. UNIT OF VOLUME

By international agreement, the unit of volume is now defined in terms of unit length, i.e. 1 m^3. The litre, previously defined as the volume occupied by 1 kilogram of pure water, should strictly no longer be used. Although the kilogram was originally intended to be the mass of 1 000 cubic centimetres of water, it is now simply the mass of the 'Kilogramme des Archives' preserved in Paris. Since this mass is not exactly equal to the value originally defined, the definition of a litre and of a cubic centimetre are no longer directly related in theory. Experimentally 1 litre = 1 000·028 cm^3, or 1 m^3 = 999·972 litre.

Volumetric apparatus is nowadays calibrated in cm^3, and although the older term 'ml' (= 0·001 l) is occasionally used, it can be taken as synonymous with 'cm^3' for all ordinary purposes.

In order to determine a given volume, one determines the weight of a liquid, generally distilled water, required to fill the volume, the weight being reduced to vacuum (section 2A).

2. CALIBRATION OF MEASURING FLASKS

Volumetric glassware can be bought in two grades in Britain. The permissible limits of error in 'Grade A' and 'Grade B' apparatus have been laid down in a scheme prepared by the Metrology Division of the National Physical Laboratory.[1] The percentage error decreases with increasing size of vessel.

Although flasks made by the best makers will, as a rule, be found sufficiently accurate, no flask should be employed for very accurate work without being tested.

3. PIPETTES

Pipettes are calibrated by weighing the water which they deliver. In carrying out the calibration, however, several precautions must be observed if an accuracy of 0·05% is to be obtained. In the first place, it must be seen that the glass of the pipette is free from all greasiness, so that the water runs from the pipette without leaving drops behind. If necessary, therefore, the pipette must be thoroughly cleaned. This is best effected by filling the pipette several times with a warm solution of potassium dichromate strongly acidified with concentrated sulphuric acid, or the pipette may be left for some time full of the acid dichromate.

The liquid must not be sucked up by mouth unless a safety tube (Fig. 2B.1) is attached to the end of the pipette. Rubber bulb devices are available which are safer and easier to manipulate.

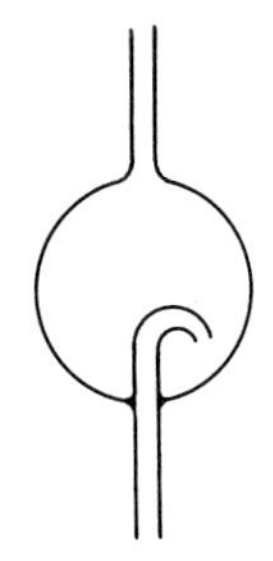

FIG. 2B.1 Safety tube for use when pipetting dangerous liquids.

Attention must be paid to the way in which the pipette is allowed to deliver, and also to the time of delivery. The pipette should be held upright or, in any case, not sloping at an angle greater than

45°, and the liquid allowed to run out freely. Immediately the liquid stops running, touch the point of the pipette momentarily against the side of the vessel, so as to remove the drop of liquid which collects at the point, and then withdraw the pipette. The pipette, also, must not be allowed to deliver too rapidly, otherwise varying amounts of liquid will be left adhering to the sides.

If a liquid other than water or aqueous solutions is to be pipetted accurately, the pipette must be calibrated specially with the liquid, as surface tension and drainage may differ.

For repetitive titrations automatic pipettes are available which deliver a fixed volume without the necessity of adjusting the level to a mark. In one design a cup contained in an outer container is filled by raising the level of liquid in the outer container above the level of the cup. The level is then lowered, and the contents of the cup discharged. This arrangement is particularly convenient when a large number of portions of the same liquid are required. Where portions are required from different liquids a syringe or piston pipette can be used. A hypodermic syringe fitted with a micrometer head to depress the piston is used to discharge very small accurately known volumes of liquid—e.g. for gas chromatography (see section 6D). An alternative—which can also be used for gases—is the precision capillary tubing syringe in which the piston is a length of closely fitting wire.

4. BURETTES

As with pipettes, so also in the case of burettes, attention must be paid to the rate at which the liquid is allowed to flow out. The minimum time of outflow advisable will depend on the volume of liquid delivered; for 30 cm^3 it should not be less than 40 seconds.

With burettes the accuracy to be attained is not in general as great as with pipettes. With a 30 cm^3 burette of internal diameter of 8–10 mm an accuracy of about 0·1% can be obtained on a reading of 10 cm^3.

Burettes can be individually calibrated if necessary by the Ostwald method: an auxiliary pipette of known volume (≈ 2 cm^3) is attached to the burette and repeatedly filled to the mark and discharged. The level of burette is noted each time. The difference between the change in volume indicated and that actually delivered can be plotted as a correction curve. However individual calibration is rarely justified. The accuracy of a titration can be increased by adding most of the titrant from a pipette so that the less accurate addition from a burette is reduced to a small fraction of the total.

Parallax. In reading a burette, it will be found that the apparent position of the liquid meniscus alters as the line of vision is raised or lowered. Owing to this '*parallax*' these readings are liable to considerable errors. These can be avoided most easily by placing a piece of mirror glass behind the burette, and then raising or lowering the line of vision until the meniscus and its image just coincide. The eye is then on the same level as the meniscus, and the error due to parallax disappears. The reading is then made with the eye in this position. Burettes can also be obtained on which the main divisions are engraved completely round the burette. With such burettes errors of parallax are readily avoided.

A related problem is the unambiguous observation of the bottom of the meniscus. The most reliable technique is to hold a white card *horizontally* behind the burette, slightly below the level of the meniscus. Light is reflected upwards from the card into the burette, and then is totally internally reflected from beneath the meniscus towards the observer.

It will be clear that the precautions to be taken in reading a burette should also be taken when using apparatus of a similar character, e.g. the barometer, eudiometer, mercury manometer, etc.

BIBLIOGRAPHY 2B: Calibration of volumetric apparatus

[1] National Physical Laboratory, *Volumetric Glassware (design, etc.)*, 1954 (H.M. Stationery Office, London).

2C Measurement of length

In addition to ordinary scales of boxwood, metal or glass (as used on mercury manometers, barometers, etc.), one frequently needs in physical chemistry instruments for measuring small lengths accurately. The following principal types cover most requirements.

Vernier calipers similar to those used in engineering workshops will measure solid objects up to about 10 cm long, with an accuracy of 0·02 mm. *Micrometer calipers* will read to 0·01 mm. Both instruments can be obtained in a range of sizes.

The vernier measuring microscope is an extremely useful instrument. The common pattern with a traverse of about 25 cm will measure objects horizontally or vertically, and can be read to 0·02 mm. The best instruments are readable to 0·01 mm. The usual working distance for a travelling microscope is about 2 in. A similar instrument furnished with a telescope instead of a microscope is called a *cathetometer*. In measuring vertical distances with it, as on a manometer, care should be taken to set the scale vertical with a plumb-line, and to set the telescope horizontal by means of the spirit-level which is attached to it.

Small objects are measured on the stage of a microscope. Most mechanical stages are fitted with a movement of about 50 mm and a vernier reading to 0·1 mm. Micrometer eye-pieces can also be obtained; these contain a cross-wire which can be moved by a micrometer screw and the position can be read to 0·01 mm.

However, the simplest method of measuring microscopic objects is by means of an eye-piece scale which is slipped into the ocular of the microscope. The effective magnification is obtained by comparing the graduations seen in the eyepiece with the fine lines engraved, or photographically recorded, on a glass graticule ('stage micrometer') which is examined through the microscope, using the same objective and ocular as used when measuring the object.

BIBLIOGRAPHY 2C: Measurement of length

See standard texts.

3
The measurement and control of temperature

3A Temperature measurement

1. THE FUNDAMENTAL SCALE OF TEMPERATURE

There are innumerable ways in which changes of temperature have been or could be measured; any physical property which is dependent on temperature and which is readily reproducible could be considered a potential basis for a 'thermometer'. However, no two substances or properties have *exactly* the same temperature-dependence, and consequently no two methods of thermometry lead to exactly the same scale of temperature. Temperature scales would therefore be entirely empirical and arbitrary if it were not for the existence of the absolute thermodynamic scale of temperature, devised by Kelvin (1851). The Kelvin scale is independent of the properties of any particular substance, being defined in terms of the efficiency of a hypothetical, thermodynamically reversible heat engine. Only the size of the degree is arbitrarily chosen by making the difference between the ice point and the steam point 100°C; the scale is defined by making the temperature of the triple point of water 273·16 K.

Practical methods of thermometry have to be related to the absolute scale to eliminate their arbitrariness. The link between the absolute and the practical scales is the gas thermometer. It can readily be proved that if a gas obeyed the ideal gas laws precisely, its volume at constant pressure (or pressure at constant volume) would be exactly proportional to its absolute temperature. Consequently, in principle it is only necessary to compare a practical thermometer—say the mercury-in-glass—with a gas thermometer to determine the magnitude of any corrections which must be applied to the former to make it conform to the absolute scale. Unfortunately, the gas thermometer itself is not convenient to use, and further, no actual gas obeys the ideal gas laws accurately. Many supplementary data are required to correct the readings of a real gas thermometer to the ideal gas or absolute temperature scale. Consequently, the work of calibrating more practicable thermometers (notably the platinum resistance thermometer) on the absolute scale has been undertaken by national laboratories for physical standards and, as a result of such work, there was set up in 1927 a list of six agreed 'basic' fixed points covering the range −183 to +1 063°C known as the International Temperature Scale. In addition, a number of 'secondary' points have been determined and recommended for calibrating instruments other than the platinum resistance thermometer. These fixed temperatures are shown in Table 3A.1, but reference must be made to the International report for the experimental procedures which must be followed to reproduce the various

TABLE 3A.1 The fixed points of the International Temperature Scale (1960 revision)

	°C (Int.)
Fundamental fixed points	
Ice point	0·01
Steam point	100·00
Primary fixed points	
B.p. liquid oxygen	−182·97
B.p. sulphur	444·6
F.p. silver	960·8
F.p. gold	1 063

temperatures accurately. It is of interest to note that above 1 063°C (the 'gold point') the gas thermometer becomes impracticable, and the optical pyrometer is taken as standard, assuming the validity of Wien's law of radiation.

2. METHOD OF MEASURING TEMPERATURE

Fortunately in physical chemistry the problems discussed above can be taken as satisfactorily solved; there are enough well-established fixed points on the International Temperature Scale to facilitate calibration of any practical thermometer with adequate accuracy. The problem in practical physical chemistry therefore reduces to the choice of the most convenient method of thermometry for the temperature range involved. Table 3A.2 summarizes some of the principal advantages and disadvantages of the chief methods of thermometry, and shows their useful range.

The platinum resistance thermometer is the most sensitive instrument available for measuring temperatures, but it is employed only in research work where the highest possible accuracy is needed. Similarly, accurate temperature measurement above 1 000°C is another specialized technique. Reference should be made to monographs on temperature measurement for further details of these methods.

In ordinary laboratory work temperatures in the range 0–1 000°C are almost always measured with either a mercury thermometer or a thermocouple. Both need calibration. Mercury thermometers can be calibrated conveniently by comparison in a well-stirred water-bath with a standard thermometer for which a certificate (e.g. from the National Physical Laboratory) is available. Similarly, a base-metal thermocouple can be calibrated against a reliable platinum thermocouple. The more general method, however, is to set up two or more suitable constant temperature baths of accurately known temperature, such as those in Table 3A.1; the readings of the thermometer in these baths are determined and a calibration chart is constructed.

3. MERCURY-IN-GLASS THERMOMETERS

These thermometers are very convenient provided absolute accuracy of about 0·1°C is sufficient. Higher accuracy can be obtained only with elaborate precautions to allow for the hysteresis of expansion or contraction of the glass, irregularity of bore, sticking of the meniscus when it is receding, and the influence of external pressure. Normally a calibration at the steam point (using a hypsometer) and the ice point (obtained using a mixture of pure ice and water in a thermos flask in which the thermometer is *suspended*) will show whether a thermometer is sufficiently reliable, and a third check at the transition temperature of sodium sulphate decahydrate (or of some other salt in Table A.2) can be added if temperatures some way from the fundamental interval are to be measured. (See section 8C for procedure.) Incidentally, the transition temperature of sodium sulphate should read 32·48°C on a perfectly accurate mercury thermometer rather than the International value of 32·38 (Table 3A.1). This is because the mercury-in-glass scale departs slightly from the absolute scale *between* 0 and 100°C, even although they are identical *at* these points; the departures in between depend on the thermometer glass, and reach a maximum of about 0·1°C at 40°C.

TABLE 3A.2

Method of thermometry	*Useful range (°C)*	*Advantages*	*Disadvantages*
Platinum resistance thermometer	−260 to +1 100	Very high accuracy can be achieved. Suitable for close temp. control	Construction difficult: auxiliary apparatus costly: size large
Liquid expansion thermometers			
(a) Mercury-in-glass (ordinary)	−30 to +350	Simple to use	Either range or accuracy limited
(b) Mercury-in-glass (hard glass, gas-filled)	−30 to +600	Simple to use	Accuracy poor at high temperatures
(c) Alcohol in glass	−110 to +50	Simple to use	Accuracy low
(d) Pentane in glass	−190 to +20	Convenient for low temperature baths	Accuracy low
(e) Mercury in steel	−30 to +500	Robust: Bourdon gauge indicator suitable for industrial use	Compensation needed for long emergent stem
Thermocouples			
(a) Copper–constantan	−250 to +400	High thermo-electric e.m.f. (10–40 μV/°C)	Deteriorates rapidly above 300°C: calibration needed
(b) Chromel–alumel	−250 to +1 100	Usable for short periods up to 1 300°C	Frequent calibration needed
(c) Platinum/13% rhodium–platinum	−100 to +1 500	Stable and reproducible: usable for short periods up to 1 700°C	High cost; low thermo-electric e.m.f. (4–12 μV/°C)
Thermistors	0–100 or more	Sensitive; physically small; useful for measuring small temperature differences	Non-linear calibration; not stable over long periods
Vapour pressure thermometers	Below room temperature	Sensitive and simple	Short range (*c.* 20°C) covered by each thermometric liquid
Radiation pyrometers			
(a) Disappearing filament optical pyrometer	+800 to $>$2 000	Covers range beyond Pt thermocouple with accuracy ±5°C	Calibration difficult: emissivity correction often needed
(b) Total-radiation pyrometer	+800 to $>$2 000	Robust, direct reading, suitable for recording	As for (a): less accurate

Special mercury thermometers with high sensitivity (large bulb, narrow capillary and short range) are used for measuring small changes of temperatures as in calorimetry (chapter 10) and for determinations of the freezing points or boiling points of dilute solutions. The Beckmann thermometer is described in chapter 7.

Emergent stem corrections. Thermometers are generally calibrated with the whole of the mercury thread immersed in a constant temperature bath, but they are often used for convenience with part of the thread exposed to the air and therefore generally at a lower temperature than that of the bulb. This introduces an error, but a correction can be applied as follows.

Suppose the temperature of the bulb is t_b, the mean temperature of the exposed part of the mercury thread is t_e, and the length of exposed mercury corresponds to n degrees on the scale. Had this exposed thread been heated to t_b also it would have expanded an additional number of scale divisions given by $n \times (t_b - t_e) \times \alpha$, where α is the apparent coefficient of expansion of mercury in glass, which is usually taken as 0·00016. In practice t_b is not known directly, but the uncorrected reading of the thermometer (t) is near enough; t_e is usually taken as the temperature registered by a thermometer with its bulb placed alongside the middle of the exposed stem. Then the corrected temperature is given by

$$t_{\text{corr.}} = t + 0{\cdot}00016n(t - t_e)$$

t = uncorrected reading: n = no. of degrees exposed: t_e = exposed thread temperature.

Emergent stem corrections may be considerable, particularly with high temperature thermometers, e.g. 20°C at 400°C. The correction should always be applied wherever it would have a significant effect on the result.

4. THERMOCOUPLES

The most widely used materials are the chromel–alumel alloys for normal work, or platinum–iridium. The latter gives a lower e.m.f. but the calibration is more stable and the metals are not attacked by corrosion. Thermocouples are readily constructed from the appropriate wires by soldering (copper–constantan), spot welding (chromel–alumel) or by melting in an oxy-coal gas flame. Figure 3A.1 shows schematically two methods of measuring temperatures with a thermocouple. In (a) the couple is simply connected by copper

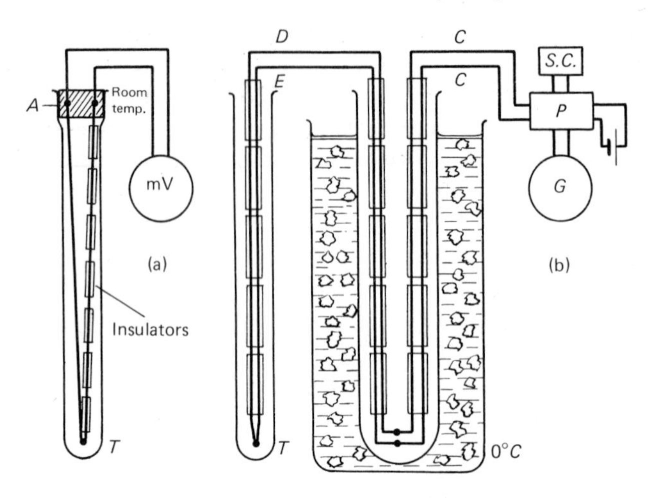

FIG. 3A.1 Measurement of temperature by thermocouple (schematic). (a) Direct method, using millivoltmeter. (b) Precision potentiometric method, with 'cold junction'. (*P*, potentiometer; *S.C.*, standard cell; *G*, galvanometer.)

leads to an accurate millivoltmeter, and the steady deflection of the needle is observed. In this arrangement the cold junction is somewhat indefinite, being at the point *A* where the copper leads are joined to the thermocouple wires. Also, the deflection of the millivoltmeter is dependent on its internal resistance and on the resistance of the leads. An accuracy of about 3°C in 1 000°C is obtainable.

Arrangement (b) is used where higher accuracy is needed. The 'cold junction' is made in a separate vessel, normally a Dewar vessel containing a mixture of ice and water. The e.m.f. generated by the thermocouple is determined by means of a sensitive potentiometer *P* similar in principle to that described for measuring the e.m.f. of voltaic cells (see chapter 4). A common type of portable thermocouple potentiometer is sensitive to about 0·01 mV and has a range up to 100 mV. This method, of course, determines the maximum (open circuit) e.m.f. of the couple, and is therefore independent of the resistance of the leads.

The calibration tables for thermocouples are generally expressed as the e.m.f. generated by the couple when the 'cold junction' is at 0°C. If, instead, the 'cold junction' is at room temperature, the e.m.f. obtained is correspondingly diminished, and a correction —the 'cold junction correction'—must be applied. Since the thermo-electric e.m.f. is not quite linear with temperature it is not accurate simply to add on the cold junction temperature. Instead, the e.m.f. generated by the cold junction is read from the tables, and this value is *added* to the observed e.m.f. to give the e.m.f. which would have been obtained had the cold junction been at 0°C. The corresponding temperature of the hot junction can then be read from the calibration tables. Table A6 in the Appendix gives typical e.m.f.–temperature tables for copper–constantan, chromel–alumel and platinum/13% rhodium–platinum couples.

5. THERMISTORS

Thermistors are small beads of a mixed oxide semiconductor to which are attached platinum leads; the beads, about 1 mm in diameter, are enclosed in a glass tube. Current is carried in semiconductors by electrons (*n* type) or by vacancies or 'holes' (*p* type). The number of carriers available varies exponentially with temperature and is proportional to $e^{-\Delta E/\boldsymbol{R}T}$ where ΔE is the energy gap between the normal energy of the carriers and the conduction band. The resistance of a thermistor thus *decreases* exponentially as the temperature is increased, typically from 100 kΩ at 0°C to 1 kΩ at 100°C. The resistance is measured in the usual way with a Wheatstone bridge (section 4C). Thermistors require fairly frequent calibration for accurate work but are very useful because of their small size and high sensitivity.

EXPERIMENT
Determine the energy gap of a thermistor.

Standard Telephone and Cables Ltd, Type *F* thermistors are suitable for this experiment. Measure the resistance as a function of temperature between 0 and 100°C using a Wheatstone bridge circuit. Control the temperature of the thermistor by a water bath, preferably thermostatted. Use ice to cool below room temperature. Measure the temperature of the bath with a mercury thermometer to which the thermistor may be attached with a rubber band. Do not allow the leads of the thermistor to become wet. Since the resistance R is (nearly) proportional to $e^{-\Delta E/\boldsymbol{R}T}$, a graph of $\log_{10} R$ against $1/T$ (where T is in K, not °C) should have a slope of $-\Delta E/2{\cdot}303\boldsymbol{R}$.

BIBLIOGRAPHY 3A: Measurement of temperature

Partington, Vol. I, section 6.
Weissberger, Vol. I, Pt 1, chapter VI.
Roberts, *Heat and Thermodynamics*, chapter 1, 4th edn, 1951 (Blackie and Son, London).

Griffiths, *Methods of Measuring Temperature*, 3rd edn, 1947 (Griffin, London).
Campbell (ed.), *High Temperature Technology*, 1956 (John Wiley, New York).

3B Temperature control—thermostats

As most of the properties studied in physical chemistry are affected by temperature, it is very important to have means whereby experiments can be carried out at constant temperature. Constant temperature baths, or *thermostats*, are therefore an essential part of the equipment of a laboratory for physical chemistry.

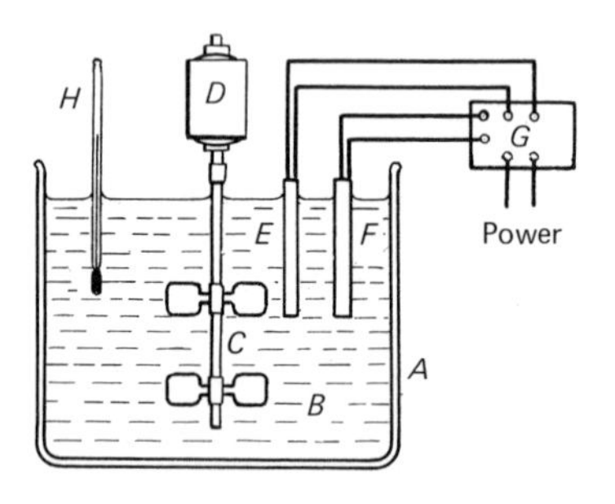

FIG. 3B.1 Components of a constant temperature bath (schematic). *A*, tank; *B*, bath liquid; *C*, stirrer; *D*, electric motor; *E*, heater; *F*, thermoregulator; *G*, relay; *H*, thermometer.

A useful range of fixed temperatures can be obtained by employing baths immersed in boiling liquids, melting solids, or salts at their cryohydric or transition temperatures (see Tables A1, A2, A3, in the Appendix). However, heated thermostat baths with automatic temperature control are more versatile and are indispensable when a constant temperature must be maintained over a long period. Innumerable methods have been devised for producing constant temperature; in this section a number of more commonly used pieces of apparatus will be described.

The usual components of a thermostat are shown diagrammatically in Fig. 3B.1.

1. THERMOSTAT BATHS

The tank *A* may be made of copper or galvanized or enamelled iron but stainless steel or Pyrex glass is better for a small thermostat. If readings must be made of immersed instruments such as manometers, one or more sides must be constructed of plate glass. Water is most often used as the bath liquid *B*; to prevent micro-biological growth a muslin bag containing mercuric chloride may be suspended in the water. For temperatures above about 40°C evaporation becomes serious, and a layer of liquid paraffin is sometimes poured on the water. A better method is to cover the surface with light hollow plastic spheres which are available commercially.

For temperatures between about 80 and 300°C a high-boiling liquid paraffin oil is generally used as bath liquid. At still higher temperatures molten salts or liquid metals (tin, lead) may be used, but wherever possible the liquid is dispensed with, and the object to be heated is simply placed in a furnace which is maintained at constant temperature (see below).

As liquids such as water are poor conductors of heat it is important to stir the bath liquid of a thermostat vigorously to maintain uniformity of temperature. A paddle rotated by an electric motor may be used; the commercial thermostat regulators for water baths described in the next section usually incorporate a stirrer.

2. TEMPERATURE REGULATION

To maintain a thermostat at a temperature above that of the surroundings, heat must be added to counteract that lost by radiation, evaporation, etc. Gas heating may be used for simplicity, but electric heating is preferable, and may be provided by metal immersion heaters (such as those used in small electric kettles), or glass heating lamps, or simply a helix of nichrome wire in a length of Pyrex glass tubing bent into the shape of a 'U'. In order to obtain automatic regulation of the temperature, a *thermoregulator* is immersed in the bath, its function being to switch on the heating when the temperature falls below the required value and switch it off again as soon as the required temperature is restored.

We shall not describe the arrangement of these devices in detail, because inexpensive commercial units are readily available capable of maintaining the temperature of a water bath constant to 0·01 K or even better (Fig. 3B.2). These often use a bimetallic helix immersed in the water as a thermoregulator: the differential expansion of the two metals causes the helix to rotate as the temperature of the water changes. The rotation is used to open or close a circuit which passes current through a relay: the contacts of the relay in turn switch the heating current. The units have a mechanical stirrer or a

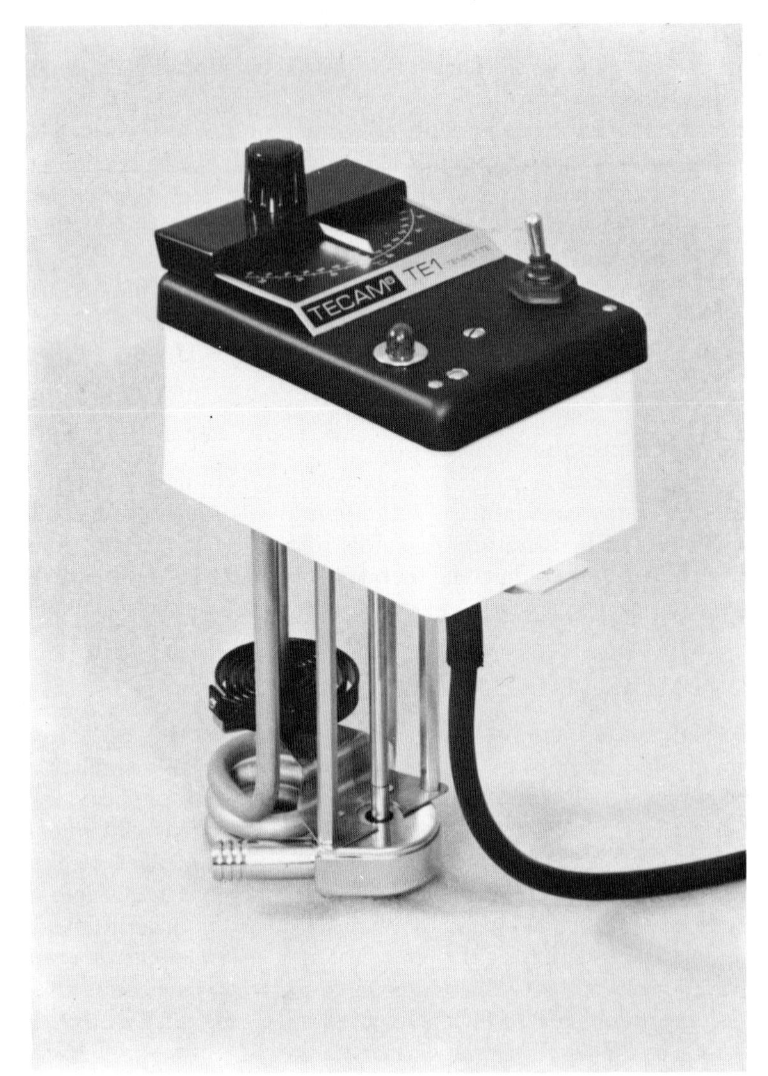

FIG. 3B.2 Water-bath temperature controller. (Courtesy of Techne (Cambridge) Ltd.)

powerful pump which circulates the water past the heating coil and mixes the contents of the tank.

The temperature of a thermostat bath should always be measured by a thermometer which is independent of the thermoregulator: the latter is not usually accurately calibrated even though it maintains a constant temperature.

3. CIRCULATION OF WATER

The maintenance of a constant temperature in an apparatus outside the thermostat (e.g. a refractometer) can be effected by circulating water from the thermostat by means of a pump. With some commercial thermostat units, the pump can be used to circulate water through external apparatus. Otherwise, the small electrically driven centrifugal pumps used for garden fountains are suitable provided the apparatus is placed on the suction side of the pump so that it receives water direct from the thermostat rather than water which has passed through the pump where its temperature may be altered. Centrifugal pumps, of course, need 'priming'. In the absence of a pump, it is possible to obtain a stream of water at constant temperature by flowing tap-water slowly through a large metal coil immersed in a thermostat of adequate heating capacity. Alternatively commercial units are available (e.g. 'Churchill') which recirculate water through an external apparatus, maintaining its temperature constant.

Again, the temperature of the apparatus should always be measured independently. It will fall below that of the circulating water if heat losses are excessive.

4. TEMPERATURES BELOW THAT OF THE SURROUNDINGS

Regulators have been designed which will switch on a source of cooling when the temperature of a thermostat bath rises above the temperature required. It is simpler, however, to provide a constant amount of cooling and to supplement this with intermittent heating controlled by a normal thermoregulator such as one of those described previously. Cooling can be effected conveniently by flowing tap-water or ice-cold water through a coil of lead or copper immersed in the thermostat, the water being run to waste afterwards. When the bath temperature falls too low, the thermoregulator switches on the supplementary heater. The flow of the cooling water can be regulated so that the heater operates only occasionally.

For temperatures somewhat below the freezing point of water, the bath liquid or liquid in the cooling coils may be a concentrated solution of calcium chloride or of ethylene glycol, and the cooling can conveniently be obtained by a small refrigerator unit.

Temperatures between 0 and $-78°C$ can be produced by adding solid carbon dioxide to a bath of alcohol, acetone or ether contained in a Dewar (Thermos) vessel. For still lower temperatures pentane cooled by liquid nitrogen can be used, and, if necessary, a low-temperature thermostat ('cryostat') can be constructed in which quantities of liquid nitrogen are delivered automatically to a pentane bath by a special thermoregulator.

5. FREEZING MIXTURES FROM ICE

The melting point of ice is lowered by addition of a solute to the liquid (chapter 7). Consequently, if powdered ice is added to say, brine, some of the ice melts and the temperature falls below 0°C. With very soluble salts it is possible to reach temperatures even below −50°C (e.g. −51°C with $CaCl_2.6H_2O$) but such mixtures, of course, have little reserve cooling power left. With moderately soluble salts, however, a useful and stable temperature is obtained by adding excess of the salt to powdered ice in order to produce the eutectic ('cryohydric') mixture of ice+salt+saturated solution. For common salt, for example, this temperature is −21°C and a number of other cryohydric temperatures for readily available substances are shown in Table A3 in the Appendix. Such freezing mixtures will hold a steady temperature as long as any ice remains unmelted, whereas ice+unsaturated brine mixtures slowly rise in temperature as the solution becomes diluted by the melting ice.

6. FREEZING MIXTURES FROM SOLID CO_2

The temperature of solid carbon dioxide when its sublimation pressure is 1 atm is −78°C, and this is therefore the normal temperature assumed by the solid, but it varies slightly, of course, with the barometric pressure. Lower temperatures can be obtained by means of reduced pressure. Low temperature baths are prepared by cooling a liquid (f.p. below −78°C) with powdered carbon dioxide; alcohol (methylated spirits) is generally used. Lumps of carbon dioxide are wrapped in a cloth and smashed with a hammer or crushed by a metal pestle and mortar. The powder is added little by little to a thermos flask half full of alcohol which is stirred with a stiff copper wire or glass rod stirrer. If too much carbon dioxide is added at a time, the effervescence causes the liquid to overflow. Liquid at −78°C will cause serious 'burns', so the initial addition should be made with caution.

Eventually a point is reached when excess carbon dioxide remains in suspension and the bath is then at −78°C. If it is to be kept at this temperature a considerable excess of powdered carbon dioxide can now be stirred in, and further quantities can be added from time to time as required. Temperatures between -78°C and room temperature can, of course, be obtained by using less carbon dioxide in the first place, but such baths will not hold a steady temperature. Instead it is better to use the carbon dioxide to freeze (partially) a pure organic liquid of suitable freezing point, as the solid-liquid equilibrium produces a constant temperature, heat conducted into the bath being taken up as the latent heat of fusion of some of the solid. Suitable liquids and their freezing points are listed in Table A1 in the Appendix.

7. TEMPERATURES IN THE RANGE 300–1 000°C

Above 300°C liquid baths become inconvenient. A number of fixed constant temperatures can be obtained by means of boiling liquids such as mercury and sulphur, but the most convenient form of heating is the electric resistance furnace. Such furnaces can be made readily in the laboratory. The heating element consists of nichrome resistance wire and this is wound on a tube or other support of ‘Pyrex’, fused silica, refractory fireclay, mullite or even iron. In the last case, the winding must be insulated from the iron by several layers of asbestos paper, applied moist. The nichrome wire should be spaced evenly to avoid local over-heating, but it is an advantage gradually to decrease the spacing of the turns towards the ends of the heated zone in order to counteract the cooling effect of the ends and thus improve the degree of uniformity of temperature along the tube. The ends of the wire are secured on the tube, and connected by treble-thickness leads to the terminals. The windings should be held in position by application of a thin layer of alundum cement (Thermal Syndicate, Ltd). Finally, the heater should be lagged by a layer of asbestos string, kieselguhr or calcined alumina powder. The thicker the lagging, the less sensitive will the furnace be to draughts, etc., but the longer it will need to reach a steady temperature. A well-lagged tube furnace with an element 1 ft long of diameter about $\frac{1}{2}$ in needs only about 0·5 kW to reach 1 000°C; for larger sizes, the power increases approximately as the area of the heated surface.

For research where the vessel temperature must be maintained absolutely constant along its length. the vessel is enclosed in a closely fitting aluminium block: the high thermal conductivity eliminates temperature gradients.

The temperature of the furnace can be adjusted by means of a large sliding rheostat connected in series with the furnace winding.

If the furnace is run from a.c. mains, the current can be regulated conveniently with a variable autotransformer such as the 'Variac' (Claud Lyons Ltd) which is less wasteful of power than a rheostat.

For accurate work, however, a furnace temperature regulator can be obtained commercially (e.g. from Chamberlain and Houkham, Motherwell, Lanarkshire, Scotland). The resistance of a platinum resistance thermometer placed in the furnace is compared with a standard resistance in a Wheatstone bridge circuit. The out-of-balance current is proportional to the difference between the temperature of the furnace and the temperature to be maintained, and is one example of an *error signal*. The current through the furnace windings is controlled by a semiconductor device known as a thyristor. This conducts current only when both (i) a positive voltage is applied and (ii) a positive trigger voltage is then applied to a control grid. These trigger signals are applied at an interval after the start of each positive cycle of the a.c. mains (Fig. 3B.3).

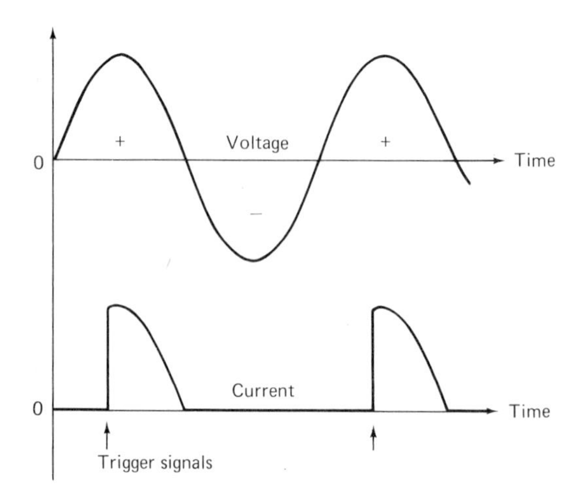

FIG. 3B.3 Voltage–current curves for a thyristor.

By varying the delay before the trigger fires the thyristor, the fraction of cycle for which current is passed, and hence the average current through the furnace, can be varied over a 4:1 range. This delay is controlled by the error or out of balance signal. The important feature of this type of control is that the correction applied to the furnace current is proportioned to the temperature error. This ensures that oscillation of the furnace temperature due to the time lag between a change of current and the corresponding change in temperature are reduced to a minimum. Also there is no dissipation or loss of power in the control circuits as there would be using a simple rheostat.

BIBLIOGRAPHY 3B: Temperature control

Partington, Vol. I, section 6 (1).
Weissberger, Vol. I, Pt 1, chapter 1.
Griffiths, *Thermostats and Temperature-regulating Instruments*, 3rd edn, 1951 (Griffin, London).
Campbell (ed.), *High Temperature Technology*, 1956 (John Wiley, New York).

4
Electrical measurements

4A Transducers

In the physical sciences many measurements of physical quantities are now made electrically. The device which converts the physical property into an electrical signal is called a transducer. We shall first consider some examples.

1. *Length.* Small movements of an object can be detected by attaching a small mirror. Light from a telescope (or better the highly collimated beam from a laser) is reflected off the mirror, partially obstructed by a knife edge and detected by a photocell or photomultiplier (see below). Movement of the mirror deflects the beam on or off the knife edge, and varies the photocurrent.

2. *Force.* Force due to weight or pressure can be detected by a strain gauge, which is a fine resistive network on a flexible support. Deformation of the support changes the geometry, and hence the electrical resistance of the network, which forms one arm of a Wheatstone bridge.

3. *Temperature.* The thermistor described above is one example of a temperature transducer; the thermocouple is another.

4. *Light.* Light can be detected by a vacuum photocell. This consists of a photocathode covered by a metal oxide and an anode enclosed in an evacuated glass tube. Light striking the photocathode causes photoemission of electrons: these are collected at the anode which is maintained at 50–100 V positive with respect to the cathode. The photocurrent is measured on a microammeter.

The spectral response of photocells is best expressed in terms of milliamperes of current passed by the cell when a flux of 1 watt of monochromatic radiation falls on the cathode. Figure 4A.1 shows that a cathode composed of a thin film of oxidized caesium on silver has a maximum of sensitivity in the near infra-red and another in the near ultra-violet. The potassium cell (curve *B*) can be used for the

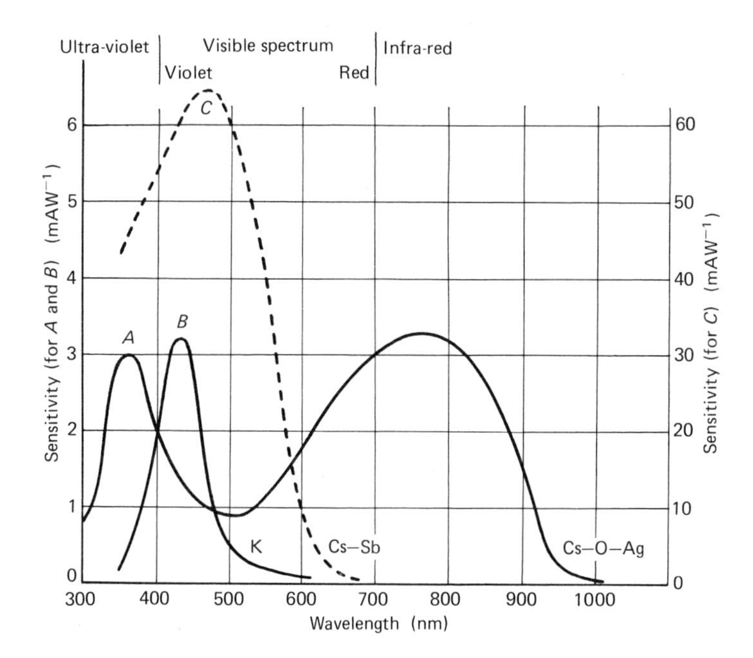

FIG. 4A.1 Approximate spectral sensitivity graphs for some emission photocells.

blue end of the spectrum, but it is now being displaced by the thin-film caesium-on-antimony cell which has a much higher sensitivity in this region (curve *C*). A sodium cell has a maximum response at about 300 nm and is therefore used in a cell with a quartz window for measurements in the ultra-violet.

The photocell suffers from a low sensitivity, and has been replaced for many applications by the photomultiplier; this is a photocell with a series of about ten additional electrodes (dynodes) maintained at about 100 V, 200 V, 300 V, . . . positive with respect to the photocathode. Photoelectrons from the cathode hit the first dynode, and more electrons are released by secondary emission. This is repeated up the chain, multiplying the original photocurrent many thousands of times. The current is collected, as before, at the anode which is kept about 100 V above the last dynode. The great advantage of the photomultiplier is that the amplification can be varied over many orders of magnitude by altering the dynode voltages. If this is not needed, an inexpensive but sensitive alternative is the selenium photoconductive cell whose resistance drops when light falls on it. It is simply connected in series with 100 V d.c. (e.g. from a battery) and a microammeter.

Further examples of transducers will be found in the remaining chapters of this book. In general transducers produce only a feeble current or voltage; electrochemical experiments also require accurate measurement of small voltages. We shall consider the measurement of current first.

BIBLIOGRAPHY 4A: Transducers

Weissburger, *Physical Methods of Chemistry*, Vol. I, Pt 1B, chapter VI, 1971 (Wiley–Interscience, New York).

4B Measurement of current

Very accurate measurements of current over a long time period can be made using a coulometer with two copper or silver electrodes immersed in a solution of their salts. The change in weight of the electrodes after a period of electrolysis multiplied by the number of ionic charges and divided by the atomic weight of the metal is equal to the charge passed in moles. Multiplication by 96 487 ($C\ mol^{-1}$) and division by the time for which current has passed gives the average current. These devices are used in very accurate electrochemical work (e.g. the determination of transport numbers), but are too cumbersome for ordinary use.

Currents above 1 μA can be measured using a normal moving coil ammeter, though the response to change in current tends to become slow for the more sensitive instruments. Convenient electronic microammeters are available commercially which will measure down to 10^{-8} A or less. The current passes through a resistance, and the voltage developed is amplified and displayed on a milliammeter. These have a rapid response, and are also fully protected against overload. The same instrument, of course, serves equally for the measurement of small voltages.

Another class of sensitive ammeter which is much used, especially for the detection of the position of balance for potentiometric work, is the mirror galvanometer. A small mirror is attached to the coil of a sensitive moving coil movement: the image of a crosswire is projected from a lamp onto the mirror and thence onto a translucent scale. By interposing repeated reflections the light path can easily be increased to 1 m, and the optical lever magnifies the sensitivity of the instrument: a deflection of up to 1 m per μA can be attained. Figure 4B.1 shows the 'Scalamp', a typical modern mirror galvanometer (Messrs Pye Instruments Ltd).

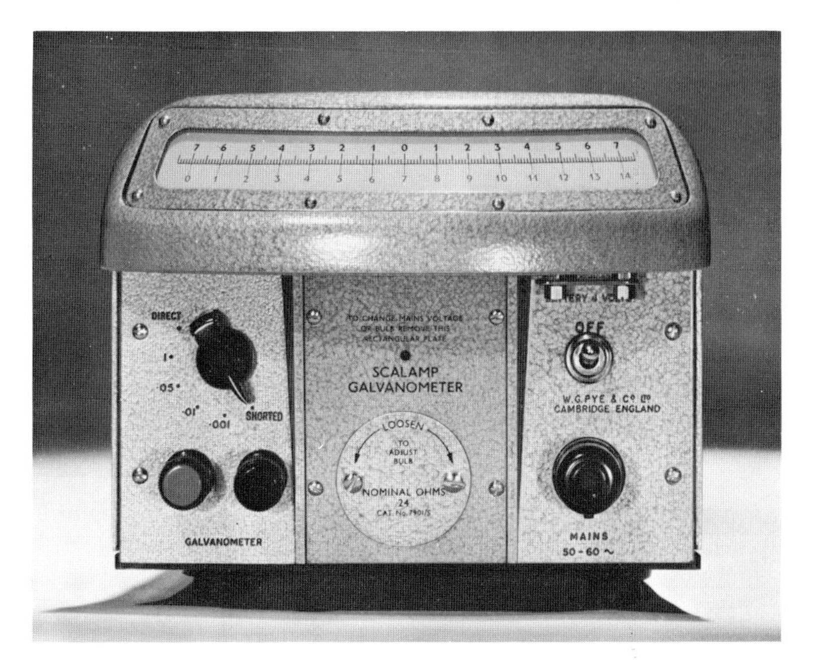

FIG. 4B.1 'Scalamp' mirror galvanometer. (Courtesy of Pye Unicam Ltd, Cambridge.)

The ultra-violet recorder is a version of the mirror galvanometer used to record a rapidly varying current for long periods of time. The source of light is a quartz lamp; the image of a pin hole is focussed onto sensitized paper instead of onto a scale. Passage of the paper records the value of the current as a function of time.

Currents can equally be converted into a voltage by passage through a resistor, and then measured by the techniques described in the next section.

BIBLIOGRAPHY 4B: Measurement of current
See bibliography to 4C.

4C Measurement of voltage

1. POTENTIOMETRIC CIRCUITS

We begin by considering the methods used to measure voltages which are not changing with time, and which must be measured without drawing an appreciable current, as for the e.m.f. of cells.

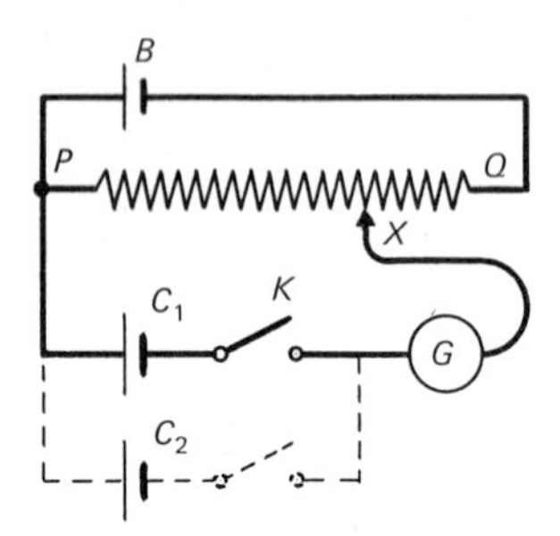

FIG. 4C.1 Basic principle of the potentiometer.

A steady external source of potential is adjusted until no current flows when it is connected in opposition to the cell. The basic principle of the potentiometer can be understood from Fig. 4C.1. A cell of large capacity, B, discharges through a calibrated resistance PQ which carries a movable contact, X. It follows from Ohm's law that, provided no current passes through the contact X, there is a uniform potential gradient (in volts per ohm) through PQ, and the potential difference across any section of it is proportional to the resistance of that section. The cell C_1 is now connected as shown, with like poles of B and C_1 connected to P. G is a sensitive galvanometer. When the key K is momentarily closed, current passes in one direction or another—as indicated by G—unless the contact X is at the precise point on PQ at which the potential equals that of C_1. Provided the e.m.f. of B exceeds that of C_1 a balance point can be found by moving X until on closing K no current passes. The potential drop along PX is then equal to the e.m.f. of the cell.

However, the e.m.f. of the working cell, B, is not accurately known, and therefore use is made of a reference or 'standard' cell, C_2, of known e.m.f. Such a cell can be preserved without change of e.m.f. because it is not called upon to provide more than an infinitesimal current. A new balance point, X', is now found for C_2 exactly as for C_1, and hence,

$$\frac{\text{e.m.f. of } C_1}{\text{e.m.f. of } C_2} = \frac{\text{resistance } PX}{\text{resistance } PX'}$$

A crude potentiometer resistance PQ can be made from a metre of resistance wire along which a sliding metal contact moves; relative resistances PX and PX' are then read from the *lengths* PX and PX'. In practice more accurate and convenient direct-reading potentiometers are used; these are described in the next section. These in turn are now being superseded by the digital voltmeters described later in section 4C.3.

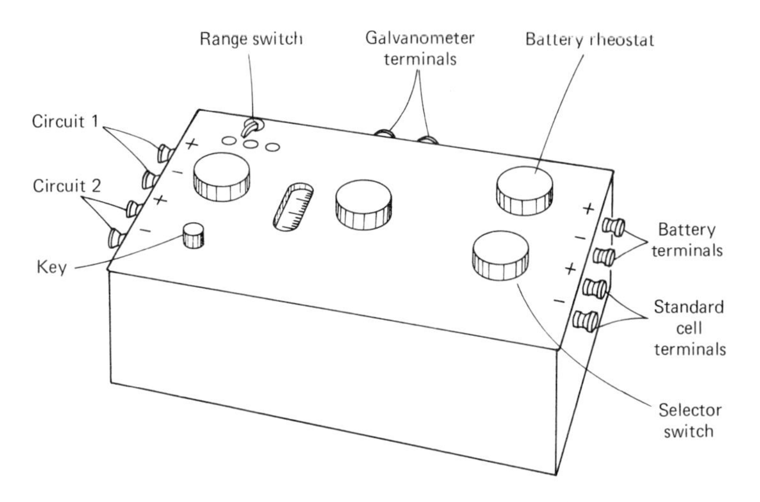

FIG. 4C.2 General purpose laboratory potentiometer.

2. THE POTENTIOMETER. Figure 4C.2 shows a typical laboratory potentiometer suitable for the measurement of the e.m.f. of cells. The methods of operation can be best understood by consideration of the circuit diagram (Fig. 4C.3).

The metre wire of the simple potentiometer (Fig. 4C.1) is replaced by a series of 18 2-ohm standard resistors R_1 and a circular 2-ohm slide-wire R_2 graduated at every 0·5 mV. The potential across $R_1 + R_2$ is first adjusted to be exactly 1·9 V. This is effected by means of the auxiliary standardization circuit (shown by the broken lines in Fig. 4C.3), by means of which the standard cell C can be balanced across the ratio arms R_3 and R_4. The rheostat R_5 is adjusted until, on closing the key S_3, no deflection is produced on the galvanometer. When this adjustment has been made, the potential across R_3 must equal that of the standard cell, 1·0186 V, and the potential across $R_1 + R_2$ is then known to be 1·9 V since $R_3 : (R_3 + R_4) = 1{\cdot}0186 : 1{\cdot}900$. In making a measurement of the e.m.f. of an external source, the selector switch S_2 is put into the position shown and a balance for no deflection is obtained by adjusting R_1 and R_2. The e.m.f. of the external source is then given by the sum of the dial readings on R_1 and R_2.

The range switch S_1 is provided to allow the instrument to measure smaller e.m.f. values with corresponding accuracy. Whereas the operation of standardizing the instrument always sets a potential of 1·9 V between the points A and B, the network of resistors R_5, R_6, R_7 and R_8 is prearranged to produce, 1·9, 0·19 or 0·019 V across

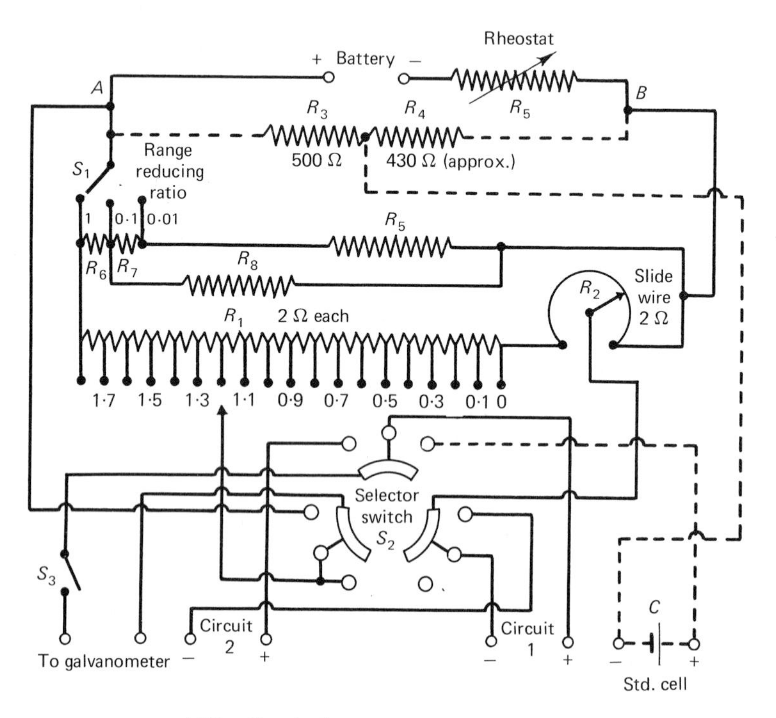

FIG. 4C.3 Circuit diagram of laboratory potentiometer.

$R_1 + R_2$ according to the setting of S_1. (The network is employed so that the total resistance shunted across AB is the same for every position of S_1; the standardization is therefore unaltered.) The greater sensitivity is useful in studying the temperature coefficient of e.m.f. of cells or the small e.m.f. produced by thermocouples (section 3A.4).

Several other potentiometers of slightly different design are also available, but the only one that will be mentioned here is the very convenient portable type in which the standard cell, galvanometer and battery are all housed in the same case. The principle of operation is similar to that of the instrument already described.

2.1. *The working cell.* The theory of the potentiometer circuit depends on the assumption that a steady source of current is available which will not alter between the operation of standardizing the instrument and that of balancing the cell on test. In practice, this requirement is most satisfactorily met by a single cell lead accumulator of capacity not less than 40 amp-hours. It is advisable to reserve a number of accumulators specifically for potentiometry. They should be kept well charged (2·1–1·9 V) and never subjected to heavy discharge. On the other hand, a freshly-charged cell is less satisfactory than one that has stood for a day after charging. The cell should be connected to the potentiometer about 10 min before final readings are required, as this makes for steadiness. Portable potentiometers generally have dry cells.

2.2 *The galvanometer.* The most satisfactory instrument for indicating the balance point in accurate potentiometry is the self-contained reflecting galvanometer unit described above, Fig. 4B.1, or an electronic micro-ammeter or null indicator. It is convenient to have ample sensitivity available for the final balance point, and yet be able to make the preliminary adjustments with a much reduced sensitivity which is instantly obtainable by rotating the contact knob of the adjustable shunt. The use of a low sensitivity at the beginning not only protects the galvanometer from damage, but also makes it easier to see whether the adjustments of potential are being made in the right direction.

2.3. *The standard cell.* The Weston standard cell[1] is used as a standard source of potential (Fig. 4C.4). It possesses the advantages not only of being easily reproduced, but also of having a small temperature coefficient of e.m.f. In this cell, mercury forms the positive pole and a cadmium amalgam containing 12·5% of cadmium, the negative pole. The electrolyte is a saturated solution of hydrated cadmium sulphate, $3CdSO_4.8H_2O$. The e.m.f. of the original, neutral, standard cells was defined as 1·01830 international volts. Commercial manufacturers of cells generally make the so-called 'acid cells' which are more stable.[2] These usually have an e.m.f. of 1·01859 V at 20°C and a temperature coefficient of -40 μV per 1°C rise of temperature. Standard cells must, of course, be kept upright.

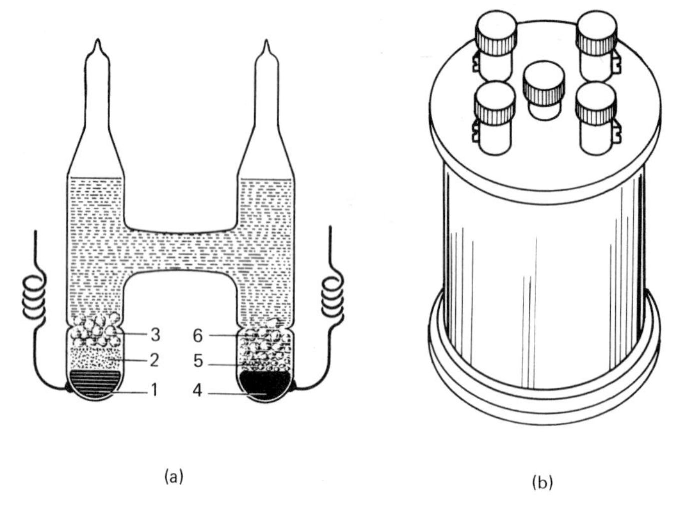

FIG. 4C.4 Weston standard cell. (a) Construction of cell. 1, Mercury; 2, paste of cadmium sulphate and mercurous sulphate; 3, 6, large crystals of cadmium sulphate; 4, cadmium amalgam; 5, small crystals of cadmium sulphate.
(b) Pair of mounted cells.

2.4. The voltage from a standard cell or indeed from any cell changes if an appreciable current is allowed to pass. For this reason the potentiometer key must only be depressed momentarily. If the opposing voltage is lower than that of the cell, the current drawn causes the cell voltage to fall, and hence move away from the correct balance point. Conversely if the opposing voltage is above that of the cell, the cell voltage rises, and again moves away from the balance point. Consequently attempts to increase the movement of the galvanometer by increasing the time the key is depressed are doomed to failure, as the cell voltage always moves away from the correct value. The movement of the potentiometer dial over which no deflection of the galvanometer spot can be observed when the key is momentarily depressed indicates the limits of the accuracy of the measurement.

3. DIGITAL VOLTMETERS (dvm)

There seems little doubt that these convenient, if expensive, instruments will replace the classical potentiometer over the next few years. An example of a digital voltmeter suitable for all but the most precise work is shown in Fig. 4C.5. Any voltage between 1·2 and 1 200 V can be measured to four significant figures, i.e. to 0·1 mV

FIG. 4C.5 Digital voltmeter. (Courtesy International Electronics Ltd, Lancashire.)

on the lowest range. For research purposes meters with up to seven significant figures are available, sensitive to 10 μV in the last digit.

The advantage of the digital voltmeter over the potentiometer are (i) digital display, avoiding reading errors, (ii) rapid balancing—about 0·2 s—so the meter can be used for varying voltage, (iii) negligible load, i.e. current drawn from the external circuit. A typical input impedence is 10 MΩ, giving only an 0·1 μA drain from a 1 V cell. Consequently the meter can be left attached to a cell to give a continuous reading, making it easy to see if the e.m.f. is drifting perhaps because of changes in the temperature of the cell.

The voltage applied to a digital voltmeter is standardized, that is to say amplified or reduced so that it falls between, say, 0 and 10 V irrespective of the range setting. The meter is basically a clock which counts the cycles of a quartz crystal oscillator. The clock is turned on at the start of the measurement, and at the same time a current is allowed to flow into a large condenser. When the voltage across the condenser is equal to the standardized input voltage the clock is turned off. The number of counts displayed is proportional to the time required to charge up the condenser, and hence to the input voltage. Most digital voltmeters calibrate themselves automatically against an internal standard cell, but it is advisable to check them from time to time against an external standard.

4. THE POTENTIOMETRIC RECORDER

In many experiments we need a graphical representation of a voltage as a function of time, e.g. the cooling curve of a liquid using a thermocouple (sections 8D.2, 4), the output from a gas chromatograph detector showing the peaks due to different species emerging from the column (section 5A), or in kinetic measurements where the concentration of a species is followed spectrophotometrically with time (e.g. section 9D.2). The laborious recording and plotting of galvanometer readings is eliminated by the use of the potentiometric recorder, which, like other electronic devices, draws negligible current from the external circuit.

An example is shown in Fig. 4C.6. This recorder accepts voltages from a fraction of 1 mV in ranges up to 100 V; these are then standardized. The pen moves from side to side as the paper is drawn underneath at a speed which can be varied from 1 to 10^{-3} cm s^{-1}.

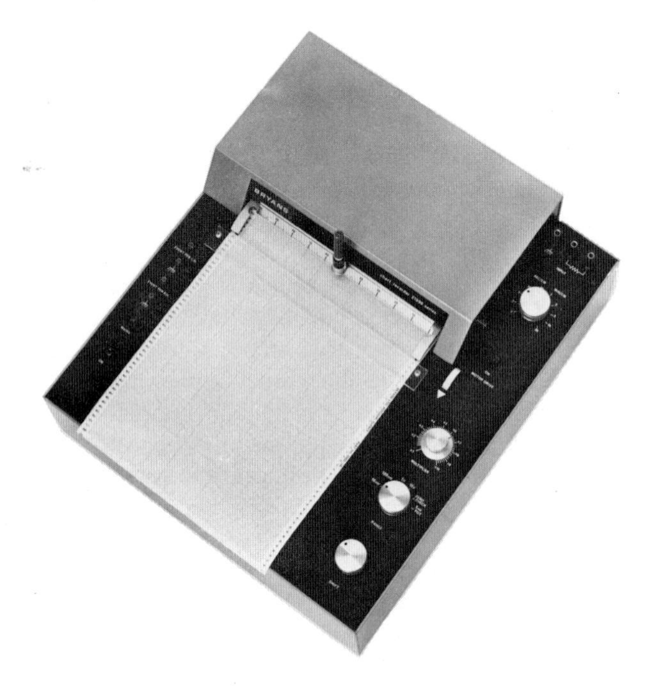

FIG. 4C.6 Potentiometric recorder. (Courtesy Bryans Ltd, Surrey.)

The pen carriage draws a contact along a slide wire which acts as a potentiometer: the difference between the voltage at the contact and the standardized input voltage is an *error signal* for the position of the pen, and is used to control the movement of the pen by an electric motor in such a way that the error signal is reduced to zero, thus correctly positioning the pen.

5. THE OSCILLOSCOPE

If the voltage observed changes too rapidly for the potentiometric recorder, i.e. in a time of 1 s or less, it can be displayed on an oscilloscope. This has a cathode-ray tube in which a narrow beam of electrons is accelerated to strike a phosphor on the face of the tube; this phosphor emits light. The beam of electrons can be deflected electrostatically in the Y direction (Fig. 4C.7) by the applied voltage (after

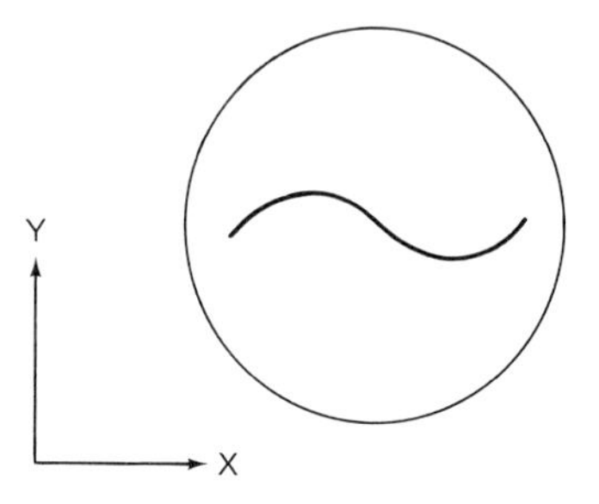

FIG. 4C.7 Cathode-ray tube display.

standardization and amplification), so the position of the spot in the Y direction indicates the applied voltage. The spot is repeatedly deflected from left to right at a constant speed by the time base; the sweep rate can be anything between 0·1 and 10^7 cm s^{-1}. The start of the sweep is initiated by the voltage change observed: either single events (single sweeps) or repetitive waveforms can be displayed. The display is normally recorded photographically; for this purpose Polaroid film which develops itself in a few seconds is particularly convenient.

BIBLIOGRAPHY 4C: Measurement of voltage

Weissberger, Vol. I, Pt 1, chapter II.
Weissberger, *Physical Methods of Chemistry*, Vol. I, Pt 1B, chapter VIII, 1971 (Wiley–Interscience, New York).
Delaney, *Electronics for the Physicist*, 1969 (Penguin, London).
Hywel White, *Elementary Electronics*, 1967 (Harper, New York).

4D Measurement of resistance and capacitance

THE WHEATSTONE BRIDGE

Ordinary ohmic resistance is measured using the Wheatstone bridge circuit (Fig. 4D.1). The unknown forms one arm of the bridge (R_1), a resistance box a second arm (R_3), and two fixed resistances the

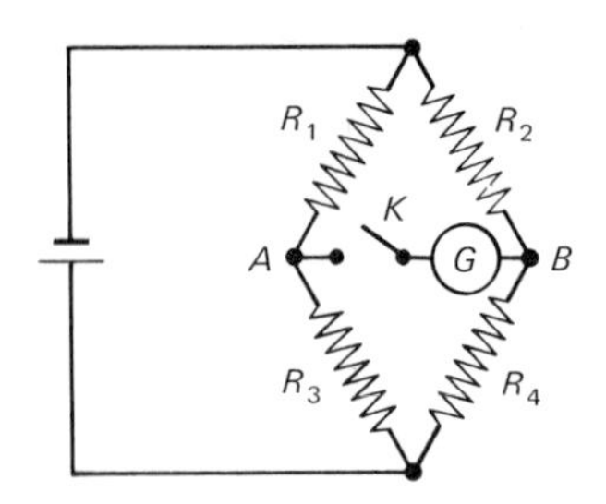

FIG. 4D.1 Schematic circuit for Wheatstone's bridge.

other two arms (R_2 and R_4). R_3 is adjusted until momentary depression of the telegraph key K produces no deflection of the sensitive galvanometer G. At this point there is no potential difference between A and B so $R_1/R_3 = R_2/R_4$. The bridge is most sensitive for $R_1 \approx R_2 \approx R_3 \approx R_4$, but this is not essential. Normally fixed resistances R_2 and R_4 are arranged so that ratios $R_2 : R_4 = 1:100$, $1:10$, $1:1$, $10:1$, $100:1$ can be switched into the bridge as required.

The resistance of conductance cells must be measured using alternating current to avoid electrolysis. In this case the accumulator is replaced by a signal generator—an a.c. source of variable frequency—and the galvanometer by earphones, or better by an electronic voltmeter or by an oscilloscope. Each arm of the bridge has an effective capacitance in parallel with each of R_1, R_2, R_3 and R_4. To observe the balance point both the resistive bridge and the capacitative bridge must be balanced simultaneously. It is thus necessary to provide variable capacitances in parallel with R_3 and also possibly fixed capacitances in parallel with R_2 and R_4. The bridge is balanced by alternately adjusting R_3 and its parallel capacitance. For a.c. work, however, the rather cumbersome procedure can be avoided by using a commercial electronic universal bridge which can be used to measure any combination of resistance, capacitance or inductance in an external circuit.

BIBLIOGRAPHY 4D: Measurement of resistance and capacitance

See standard textbooks on electricity.

5
Density of gases and vapours

Determinations of the density of gases and vapours have played an important part in the history of chemistry because the molecular weight of a substance in the gaseous state can be calculated from its density. This method of determining molecular weight is rarely needed nowadays because the molecular weights of stable molecules are well established. The determination of molecular weight has been supplanted by other forms of analysis as a method of identifying molecular species. However, the determination is still useful for less stable species, and for analysis of gas mixtures, e.g. of isotopic mixtures.

5A Gas density

1. The *absolute* density of a gas is the mass of unit volume, at the stated temperature and pressure. The *standard* density is the absolute density at 0°C and 1 atmosphere pressure (i.e. the pressure exerted by a column of mercury 760 mm high at 0°C and at sea-level in latitude 45°: $1{\cdot}0133 \times 10^5$ N m^{-2}). In connection with molecular weight determinations, *relative* gas densities are often used; these are the ratio of the density of the gas to that of a reference gas—usually hydrogen or oxygen—at the same temperature and pressure. Since Avogadro's hypothesis states that equal volumes of gases at the same temperature and pressure contain equal numbers of molecules, the relative gas density is equal to the ratio of the molecular weights of the two gases.

The ideal gas law is:

$$PV = \frac{m}{M} RT,$$

where P is the pressure, V the volume, m the mass, M the molecular weight, $\boldsymbol{R}$ the gas constant and T the absolute temperature. The introduction of the kilogram instead of the gram as the SI unit of mass has led to a slight complication. The gram molecular weight, the mass of 1 mole in grams, and the relative molecular mass, the mass of 1 molecule relative to $C^{12} = 12$, are all numerically equal. However, the molecular weight in SI units, i.e. in kg, is reduced by a factor of 1 000; m of course is also expressed in kg.

All gases deviate to some extent from the ideal gas law, so Avogadro's hypothesis is not *accurately* correct under ordinary conditions, nor would 1 mole of every gas occupy at s.t.p. *exactly* 0·022414 m^3 which is the volume occupied by an ideal gas under these conditions. Ordinary gases such as O_2, N_2, H_2, etc., deviate by less than 1% from the ideal gas law at 1 atmosphere pressure, but easily condensible gases such as CO_2 and NO_2, and, above all, *vapours* (i.e. gases at temperatures below their critical temperature) show considerable deviations. Consequently, normal gas or vapour densities give only approximate values of molecular weights. Where the highest accuracy is required, as in atomic weight work, either a more accurate equation of state is employed (for example, the van der Waals equation) or, better, use is made of the fact that deviations from the ideal gas law diminish as the pressure is reduced. Thus, in the 'method of limiting densities', the relative density of the gas compared with oxygen is determined at a series of different pressures, and the value for zero pressure is obtained by extrapolation. This method eliminates all errors due to non-ideality of the gases.

2. METHODS OF DETERMINATION OF GAS DENSITY

The oldest and still the most convenient method of determining the density of a gas is to weigh it in a globe of known volume (Regnault, 1847). With proper precautions a very high accuracy can be achieved, as in the classic work of Lord Rayleigh[1] which led to the discovery of the inert gases.

The chief difficulties encountered in Regnault's method arise from the necessity of weighing large glass vessels (see below). This operation can be avoided if the gas can be weighed in some condensed form either before or after its pressure is measured at controlled temperature in a large vessel of known volume. For example, the weight of a considerable volume of CO_2 could be obtained from the loss of weight of a sample of pure calcite from which the gas was obtained by heating. Alternatively, the CO_2 could be absorbed and weighed in a soda-lime tube. In the hands of the Spanish chemist, Moles, an accuracy approaching 1 in 10^5 has been achieved with this '*volumeter method*'.

The third method, now the most often used, is the *gas buoyancy micro-balance*, which depends on the principle of Archimedes. The apparatus usually takes the form of a small closed silica bulb which is attached to one arm of a light torsion- or other micro-balance, the whole apparatus being enclosed in a vessel to which the gas can be admitted. The buoyancy of the bulb depends on the density, and hence pressure, of the gas. The pressure (p_A) is found at which the pointer of the micro-balance coincides with some arbitrary mark, and the gas is then replaced by a reference gas (usually oxygen); the pressure is again adjusted until the same position of the balance is restored. A different pressure (p_B) is, of course, required, but the density of the first gas at pressure p_A must be equal to the density of the reference gas at pressure p_B. This gives the required gas density, assuming the ideal gas law to be applicable since

$$\frac{\text{standard density of gas } A}{\text{standard density of gas } B} = \frac{\text{pressure of gas } B}{\text{pressure of gas } A}$$

Alternatively, the method of limiting densities may be employed for accurate determination of molecular weights.

EXPERIMENT

Determine the absolute density of carbon dioxide (or dry air) by the globe method.

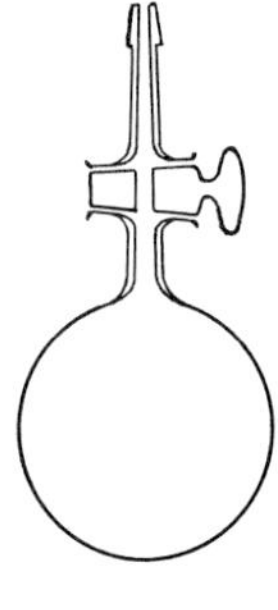

FIG. 5A.1 Gas density globe.

The glass globe for this determination (Fig. 5A.1) should have a volume of about 500 cm^3, and carry a glass stopcock preferably of the hollow-key type used for vacuum work. The stopcock should be close to the bulb and attached by fairly stout-walled tubing to reduce danger of breakage; also, the end of the tube may be drawn out a little to facilitate attachment and removal of rubber pressure tubing.

The first step is to determine the volume of the globe. This is obtained by weighing it evacuated and then full of water. It is first evacuated by means of an oil or water pump and then weighed. The tap is then opened while the end of the tube dips under the surface of a quantity of recently boiled distilled water. When the globe is full, it is placed in a water-bath, the tap being left open until the globe has taken the temperature of the bath, which is noted. The tap is then closed so that its bore remains full of water, and the tube above the tap is dried by means of filter paper. The globe is weighed to 10 mg on a large balance. The difference between the weight of the exhausted globe and that of the globe full of water gives the true weight of the water since the buoyancy acting on the globe is the same for both weighings and the buoyancy correction for the weights is negligible in the present instance. The density of water at the temperature at which the globe was filled is read from Table A4, and hence the volume of the globe is calculated.

Next, the globe is emptied by taking out the key of the tap and removing the water by a narrow glass jet connected with a filter pump. It is a good plan to remove grease from the tap at this stage and scrupulously clean the globe inside and outside with chromic acid mixture followed by plenty of distilled water. Thereafter the globe itself should not be handled by the fingers, but always with a clean chamois leather cloth or held by a piece of stout copper wire which is attached to the tap and bent to form a hook by which the globe can be hung on the balance. The inside of the globe is dried by rinsing several times with alcohol and then with ether, the last drops of ether being removed by repeated evacuation at a water-pump. The key of the tap is given a *thin* smear of Apiezon vacuum grease, and the globe is then ready for the main determination.

An arrangement such as that shown in Fig. 5A.2 is convenient for filling the globe with carbon dioxide. The mercury manometer H is intended to verify the degree of vacuum produced by the pump and to act as a safety valve from which excess pressure of gas can escape. It also serves as an indicator of the presence of any leakages in the apparatus. If present, these must be located and dealt with before proceeding with the determination.

The globe A is evacuated to the best vacuum obtainable and then weighed as accurately as possible after it has been wiped with the chamois leather and has stood long enough in the balance case to reach thermal equilibrium. (Any leakage of the stopcock will be

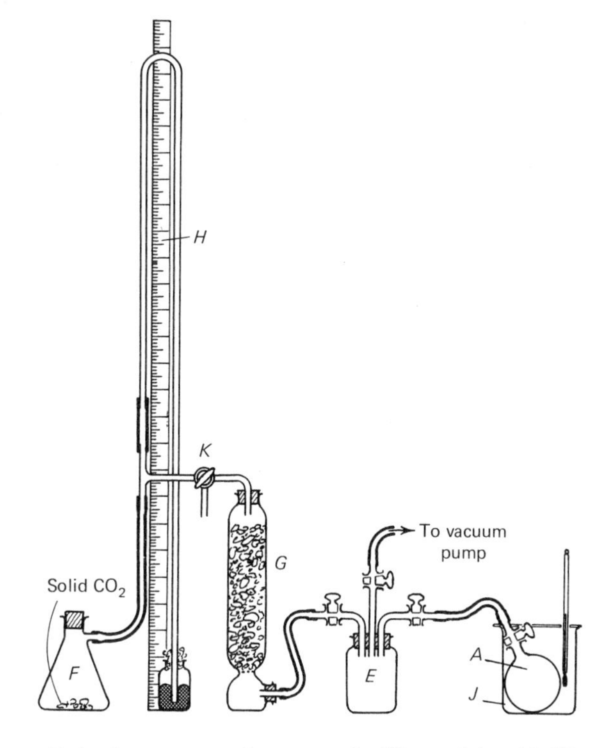

FIG. 5A.2 Arrangement of apparatus for filling a globe with CO_2.

detected by a gradual increase of weight.) If the density of the atmosphere were to remain constant throughout the determinations, clearly no buoyancy corrections would be needed since the buoyancy of the globe would remain the same whether the latter were empty or contained gas. However, since it may be necessary to use weighings made on different days and therefore possibly subject to different buoyancy corrections and since the correction is not very accurate because the humidity as well as the pressure influences the atmospheric density, it is usual in this type of work to eliminate buoyancy corrections by using a counterpoise globe of about the same size as the one in which the gas is to be weighed. If the counterpoise globe is subjected to the same handling and wiping it also helps to reduce errors which might be introduced by variable amounts of adsorbed moisture on the surface of the glass.

The next step is to fill the globe with the gas at atmospheric pressure. If the density of dry air is to be determined, air is simply admitted slowly through a calcium chloride tube. If CO_2 is to be used, it may be prepared by a Kipp's apparatus, but is more conveniently obtained from a few pieces of solid CO_2 placed in a filter flask F which can be evacuated once or twice to remove air. CO_2 can also be obtained from a cylinder. However prepared, the gas should be dried by a small tower G of calcium chloride or silica gel, and the whole apparatus should be flushed through with CO_2 several times to remove air.

While the globe is being filled it can be kept in a cardboard or wooden box J, loosely wrapped in the chamois leather and with a thermometer in contact with it. This plan obviates difficulties of drying and temperature change which are incurred if the globe is placed in a water-bath, and it also saves time.

The globe is filled initially to a pressure slightly greater than atmospheric, but when sufficient time has been allowed for the temperature to become steady, the tap K is opened to the air for a few moments to allow the pressure in the globe to become equal to that of the atmosphere and the tap on the globe is then closed. The barometer is read. The globe is finally weighed again with the same precautions as before.

In order to determine the *reproducibility* of the measurements, the globe should be evacuated and weighed again; then filled with gas and weighed once more. These operations are probably the factors which limit the accuracy of the determination of gas density. As a check on the purity of the CO_2, a test may be made to see whether it is entirely absorbed by air-free potassium hydroxide solution. The bore of the tube beyond the tap should first be filled with the solution and then the globe inverted and stood in the solution and the tap opened. If more than a small bubble of gas remains unabsorbed its volume can be estimated (e.g. by weighing the nearly full globe, filling and weighing again) and a correction applied on the assumption that it consists of air.

Calculation. If W kg is the weight of gas filling the globe, the volume of which has been calculated to be V m^3, and if t(°C) is the temperature and p the barometric pressure in mm Hg (corrected for expansion of the scale and gravitational value (see Appendix) at which the globe was filled, then the volume (V_0) of the gas at standard temperature and pressure would be

$$V_0 = \frac{V \times 273 \times p}{(273+t) \times 760} \text{ m}^3 \text{ and its standard density } d = W/V_0 \text{ kg m}^{-3}$$

and approximate molecular weight $M' = 0{\cdot}022415\ W/V_0$ kg, since 1 mole of an ideal gas occupies $0{\cdot}022415\ m^3$ at s.t.p.

It is of interest to calculate the effect on M of assuming the van der Waals equation of state to apply instead of the ideal gas equation, namely, for 1 mole of gas

$$(p + a/V^2)(V - b) = RT$$

The full equation, being cubic in V, is awkward to use, but on expansion, neglecting the very small term ab/V^2 and putting $a/V \approx ap/RT$, the equation simplifies to

$$pV = RT - p(a/RT - b)$$

This shows that for moderate pressures the deviation from Boyle's law is linear, the term $(a/RT - b)$ being called the 'compressibility coefficient' of the gas. The van der Waals constants, a and b, may be calculated approximately from the critical constants of the gas. For CO_2; $a = 0{\cdot}364\ N\ m^4\ mol^{-2}$, $b = 4{\cdot}27 \times 10^{-5}\ m^3\ mol^{-1}$. ($p$ in $N\ m^{-2}$, V in $m^3\ mol^{-1}$, $R = 8{\cdot}314\ J\ K^{-1}\ mol^{-1}$.) Hence one can calculate the volume which 1 mole of CO_2 would occupy at the temperature and pressure of the experiment. The ratio of this calculated molar volume to the observed value must be equal to the ratio of the weight taken (W) to the true molecular weight, which can therefore be calculated.

Results. The results obtained for standard gas density should not differ by more than about 0·5% from the accepted values: viz. dry air, $1{\cdot}2928\ kg\ m^{-3}$; CO_2, $1{\cdot}9768\ kg\ m^{-3}$.

BIBLIOGRAPHY 5A: Gas density

Partington, Vol. I, section 7 (D).
[1] Rayleigh, *Proc. Roy. Soc.*, 1899, **55**, 340.

5B Vapour density

1. The two well-known classic methods of determining the density of vapour of a substance which is liquid at normal temperatures are those of Dumas (1826) and of Victor Meyer (1882).

Dumas placed a sample of a volatile liquid in a bulb which was placed in a water-bath. The liquid evaporated displacing the air in the bulb; when this was complete and the bulb filled with vapour the bulb was sealed and weighed.

In Victor Meyer's method the vaporized liquid displaces air from a vessel maintained at a high constant temperature, but the volume of air is measured at ordinary temperatures. In some modifications of the method the volume is maintained constant and the increase of pressure is determined instead. The gas buoyancy micro-balance (suitably heated) can also be adapted to measure vapour densities, and has superseded these older methods.

2. ANALYSIS OF BINARY MIXTURES

Vapour density determinations can be employed to analyse binary mixtures of volatile liquids, provided the molecular weights of the components differ considerably. If the pure components give vapours of density d_1 and d_2, and a mixture containing w_1 g and w_2 g of the two components gives a vapour density d, then

$$(w_1+w_2)/d = w_1/d_1+w_2/d_2,$$

or,

$$w_1/w_2(1/d-1/d_1) = 1/d_2-1/d$$

Hence the ratio w_1/w_2 can be calculated.

BIBLIOGRAPHY 5B: Vapour density

Partington, Vol. I, section 7 (D).
Weissberger, Vol. I, Pt 1, chapter IV, section V.

6 Properties of liquids

6A Density

1. The density or specific gravity of a liquid is the mass of unit volume of the liquid. The relative density is the ratio of the weight of a given volume of the substance to the weight of an equal volume of water at the same temperature ($d_{t^\circ}^{t^\circ}$); and the density of the substance at temperature t° is equal to the relative density multiplied by the density of water at that temperature.

Densities of liquids are generally measured either by weighing a definite volume of the liquid in a density bottle or pyknometer or by determining the buoyancy acting on a 'sinker' immersed in a liquid (Principle of Archimedes). Small *changes* of density are sometimes determined by measuring the *rate* of rise or fall of a small immersed quartz 'float' of pre-arranged overall density. Where sufficient liquid is available, the density can be determined, approximately, by means of hydrometers.

2. THE PYKNOMETER

When only small quantities of liquid are available, or where greater accuracy is required, the density of liquids is best determined by means of vessels of accurately defined volume, called pyknometers. These are made in very varying shapes, but the simplest and most generally useful form is the Ostwald modification of the Sprengel pyknometer (Fig. 6A.1(a)). The volume of the pyknometer should be about 5–15 cm^3. This will allow of an accuracy of about 1 unit in the fourth place of decimals, which is quite sufficient for general purposes.

In carrying out a determination of the density of a liquid, the pyknometer must first of all be cleaned and dried, by washing well

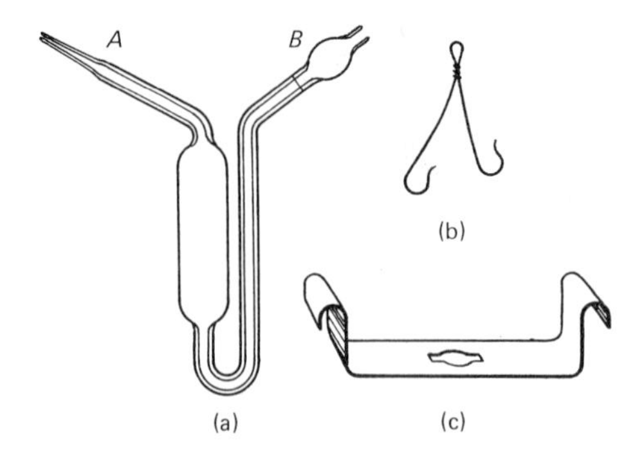

FIG. 6A.1 Pyknometer for liquids. (a) Pyknometer vessel. (b) Wire support for weighing. (c) Bracket for supporting pyknometer in thermostat.

with distilled water (if necessary, with other solvents first, then with water), and then, successively, with a small quantity of alcohol (redistilled methylated spirit) and of ether. A current of clean air is then drawn through the tube.

The pyknometer, cleaned and dried, is first weighed empty. For this purpose it is suspended from the end of the balance beam by means of a double hook (Fig. 6A.1(b)) made of glass or of platinum or copper wire. (It is not advisable to attach a wire permanently to the pyknometer, because of the greater difficulty in removing all moisture after the pyknometer has been immersed in water.) The pyknometer is then filled with distilled water by attaching a piece of rubber tubing to the end *B*, and sucking gently while end *A* dips in the water. The pyknometer is then suspended in a large beaker of water (or, better, a thermostat, chapter 3) whose temperature must be kept constant to within 0·1°C, as shown by a thermometer immersed in the water.

The pyknometer may be suspended in the bath by means of a wire hook placed over a glass rod laid across the top of the beaker; but it is better to use a holder cut from sheet zinc or copper (Fig. 6A.1(c)), and furnished with lugs which can be hooked over the edge of the bath. In the sheet of metal a hole is cut, which allows the body of the pyknometer to pass through, while the arms rest against the ends of the hole. The length of the opening should be such as to allow the pyknometer to pass so far through that the mark on the tube *B* of the pyknometer is just above the metal plate; and the water in the bath should be of such a height that it just touches the underside of the plate. By means of this arrangement, the danger of water getting into the ends of the pyknometer tubes is avoided, and the pyknometer is held in position more securely than by hooks.

When the pyknometer and its contents have taken the temperature of the bath (say after 15–20 min), the amount of water must be adjusted so that it fills the pyknometer from the point of the tube A to mark on B (Fig. 6A.1(a)). If there is too little water, a rod or tube carrying a drop of water is placed against the end of the tube A, when water will be drawn into the pyknometer by capillarity. If there is too much water, a piece of filter paper is carefully placed against the end of A, whereby water can be drawn from the pyknometer until the meniscus stands exactly opposite the mark on B. Instead of using filter paper, the adjustment can be made by gently compressing a short length of rubber tubing attached to the end of B until the meniscus is driven down to the mark. Before releasing the tube any drop of water which may have collected at the point of A is removed by means of a glass rod.

The pyknometer is now removed from the bath, and the outside carefully dried by means of a cloth; care must be taken that none of the water is expelled from the pyknometer by the heat of the hand or by the natural expansion of the liquid when the density is being determined at temperatures below that of the room. When the pyknometer has taken the temperature of the balance case, it is weighed. If concordant and accurate weighings are to be obtained, it is essential that the outside of the pyknometer shall always be dried and treated in exactly the same way, since otherwise the amount of moisture which remains on the surface will vary, and may cause an appreciable error.

When the weight of the pyknometer filled with water has been determined, the pyknometer is emptied and dried and filled with the liquid the density of which is required. It is placed as before in the bath at constant temperature, the liquid is adjusted to the mark, the pyknometer dried with a cloth as before and weighed.

Calculation of the density. If the temperature at which the pyknometer is filled with water and with the other liquid is the same, then the ratio of the weight of liquid (W') to the weight of water (W) gives the relative density (uncorrected for the buoyancy of the air) of the liquid compared with that of water at the same temperature. This is represented by $d_t^t = W'/W$. For certain purposes, as in the determination of the relative viscosity (section 6C), this ratio is all that is required; but in all cases where the specific gravity of the liquid is desired, we must multiply d_t^t by the density of water at t°C, $D_t\,(H_2O)$.

The values of W' and W must be corrected for the buoyancy of the air (section 3A(3)). After some manipulation we obtain:

$$D_{t^\circ}\ (\text{liquid}) = \frac{W'}{W} D_{t^\circ}\ (H_2O) + 1{\cdot}2\left(1 - \frac{W'}{W}\right) \text{kg m}^{-3}$$

The density of air is taken as 1·2 kg m^{-3}.

If the temperatures t at which the pyknometer is filled with water and with the other liquid are not the same, a further correction is necessary for the expansion of the glass, and one obtains as the general expression for the specific gravity of a liquid:

$$D_{t^\circ}\ (\text{liquid}) = \frac{W'}{W} D_{t^\circ}\ (H_2O)\ (1 + 0{\cdot}024\ (t_{H_2O} - t_{\text{liquid}})) + 1{\cdot}2\left(1 - \frac{W'}{W}\right) \text{kg m}^{-3}$$

Note: The use of either of the two formulae given above for the calculation of the density may introduce errors in the fifth decimal place of the density value, owing to variations in the value of the density of the air. In the following table are given the values of the density of air half saturated with moisture, at different barometric pressures (corrected):

Temperature (°C)	*Pressure in mm Hg; density in kg m^{-3}*				
	740	750	760	770	780
10	1·211	1·228	1·244	1·261	1·278
15	1·190	1·206	1·222	1·238	1·254
20	1·168	1·184	1·199	1·215	1·230
25	1·146	1·162	1·177	1·193	1·208

(For a discussion of various formulae for correcting for the buoyancy of the air, the papers quoted in Bibliography 6A[1] should be consulted.)

3. PARTIAL MOLAR QUANTITIES

Partial molar quantities are important in the theory of solutions and elsewhere. They are introduced because the extensive properties of a system are not precisely additive when a solution is made up from the pure components. However, if the property is X and there are n_i moles of the ith component

$$dX = \left(\frac{\partial X}{\partial T}\right)_{P, n_i} dt + \left(\frac{\partial X}{\partial P}\right)_{T, n_i} dP + \left(\frac{\partial X}{\partial n_1}\right)_{T, P, n_2, n_3, \ldots} dn_1 + \left(\frac{\partial X}{\partial n_2}\right)_{T, P, n_1, n_3, \ldots} dn_2$$

The partial molar quantity for the ith component is defined as

$$\bar{X}_i = \left(\frac{\partial X}{\partial n_i}\right)_{T, P, n_j}$$

so that at constant temperature and pressure

$$dX_{T, P} = \bar{X}_1 \, dn_1 + \bar{X}_2 \, dn_2 + \cdots$$

Integrating

$$X = \bar{X}_1 n_1 + \bar{X}_2 n_2 + \cdots$$

The total X can be regarded as being made up of the sum of the number of moles of each component multiplied by the partial molar quantities; the latter are, of course, in general a function of the mixture composition.

For a two-component system the molarity m is the number of moles of component 1 per kg of 2. If the total value of X per kg of 2 is X_T (the 'integral X of mixing').

$$\left(\frac{\partial X_T}{\partial m}\right)_{T, P} = \left(\frac{\partial X}{\partial n_1}\right)_{T, P, n_2} = \bar{X}_1.$$

The partial molar quantity $\bar{X}_1$ at any value of m is thus the slope of a plot of X_T against m. If the molecular weight of 2 is M_2 kg mol^{-1}, $n_2 = 1/M_2$ and so the corresponding value of $\bar{X}_2$ is $M_2(X_T - m\bar{X}_1)$. $\bar{X}_1$ and $\bar{X}_2$ are normally measured over the whole concentration range using two plots: X_T against the molality of 1 for mixtures in which 1 is dilute and X_T against the molality of 2 for those in which 2 is dilute. In this last case the slope of the plot is $\bar{X}_2$ and $\bar{X}_1 = M_1(X_T - m\bar{X}_2)$.

EXPERIMENT

Determine the kilogram molecular volume of pure ethanol at 25·0°C and its partial molal volume in dilute aqueous solution.

Molar and partial molar volumes can be calculated from accurate density data (see below). The experiment therefore consists in measuring accurately the density of pure ethanol and of a number of ethanol–water mixtures by the pyknometer method as described above. At least four solutions should be made up accurately *by weight* in stoppered flasks—for example, approximately, 5, 10, 15 and 20 weight per cent—and their densities determined. The work can be expedited by using two pyknometers.

Calculation. The density of water at 25°C is 997·07 kg m^{-3}. Hence, the density of ethanol and the ethanol–water solutions can be calculated. The specific volume is 1/density (m^3 kg^{-1}). The molecular weight of ethanol is 0·04607 kg and hence its molecular volume V_m is (0·04607/density) m^3 mol^{-1}.

If a solution contains W_1 kg ethanol and W_2 kg water, the weight fraction F_2 of water is $W_2/(W_1+W_2)$. If its density is D, the volume is $1/D$ m^3 (kg soln)$^{-1}$, so $V_T=(1/F_2D)$ m^3 (kg H_2O)$^{-1}$. The solution contains (W_1/F) kg ethanol (kg H_2O)$^{-1}$ so the molality m_1 is $W_1/0{\cdot}04607F_2$ mol (kg H_2O)$^{-1}$. Since $n_1=m_1$ and $n_2=1/0{\cdot}018015=55{\cdot}51$, the mole fraction x_1 of ethanol is $m_1/(m_1+55{\cdot}51)$.

Plot V_T against m_1, remembering you have a value for $m_1=0$. Draw a smooth curve through the points, and by drawing tangents obtain values of $\overline{V}_1$ and hence of $\overline{V}_2$. Plot $\overline{V}_1$ and $\overline{V}_2$ against x_1.

If there were no change of volume on mixing, i.e. if the solution were ideal, each addition of 1 mole of ethanol would increase the total volume by its molar volume V_0, so $\overline{V}_1$ would be equal to V_0. Show the value of V_0 on your plot.

This experiment can be extended by covering the whole concentration range, plotting V_T against the molality of water for the mixtures rich in ethanol.

BIBLIOGRAPHY 6A: Density of liquids

Partington, Vol. II, section 8 (B).
Weissberger, Vol. I, Pt 1, chapter VI.

6B Vapour pressure

1. THEORY

When a liquid is introduced into an evacuated vessel, some of it evaporates (with absorption of latent heat) until a certain *saturation vapour pressure* is set up which is in equilibrium with the liquid. This pressure p has a definite value at a given temperature, T (K), for a given pure liquid. At equilibrium the rate of evaporation of molecules from the liquid is equal to the rate of condensation of molecules from the vapour into the liquid, and the vapour–liquid equilibrium can be treated either by the kinetic theory or by thermodynamics. The relationship between p and T obtained from either method is

$$\frac{\mathrm{d}\log_e p}{\mathrm{d}T} = \frac{L}{\boldsymbol{R}T^2}$$

in which L is the molar latent heat at T and $\boldsymbol{R}$ is the gas constant, $8{\cdot}314\ \mathrm{J\ mol^{-1}\ K^{-1}}$. This equation is a form of the Clapeyron–Clausius equation and is valid provided the molar volume of the liquid is negligible compared with that of the vapour, and the vapour obeys the ideal gas laws. If the further assumption is made that L is a constant for the temperature range T_1 to T_2 the equation can be integrated, and gives

$$\log_e\left(\frac{p_2}{p_1}\right) = \frac{L}{\boldsymbol{R}}\left(\frac{1}{T_1} - \frac{1}{T_2}\right)$$

or, in terms of decadic logarithms,

$$\log_{10}\left(\frac{p_2}{p_1}\right) = \frac{L}{2{\cdot}303\boldsymbol{R}}\left(\frac{1}{T_1} - \frac{1}{T_2}\right)$$

This vapour pressure equation holds fairly accurately for liquids (or solids) over a limited range of temperature, and a graph of $\log_{10} p$ against $1/T$ gives a straight line of slope $-(L/2{\cdot}303\boldsymbol{R})$. This graph, however, shows appreciable curvature if plotted for a wide range of temperature, because L decreases with rise of temperature, according to the Kirchhoff equation.

The above relationships apply to vapour pressure of all liquids and solids, whether at normal, high or low temperatures, but, naturally, different experimental techniques are needed according to the pressure range concerned. In addition, several distinct methods are in use; four typical ones are described below.

2. METHODS

In *static methods*, the pressure exerted when excess liquid is introduced into an evacuated vessel is measured directly or indirectly. For temperatures below that of the surroundings, the pressure can be measured directly (as in Method 1) on a mercury or other manometer of suitable sensitivity. For higher temperatures it is necessary

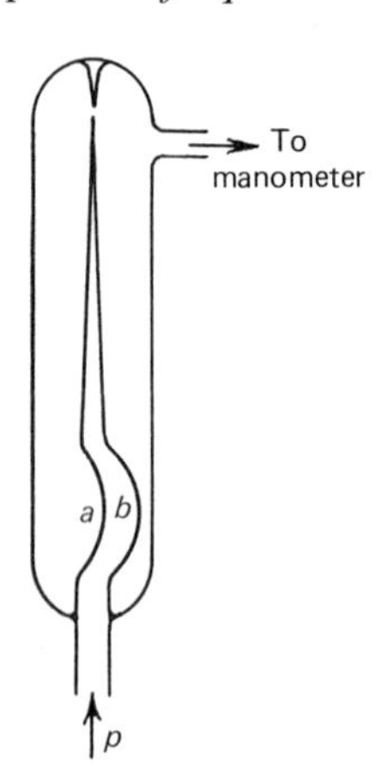

FIG. 6B.1 Principle of the glass 'spoon' gauge. (*a*, *b* are flexible, being very thin.)

to interpose a pressure-sensitive glass membrane (e.g. 'spoon' gauge, Fig. 6B.1, or 'spiral' gauge in which a thin hollow glass spiral acts as membrane) or a liquid barrier (as in the isoteniscope, Method 2) between the vapour and the manometer, since otherwise liquid distils from the warmer to colder parts of the apparatus. The pressure of air needed to counterbalance the vapour pressure is then measured. In static methods it is essential to remove dissolved air from the liquid (see below).

There are two principal *dynamic methods*. The first is equivalent to determining the boiling point of the liquid at a series of different pressures; clearly a liquid boils when its vapour pressure equals the pressure in the vessel containing it. There are several procedures for measuring boiling points under reduced pressure—for example, the method of Ramsay and Young (Method 3) and various micro methods.[1] The standard method of measuring boiling points accurately (chapter 7) can readily be adapted to work at different pressures by means of a manostat or simple pressure-regulating apparatus such as that shown in Figs 6B.3 and 6B.4.

The other dynamic method, sometimes called the gas-saturation or transpiration method, consists in passing through the liquid a measured volume of a gas that is chemically inert to it, and determining the quantity of liquid carried away by the saturated gas. This method, being cumulative, is particularly suitable for liquids of rather low vapour pressure. *Very* low vapour pressures such as those given by high-boiling oils, liquid or solid metals, etc., are best measured by one of the *high vacuum methods*—for example, by determining rates of evaporation or effusion, or by the molecular bombardment pressure exerted by the vapour.[1] Methods of determining *partial* vapour pressures, e.g. of solutions, by comparison with solutions of known vapour pressure (the 'isopiestic' method) are mentioned in chapter 7. The vapour pressure over salt hydrates is considered in chapter 8.

3. METHOD 1. *The direct, static method* (*low temperatures*)

Figure 6B.2 shows an apparatus suitable for measuring vapour pressures in the range 1–150 kN m^{-2} and for temperatures between −80°C and that of the laboratory. The essential parts are the bulb *A* containing the liquid, the manometer and a pump for removing air from the apparatus. The bulb *A* is almost filled with the liquid and is then connected by a ground glass ball joint *B* to the rest of the apparatus. *A* is cooled in a liquid bath which is preferably contained in a small Thermos flask *F*.

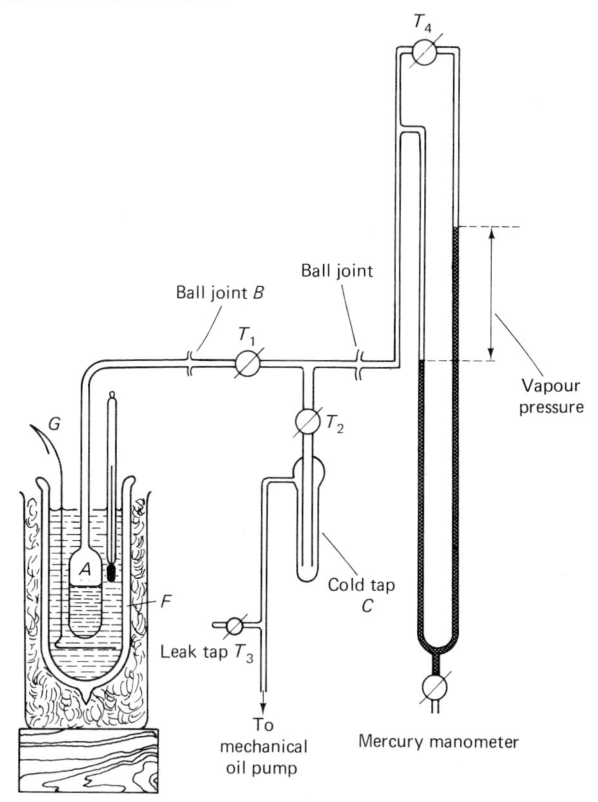

FIG. 6B.2 Apparatus for determination of vapour pressure of liquids by the static method at temperatures below that of the laboratory.

Temperatures should be measured with a low temperature mercury, alcohol or pentane thermometer graduated in degrees or tenths. The thermometer should be checked at two or three known temperatures (chapter 3).

EXPERIMENT

To measure the vapour pressure of diethyl ether

Bulb A is filled with ether: (a) the ground glass ball joint B is cleaned with acetone until it is free from grease, (b) bulb A is cleaned thoroughly with acetone and dried, (c) A is rinsed out twice with a few cm^3 of ether. (Remember ether is inflammable and explosive.) (d) A is half filled with sodium-dried diethyl ether, (e) joint B is allowed to dry, re-greased lightly with vacuum grease and attached to the apparatus with the clips.

Taps T_2 and T_3 are closed, and the vacuum pump turned on. After a minute or so, the cold-trap C is cooled with a methylated spirit—'dry ice' (solid CO_2) mixture in a Dewar flask so as to protect the pump from ether vapour. Dry ice is added *bit by bit* to the acetone and the Dewar then raised slowly around the cold trap so as to cool it gradually. If this is done too rapidly, there will be violent effervescence of CO_2; serious burns result from splashes of the cold liquid.

T_4 and then T_2 are opened, and the manometer is pumped for half a minute. T_1 is partly opened to allow the ether to boil gently under reduced pressure for a few seconds to sweep out the air. T_1 is closed again and the ether in bulb A frozen with acetone—dry ice mixture, carefully observing the precautions already mentioned. T_1 is opened again and bulb A is pumped until as good a vacuum as possible is attained. T_4 and T_2 are closed to check for the absence of leaks: the manometer levels should remain steady for at least 1 min. The vapour pressure of ether at the methylated spirit–dry ice temperature of -78°C is negligible.

The -78°C bath around bulb A is replaced by a series of temperature baths, measuring the vapour pressure of the ether at each temperature in turn. Each bath is *partially* frozen by adding dry ice. The liquid must be stirred continuously with a thick glass rod during the addition. The temperature of all the low temperature baths should be measured with a pentane thermometer calibrated at the ice-point (0°C) and the carbon dioxide-point (-78°C); mercury-in-glass thermometer is used above -6°C. The vapour pressure at each

Temperature (°C)	*How obtained*
About −63·5	f.p. $CHCl_3$
−45·2	f.p. C_6H_5Cl
−23·0	f.p. CCl_4
−6·1	f.p. $C_6H_5NH_2$
0	ice + water
12	water + a little ice
Room temperature	Empty Dewar

temperature is the difference between the heights of the mercury columns.

At the end of the experiment, the ether is frozen once more with the dry ice bath: the mercury levels should be equal again if there have been no leaks during the experiment. The apparatus is closed down by closing taps T_1 and T_2, opening T_4, switching off the pump and opening the leak tap T_3. Finally the two cold baths (Dewars) are removed.

Calculation. A graph of $\log_{10} p$ (in mm Hg) against T^{-1} (in K) is plotted. L is obtained from the slope which is $L/2{\cdot}303\boldsymbol{R}$. The normal boiling point, i.e. T when $p = 760$ mm Hg, should also be obtained by extrapolation.

4. METHOD 2. *The isoteniscope of Smith and Menzies*[2]

The isoteniscope (Fig. 6B.3) consists of a glass bulb A of about 2 cm diameter connected to a small U-tube B with limbs about 3–4 cm long. The principle of the device is that the pressure of the vapour in A is balanced by an adjustable pressure of air in the external apparatus so that the liquid stands at the same level in the limbs of the U-tube B. The pressure registered by the manometer M is then equal to the barometric pressure *minus* the vapour pressure. The temperature of A is that of the bath in which it is immersed; this is preferably a small glass-sided thermostat with control gear of a type providing rapid adjustment of temperature. (Section 3B.)

The air pressure is regulated by tap C, which connects to a filter-pump, and tap D which admits air through a capillary leak. (Alternatively small 'doses' of air may be admitted by having two taps with a small volume between them.) The large bottle E acts as a buffer to stabilize the pressure; it should be covered by felt and wire gauze to minimize temperature fluctuations and to guard against flying glass in case it collapses under vacuum.

To make a determination the apparatus is first assembled and tested for leaks. Liquid is then introduced until *A* and *B* are about two-thirds full. The pressure is reduced to make the liquid in *A* boil so that air is displaced from the isoteniscope. The pressure is then

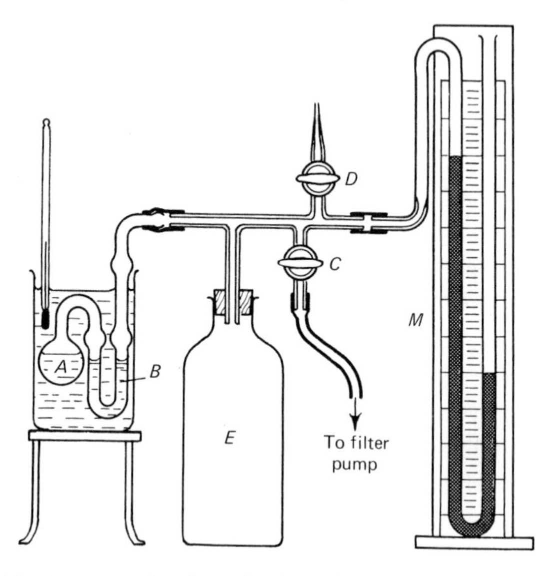

FIG. 6B.3 Apparatus for determination of vapour pressure by the isoteniscope method.

cautiously restored by opening *D* until the liquid in *B* stands level, and this balance is maintained while the isoteniscope attains the temperature of the thermostat. The final adjustment of pressure is carried out carefully to bring the columns of liquid in *B* to the same level as closely as can be judged by the eye. The thermometer, manometer and barometer are then read. The evacuation process and vapour pressure measurement should be repeated to ensure that air has been entirely removed. If the result is duplicated, further measurements at other temperatures can be proceeded with, care being taken that no air gets back into *A*. Sufficient time must be given at each temperature to allow the liquid to take the temperature of the bath; the vapour pressure should then remain steady indefinitely.

The isoteniscope is particularly suitable for the temperature range 20–150°C and for vapour pressures of 1–100 kN m^{-2}.

EXPERIMENT
Vapour pressure of carbon tetrachloride.

The vapour pressure of CCl_4 is measured at ten temperatures between 20 and 45°C. The liquid must be thoroughly degassed before use otherwise the measurements will be vitiated by the slow release of dissolved air: the liquid is tipped into the bulb and outgassed by pumping and shaking; finally some of the liquid is tipped back into the U-tube.

The temperatures of the bath are measured by a mercury-in-glass thermometer; the readings should be corrected for emergent stem error (section 3A). Manometer readings should only be taken after the temperature is steady for at least 2 min; manometer and barometer readings should be corrected for temperature and latitude (see Appendix, Tables A11 and A12). The calculations are similar to those for the previous experiment.

5. METHOD 3. *The method of Ramsay and Young*[3]

This is an ingenious method of obtaining accurate boiling points of a liquid under reduced pressure. Super-heating is avoided, and comparatively little liquid is needed. Figure 6B.4 shows a convenient apparatus.

The bulb of a thermometer, *A*, is *thinly* covered with a cotton-wool wick which is kept moist with the liquid to be studied, the tap-funnel

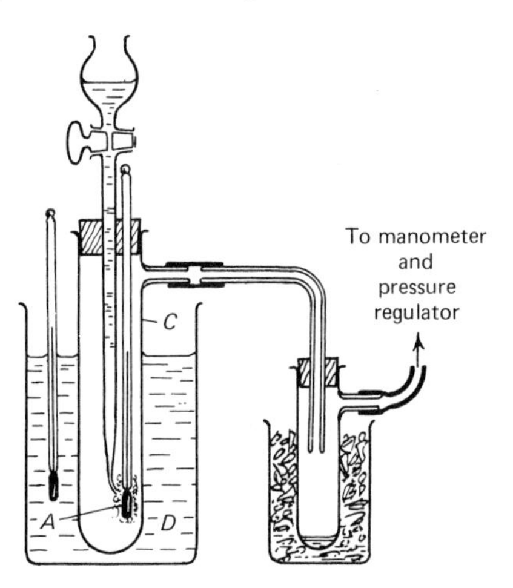

FIG. 6B.4 Apparatus for determination of vapour pressure by the method of Ramsay and Young.

being adjusted to give a very slow flow of liquid sufficient merely to replace that evaporated. The film of liquid around *A* is heated only by radiation from the outer tube which is immersed in liquid bath *D*. The temperature of *D* is maintained at about 10–20°C above that indicated by *A*. *D* need not therefore be accurately thermostated, but may consist simply of a large beaker of water or oil heated by a Bunsen burner.

A low pressure is maintained in vessel *C* either by an arrangement similar to that used with the isoteniscope and shown in Fig. 6B.5 or by some form of automatic pressure regulator ('manostat') of which many forms have been devised.[4]

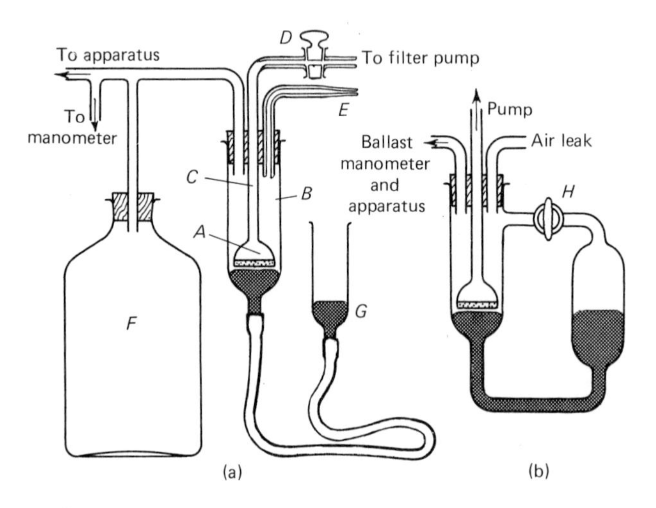

FIG. 6B.5 Manostat employing sintered-glass disc and mercury cut-off.

Manostat. Figure 6B.5(a) shows a simple reduced-pressure manostat suitable for the pressure range 1–100 kN m^{-2}. The key component is a sintered glass filter (*A*) of porosity No. 4—a size which will permit passage of air but not of mercury, even under vacuum. The tube *C* is kept at low pressure by means of a filter pump or oil pump, and the rate of evacuation is adjusted by tap *D* so that it slightly exceeds the rate of leakage of air into *B* through a capillary leak *E*. (Easier control is obtained with needle valves at *D* and *E*.) The mercury level rises until it reaches the sintered disc and cuts off the pump. It is important to have a ballast volume *F* connected to *B* to minimize fluctuations. If the disc *A* is exactly horizontal the manostat will work with an on–off action and an oscillation of pressure, but if the disc is slightly tilted a gradual cut-off is obtained, and a pressure steady to better than 0·1 mm can be obtained. The operating pressure of the manostat is readily altered by raising or lowering the reservoir *G*. The manostat can be made much more compact and the inconvenience of a movable mercury reservoir can be avoided by the modified form of apparatus shown in Fig. 6B.5(b). The vacuum stopcock *H* is kept open at first until the pressure, as indicated by the manometer, is reduced to the required value. *H* is then closed, and any further fall of pressure now causes the manostat to operate.

The same manostat can be used to maintain constant positive pressures somewhat above atmospheric pressure by feeding a slow stream of compressed air in through the tube *D* and allowing the excess to escape to the atmosphere through the sintered disc *A*.

Another convenient form of manostat employs a 'Cartesian diver' as the pressure-sensitive element.[4]

In making a measurement by the method of Ramsay and Young, the pressure is first reduced to about 4 kN m^{-2}. Evaporation occurs from the liquid surrounding the thermometer bulb, and the temperature falls. Providing the bath temperature is kept within the limits stated, a steady distillation can be achieved, the rate of supply of heat by radiation being balanced by the rate of cooling by evaporation, and a steady temperature is registered by the thermometer. This temperature and also the manometer reading are recorded. At this temperature the liquid has a vapour pressure equal to the controlled pressure in the apparatus, since the liquid is surrounded by and is in equilibrium with a film of air-free vapour in the interstices of the cotton-wool. Further readings are taken with higher pressures and temperatures until a vapour pressure of 1 atmosphere is reached. If the thermometer thread is entirely in the heated zone no emergent stem correction is needed, but a correction for the effect of reduced pressure—probably 0·1–0·2°C at the lowest pressures—may be found necessary. This correction is determined, of course, with no liquid on the wick, and may be assumed proportional to pressure change and independent of temperature, except for high temperatures.

6. METHOD 4. *The transpiration (gas-saturation) method*

The essential measurements in this method are (a) the quantity of gas passed, (b) the quantity of liquid (or solid) evaporated, (c) the temperature and (d) the pressure at the saturator. The general arrangement of apparatus is shown schematically in Fig. 6B.6(a).

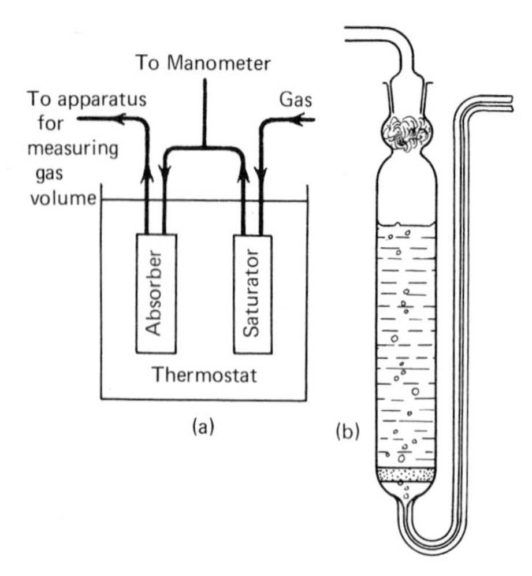

FIG. 6B.6 Determination of vapour pressure by the transpiration method. (a) Arrangement of apparatus (schematic). (b) Vessel for saturating a gas with a liquid.

There are so many ways in which (a) and (b) may be determined, depending on the substance to be studied, that only general methods will be indicated here.

If the vapour pressure is very low, a large volume of gas will be needed to transport measurable quantities of vapour. Volumes of say 5 dm^3 or more are best measured by a laboratory gas-meter.

6.1. *Wet gas meters.* The pressure of gas in the meter (P_m) and also the temperature (t_m) must be determined. The saturation water-vapour pressure at t_m is read off from tables (Appendix Table A5)—say p_w. Then if V_m is the volume of (wet) gas indicated by the meter, the content of dry gas is equivalent to
$V_m \times [(P_m - p_w)/P_m]$ at t_m; or a volume of

$$V_0 = V_m \frac{(P_m - P_w)}{P_m} \times \frac{273{\cdot}1}{(273{\cdot}1 + t_m)} \times \frac{(P_m - p_w)}{P_0}$$

at s.t.p., where P_0 is 1 atm. and P_m, p_w and P_0 are expressed in the same units. Wet gas meters must be carefully levelled and filled to the correct level with water before use. They may be tested against a calibrated aspirator bottle.

6.2. *Gas flow meters.* Another method of measuring the volume of gas passed through the system is to arrange a constant rate of flow of gas by means of a manostat (above, this section) or a constant pressure head with 'blow-off' (Fig. 6B.7) and to determine the time of flow and the *rate* of flow of gas. The latter can be measured by a capillary flow-meter *B*, or soap-bubble flow-meter *C*.

Capillary flow-meters are particularly suitable for indicating moderate rates of gas flow; the fall of pressure across the ends of a capillary resistance *a*, is indicated on a U-tube manometer *b*, containing, for instance, butyl phthalate (density $\approx$ 1 050 kg m^3). Since

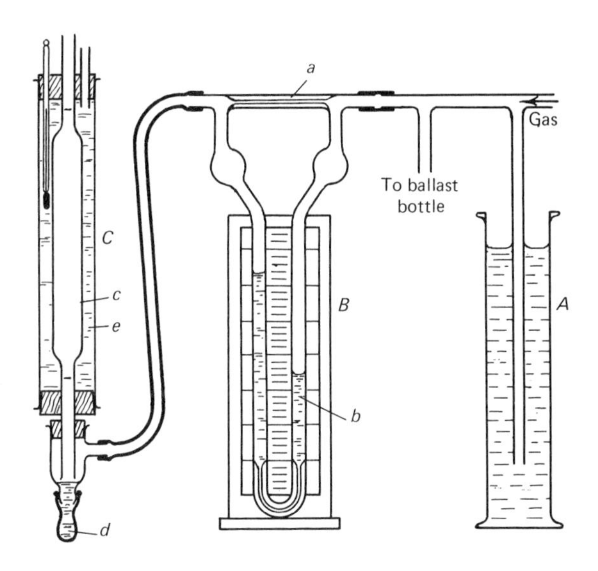

FIG. 6B.7 Determination of the rate of flow of a gas by means of a capillary flow-meter *B* or soap-bubble flow-meter *C*.

they depend on the viscosity of the gas, capillary flow-meters need calibration with each gas used. Figure 6B.7 shows a suitable method of calibration, using a soap-bubble flow-meter as standard. This device consists of a vertical tube *c*, which may be a graduated burette or simply a calibrated volume defined by marks at top and bottom. The lower end of *c* stands a little above a small reservoir of soap solution, the level of which can be made to rise and touch the tube *c* momentarily when required by compressing the small rubber bulb *d*. If gas is passed steadily through *c*, it carries a horizontal soap film upwards, and if the time taken for the film to travel between the two marks is noted, the rate of gas flow can be calculated. The tube *c* should be surrounded by a water-jacket *e* to define its temperature. The gas is presumed to be saturated with water vapour, and the volume is therefore corrected as above. Such flow-meters can be made with capacities ranging from 1 to 100 cm^3 s^{-1}.

If a sensitive method is available for determining the quantity of liquid transported in the gas-saturation method, it may suffice to pass only a few litres of gas, and this quantity can be most easily measured by running water from an aspirator bottle, measuring it subsequently by volume or weight. The usual temperature, pressure and humidity corrections are needed.

Inexpensive commercial flow-meters are available in which a steel or glass ball is supported in a vertical tube by the flow of gas up the tube: the tube is tapered, and the height of the ball indicates the flow rate.

6.3. *Saturators.* If a gas is bubbled through a liquid by a simple immersed jet bubbler, it is unlikely to reach saturation at once, and several such bubblers must be used in series. Much better contact between gas and liquid is achieved if the gas has to pass up a column of glass beads or short pieces of glass tubing wet with the liquid. Alternatively the gas stream may be broken into very small bubbles by means of a sintered glass disc, as in the saturator shown in Fig. 6B.6(b). The small bulb above the liquid is packed with glass wool to trap spray which otherwise is easily carried away in the gas stream.

The quantity of liquid evaporated during the experiment can be measured in several ways, namely (i) by loss of weight of the saturator, or (ii) by absorbing or condensing the vapour in another vessel and determining the increase of weight of the absorber, or (iii) by absorbing the vapour and determining the quantity by chemical or physical methods of analysis.

Calculation. The ideal gas laws and Dalton's law are assumed to hold. The measurements of gas flow lead to a value for V_0 m^3, the s.t.p. volume of dry gas passed through the saturator (see above). Since 1 mole of gas at s.t.p. occupies 0·022414 m^3, the number of moles of carrier gas is $V_0/0{\cdot}022414$. Similarly, the number of moles of liquid vaporized is the mass w divided by the molecular weight of the substance, M. The gas-vapour mixture emerges from the saturator at a total pressure P_s, the vapour being at its saturation vapour pressure p_s and the carrier gas making up the balance, i.e. $(P_s - p_s)$. Since gas and vapour occupy the same volume, namely, the volume of mixture passing out of the saturator, their partial pressures in the mixture must be in the ratio of the number of moles of each, hence

$$\frac{p_s}{(P_s - p_s)} = \frac{w/M}{V_0/0{\cdot}022414}$$

In this equation P_s, w and V_0 are determined, M is known and therefore the vapour pressure p_s can be calculated.

7. FURTHER EXPERIMENTS

Detailed experiments for methods 1 and 2 are given above.

If time permits the use of more than one method, the same liquid can be studied by two methods over a very wide range of temperature. For example, volatile liquids such as ethyl ether, ethyl bromide or methyl formate may be studied by method 1 for temperatures below room temperature and methods 2 or 3 from room temperature to their boiling points. Less volatile liquids (e.g. carbon tetrachloride, water, glacial acetic acid, toluene, xylene, pyridine, chlorobenzene, n-heptane, n-octane) may be studied by a combination of method 3 (or 2) (for high pressures) with method 4 for low vapour pressures.

BIBLIOGRAPHY 6B: Vapour pressure of liquids

Partington, Vol. II, section 8 (J).
Weissberger, Vol. I, Pt 1, chapter IX.
[1] See Partington, *loc. cit.*, for further details and references.
[2] Smith and Menzies, *J. Amer. Chem. Soc.*, 1910, **32**, 1412.
[3] Ramsay and Young, *J. Chem. Soc.*, 1885, **47**, 42.
[4] Thomson in Weissberger, *op. cit.*; Partington, *op. cit.*, Vol. I, 7 (A), 8.

6C Viscosity

1. THEORY

When a liquid flows through a tube, the layer of liquid in contact with the wall of the tube is stationary whereas the liquid in the centre has the highest velocity; intermediate layers move with a gradation of velocities. The flowing liquid may therefore be regarded as composed of a number of concentric tubes sliding past one another like the tubes of a telescope. Each layer exerts a drag on the next, and work must be done to maintain the flow. Newton deduced that the internal friction or *viscosity* would produce retarding forces proportional to the *velocity gradient* (dV/dx) normal to the direction of flow and to the area of contact (A) between the moving sheets of liquid, i.e. $F \propto A\, dV/dx$, or $F = \eta A\, dV/dx$, where η is a constant. This law holds for all homogeneous liquids, but not for suspensions or colloidal solutions, which are therefore called non-Newtonian fluids (chapter 15). The proportionality constant η is the *coefficient of viscosity*; in SI units it has the dimensions of $kg\ m^{-1}\ s^{-1}$. Common liquids range in viscosity from 2×10^{-4} (ether) to 0·8 (glycerol), water being about 0·1, all in $kg\ m^{-1}\ s^{-1}$. Older data will be found expressed in the c.g.s. unit of viscosity, the poise. 1 poise = $1\ g\ cm^{-1}\ s^{-1} = 0{\cdot}1\ kg\ m^{-1}\ s^{-1}$. Other quantities sometimes employed are the *fluidity* ϕ defined as the reciprocal of viscosity, i.e. $\phi = 1/\eta$, and the *kinematic viscosity* ν defined as viscosity divided by density, i.e. $\nu = \eta/\rho$.

Kinematic viscosity is more significant than absolute viscosity in problems of hydrodynamics because the flow properties of a liquid depend on inertia as well as internal friction.

Viscosity *decreases* considerably with rise of temperature (roughly about 2% per degree), and follows fairly closely the Andrade equation: namely,

$$\eta = A\, e^{B/RT}$$

in which A and B are constants for a given liquid. By analogy with the Arrhenius theory of reaction velocities (cf. chapter 14), B, which has the dimensions of work, can be regarded as the 'activation energy for viscous flow', although the structural interpretation of the quantity is not quite clear. It is probably related to the work needed to form 'holes' in the liquid into which molecules can move, thus permitting relative motion to take place.[1] Viscosity depends on molecular size and, particularly, *length* and also on the magnitude of the intermolecular forces; non-polar organic liquids (e.g. benzene) generally have low viscosities, whereas liquids in which directed bonding can occur between the molecules (e.g. hydrogen bonds in glycerol) have relatively high viscosities. The viscosity of colloidal solutions is discussed in chapter 15.

2. METHODS OF MEASURING VISCOSITY

Because of the importance of viscosity in many sciences and in technology, numerous instruments for measuring it (*viscometers*) have been developed to suit different needs. A viscometer must provide means of measuring the rate of flow in the liquid and the force exerted or work done in producing the flow. The fundamental equation (above) is applied directly in *rotating cylinder viscometers*. In these a thin film of liquid is sheared between inner and outer concentric cylinders. In the Couette viscometer the outer cylinder is rotated at constant speed and the force exerted on the inner cylinder is determined by the steady deflection of the torsion wire on which the inner cylinder is suspended.

All other methods depend on flow relationships derived from the fundamental equation. Thus, many forms of apparatus employ flow through *capillary tubes*, and then rely on Poiseuille's equation for such flow, namely:

$$V = \frac{p\pi r^4 t}{8\eta l}$$

V = volume of fluid (viscosity η) passing through a tube of length l, radius r, in time t, when a pressure difference p is maintained between the ends of the tube. This equation holds accurately for 'streamline flow', but not for the 'turbulent flow' which sets in at high velocities (namely, when the mean velocity of flow along the tube exceeds $10\ \eta/r\rho\ \mathrm{m\ s^{-1}}$). A small error arises in practice because the liquid emerging from a capillary tube possesses appreciable kinetic energy, and since this is not allowed for in Poiseuille's equation a kinetic energy correction is needed. Consequently, accurate *absolute* measurement of viscosity is difficult. As often happens, however, in physical chemistry comparative measurements are entirely satisfactory and much more convenient. In fact, simple U-tube capillary viscometers such as Ostwald's viscometer (see below) are used more than any other type.

Another well-established accurate hydrodynamic law is Stokes's law for the terminal rate of fall (v) of a sphere (radius r, density ρ_1) through a liquid (density ρ_2), namely,

$$v = \frac{2}{9}\frac{(\rho_1 - \rho_2)r^2 \boldsymbol{g}}{\eta}$$

In the *falling-sphere viscometer* a ball-bearing is timed in falling a measured distance through a cylindrical tube of liquid. Corrections are needed, however, for the influence of the walls and bottom of the tube. The former is usually quite important; it can be allowed for approximately by introducing a factor of $(1 + 2{\cdot}1r/R)$ below the line of the r.h.s. of Stokes's law (R = radius of tube).[2] Here again relative determinations with the same tube and ball largely eliminate these complications. The method is particularly suitable for viscous oils which are available in quantity. The simple apparatus of Gibson and Jacobs[3,7] (Fig. 6C.1), is suitable.

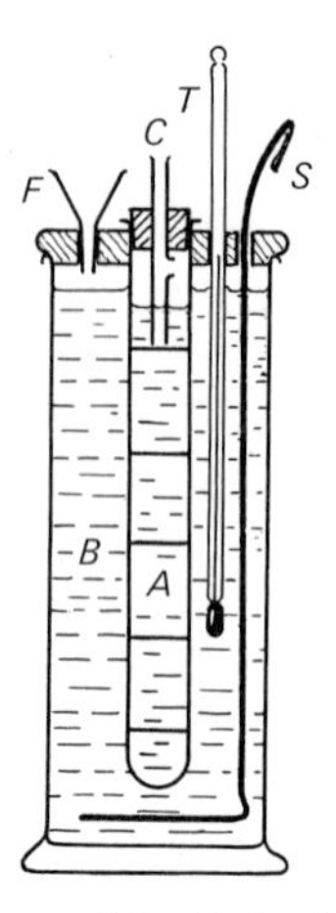

FIG. 6C.1 Apparatus of Gibson and Jacobs for determination of viscosity of a liquid by the falling-sphere method.[3,7]

Other methods of measuring viscosity depend on more complicated systems of flow—for example, the damping of torsional oscillations of a disc suspended in the liquid on a torsion wire[4] or of a hollow sphere[5] or hollow cylinder[6] containing the liquid. These methods have been used for liquid metals at high temperatures. In addition, many technical viscometers are in use for comparing empirically the flow characteristics ('rheological properties') of materials such as oils, paints, clays, etc. (section 15C).

3. THE OSTWALD VISCOMETER

The usual form of this simple, yet accurate, apparatus for comparing viscosities of different liquids is shown in Fig. 6C.2(a). The left-hand limb of the U-tube is essentially a pipette with two defining marks *A* and *B*, and a capillary resistance *C* through which the liquid contained in bulb *D* flows under gravity back into the bulb *E* in the right-hand limb. A definite volume of liquid is employed, and is delivered into tube *F* from a calibrated pipette; the quantity should be such that when the liquid is sucked up into the left-hand limb until the meniscus stands above the mark *A* then the meniscus on the right stands at the bottom of bulb *E*. The liquid is released from this position and allowed to flow back. When the meniscus passes mark *A* a stop-watch is started and when it reaches mark *B* the watch is stopped, and the time of outflow is noted.

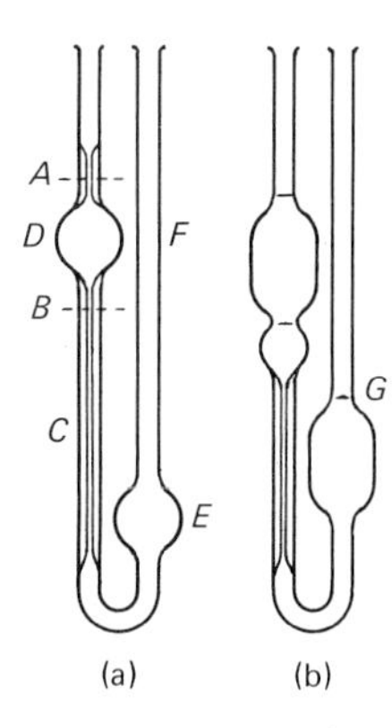

FIG. 6C.2 Ostwald viscometers. (a) Traditional form. (b) British Standards Institution form.[7]

The force driving the liquid through the capillary is equal to $h \times \rho_1 \times \boldsymbol{g}$, where h is the mean difference of level of liquid in the two limbs of the tube, ρ_1 is the density of the liquid and $\boldsymbol{g}$ the gravitational constant. The resistance to flow depends on the dimensions of the capillary (which are constant) and on the viscosity of the liquid. If now, the *same volume* of a second liquid is introduced into the tube, the mean difference of level of the two liquid surfaces will also be h, so that the driving force is now $h \times \rho_2 \times \boldsymbol{g}$. Thus, the driving force is proportional to the densities of the liquids while the resistance is proportional to their viscosities. Since the *rate* of flow is proportional to force/resistance, the *times* of outflow (t_1 and t_2) for the same volume of the two liquids are in the inverse ratio, i.e.

$$\frac{t_1}{t_2} = \frac{\eta_1/\rho_1}{\eta_2/\rho_2}, \quad \text{or} \quad \frac{\eta_1}{\eta_2} = \frac{\rho_1 t_1}{\rho_2 t_2}$$

This expression gives the relative viscosities of the liquids; if the absolute viscosity of one of them is known, that of the other can be calculated.

Since the rate of flow through a capillary tube depends on r^4, and r (the internal radius) can be varied from, say, 0·2 to 2 mm, Ostwald viscometers can be made to cover a range of 10^4 in viscosity. Suitable dimensions for a set of five viscometers of the 'standard' pattern shown in Fig. 6C.2(b) have been recommended by the British Standards Institution.[7] The No. 0 and No. 1 sizes, which together will suit most ordinary liquids, have capillaries 12 cm long and 0·038 and 0·060 cm internal diameter respectively.

EXPERIMENT

Determine the absolute viscosity of a pure liquid (such as benzene), and the influence of temperature on its viscosity.

Practical details. A viscometer should be selected having a flow time between 1 and 10 min with the given liquid. It must first be thoroughly cleaned with warm chromic acid so that there are no obstructions in the capillary and the liquid runs cleanly without leaving drops behind. It is then thoroughly washed by drawing distilled water through it with the aid of a filter pump, rinsed with alcohol and ether and then dried with a stream of air *filtered through cotton-wool* to exclude dust.

The viscometer is fastened, accurately vertical, in a glass-sided thermostat (or, for temperatures near that of the laboratory, in a large beaker of water). The temperature should be controlled within 0·1°C. The mark A should be well below the surface. A piece of rubber tubing, cleaned internally to remove dust, may be attached to the tube above A and used when sucking up the liquid into the left-hand limb. A suitable quantity of liquid, usually about 10 cm^3, is measured exactly into the viscometer with a pipette, and allowed 10–15 min to reach the temperature of the thermostat. The liquid is then sucked up and released, and the time of outflow between the marks is determined with a stop-watch reading to 0·2 s. The determination is repeated a number of times. If the time of outflow is about 100 s, the different readings should not deviate from the mean by more than 0·1–0·3 s. Greater deviations point to dust in the capillary tube. To determine the influence of temperature on the viscosity, the time of outflow should be measured at intervals of 5° between, say, 25° and 50°C. A small error arises from the change of volume of the liquid owing to expansion, but this may be neglected provided that during the outflow period the lower meniscus lies inside the bulb E, so that the change of level is small. (In the 'standard' viscometer, Fig. 6C.2(b), a mark G is provided for adjustment of the level to allow for expansion and also change of surface tension.) The density of the liquid at each temperature will be required; this may be determined by means of a pyknometer or taken from standard tables. The viscometer must be calibrated separately at one temperature with exactly the same volume of a liquid of known viscosity and density, usually water, the viscosity of which may be taken as $8{\cdot}95 \times 10^{-4}$ kg m^{-1} s^{-1} at 25·0°C. Pure aniline ($\eta_{25} = 3{\cdot}61 \times 10^{-3}$ kg m^{-1} s^{-1}, $\rho_{25} = 1\,017$ kg m^{-3}) could be used for calibrating No. 1 size viscometers, and 60% (by wt) sucrose solution ($\eta_{25} = 4{\cdot}35 \times 10^{-2}$

$kg\,m^{-1}\,s^{-1}$, $\rho_{25} = 1\,280\ kg\,m^{-3}$) for higher range viscometers.[8] Special oils can also be obtained for calibrating viscometers.

Treatment of results. From the densities and times of flow calculate the absolute viscosity of the liquid at each temperature. By plotting the fluidities ($= 1/\eta$) against the temperature, a graph of gentle curvature is obtained, which is convenient for interpolation of values. Compare the results with published data. Plot also $\log_{10} \eta$ against $1/T(\mathrm{K})$; a straight line graph should be obtained if Andrade's equation is followed, and the slope is equal to $E/2{\cdot}303\boldsymbol{R}$ (see section 6C). It is instructive to plot also a graph of fluidity against specific volume ($= 1/\rho$) of the liquid at each temperature. The approximate straight line which results suggests a close connection between the ease of flow and the 'free volume' in the interstices of the liquid.

4. VISCOSITY OF POLYMER SOLUTIONS

The viscosities of solutions of high polymers depend on the sizes and shapes of the molecules in solution and thus the measurement of viscosity is a useful method in the study of polymer configurations. Einstein, in 1906, showed that for a dilute solution in which the solute particles behaved as rigid spheres,

$$\lim_{\phi \to 0} \left[\frac{(n/n_0) - 1}{\phi} \right] = 2{\cdot}5 \tag{1}$$

In this equation, ϕ is the volume fraction of the solution occupied by the solute particles, n the viscosity of the solution and n_0 that of the pure solvent. n/n_0 is called the *viscosity ratio*: clearly $(n/n_0) - 1$, formerly called the *specific viscosity*, represents the relative increase in viscosity due to the solute. Equation (1) may be generalized for particles of other shapes by replacing the number 2·5 on the right-hand side of the equation by a parameter f which depends on the configuration of the solute particles.

As

$$\phi = w\bar{v}/V = c\bar{v} \tag{2}$$

where w is the mass of solute, V the volume of solution and $\bar{v}$ the partial specific volume of solute, we may write Eqn (1) in the form

$$[\eta] = \lim_{c \to 0} \left[\frac{(\eta/\eta_0) - 1}{c} \right] = f\bar{v}_0 \tag{3}$$

where $\bar{v}_0$ is the value of $\bar{v}$ at infinite dilution. The quantity $\{(\eta/\eta_0) - 1\}/c$ is called the *viscosity number* of the solution and $[\eta]$ is the *limiting viscosity number*. The value of f depends on the shape of the polymer molecules; for example for ellipsoidal particles, f is related to the axial ratio of the ellipsoid.

The viscosity number of a polymer solution depends on the size and shape of the polymer molecules and on their concentration. Huggins (1942) suggested that the concentration dependence of the viscosity number should be expressible by a virial expansion in powers of c. To terms of order c, the Huggins equation is

$$\frac{(\eta/\eta_0)-1}{c} = [\eta]+k[\eta]^2c \tag{4}$$

EXPERIMENT

Two grams of polystyrene are dissolved in 100 cm^3 of benzene. 50 cm^3 of this solution is diluted to 100 cm^3 and, by successive dilution of half of this solution, solutions containing 2, 1, 0·5, 0·25 and 0·125% (wt/vol.) of polystyrene are prepared. *Benzene solution must not be sucked into a pipette by mouth.* The viscosities of pure benzene and of each of the above solutions are measured at 25°C. The viscometers should be calibrated with water, or, if necessary, aniline. The density of benzene at 25°C is 873·63 kg m^{-3}.

A graph of η is plotted against c. The viscosity number is calculated for each solution and the Huggins equation tested by plotting viscosity number against concentration. $[\eta]$ and k can be obtained from the scope and intercept of the plot.

When the experiment is completed, the viscometers should be washed out thoroughly with benzene.

BIBLIOGRAPHY 6C: Viscosity

Partington, Vol. II, section 8 (E).
Weissberger, Vol. I, Pt 1, chapter 8.
Merrington, *Viscometry*, 1949 (Arnold, London).
Dinsdale and Moore, *Viscosity and its Measurement*, 1962 (Institute of Physics and Physical Society Monograph for Students, Chapman and Hall–Reinholt, London).
[1] Partington, *op. cit.*; Glasstone, Laidler and Eyring, *Theory of Rate Processes*, 1941 (McGraw-Hill, New York).
[2] For further details see Merrington, *op. cit.*
[3] Gibson and Jacobs, *J. Chem. Soc.*, 1920, **117**, 473.
[4] Fawsitt, *Proc. Roy. Soc.*, 1908, **80**(A), 290.
[5] Andrade and Chiong, *Proc. Phys. Soc.*, 1936, **48**, 247.
[6] Hopkins and Toye, *Proc. Phys. Soc.*, 1950, **63** (B), 773.
[7] British Standards Institution (London), Specification No. 188, 1937, 'Determination of Viscosity of Liquids in Absolute (c.g.s.) Units'.
[8] See *Int. Crit. Tables*, Vol. 5, p. 23.

6D Surface tension

1. THEORY

Surface tension is a manifestation of the forces of attraction that hold the molecules together in the liquid (or solid) state; thus, liquid droplets tend to become spheres—the form of least surface area—because of the mutual cohesion of the molecules. Conversely, work must be expended to increase the surface area of a liquid, as in expanding a soap bubble. The reality of surface tension (γ) is convincingly demonstrated in the Dupré frame experiment (Fig. 6D.1(a))

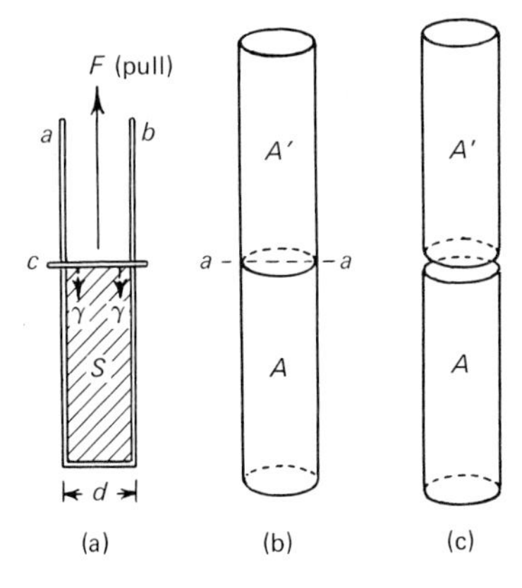

FIG. 6D.1 Significance of surface tension. (a) Dupré frame experiment—force exerted by a soap film. (b), (c) Relation of surface energy to cohesion (see text).

in which a smooth wire c slides on two parallel wires a and b and stretches a soap film S. If a force F N, is needed to balance the pull of the film (width d m), then $F=2d\gamma$, since γ is the force per unit length of the surface, and in this experiment there are two surfaces of liquid. If c is advanced by 1 m, the work done is $F\times 1$ J, while the surface area increases by $2\times d\times 1$ m^2. Therefore, the work required to form 1 m^2 of new surface is $F/2d=2d\gamma/2d=\gamma$ J. This shows that γ can be regarded either as a *tension* (N m^{-1}) acting along the surface or as surface (free) *energy* (in J m^{-2}), the two concepts being entirely equivalent. Older surface tension data will be given in the c.g.s. units of dyne cm^{-1} (1 dyne cm$^{-1}=10^{-3}$ N m^{-1}) or erg cm^{-2} (10^{-3} J m^{-2}).

The relation between surface energy and intermolecular forces can be readily appreciated by an imaginary experiment illustrated in Fig. 6D.1(b) and (c). Imagine a column of liquid AA' to be torn apart across the plane aa (area 1 m^2). Work is required to separate the two parts against the cohesive forces ('work of cohesion'); this work is equal to twice the surface energy since two m^2 of new surface are thus formed. It is understandable, therefore, that high surface tensions are found with liquids which have strong cohesive forces and consequently high latent heats of vaporization and high boiling points (e.g. liquid metals), whereas volatile organic liquids (e.g. ether) have low surface tensions. Similarly, for a given liquid, the surface tension, like the latent heat, decreases with rise of temperature and becomes zero at the critical point.

At temperatures not too close to the critical temperature, γ decreases linearly with temperature, but a relationship more accurate for a wide range of temperature is that discovered by MacLeod (1923), namely, $\gamma = k(\rho_l - \rho_v)^4$, where ρ_l and ρ_v are the densities of the liquid and the vapour and k is a constant. (ρ_v is generally negligible compared with ρ_l.) Sugden (1924) found that the k values of different substances can be correlated by use of a function called the *parachor*, which is defined by $[P] = M\gamma^{\frac{1}{4}}/\rho_l$. ($M$=mol. wt.) The parachor is approximately an additive property so that the value for a given molecule can be estimated as the sum of a series of atomic and structural constants; it is not a reliable guide to molecular structure as was originally thought. It is useful, however, for estimating values of surface tension of liquids of known structure and density.

If the imaginary experiment of Fig. 6D.1(b, c) were conducted with columns A and A' of *different*, immiscible, liquids, the work required would give information about the *work of adhesion* of the liquids. Clearly, an interface between any two phases has surface energy (*interfacial energy*) analogous to that of a liquid–gas surface, and the corresponding force required to extend the interface is called the *interfacial tension*. This quantity is of interest in connection with wetting phenomena, detergency and emulsification (chapter 15). The relation between surface (or interfacial) tension and adsorption is also considered in chapter 15.

2. METHODS OF MEASURING SURFACE TENSION

The innumerable methods that have been described can be classified into six groups: (1) direct measurement of capillary pull (Wilhelmy, du Nouy), (2) capillary rise (in single tubes, differential tubes, parallel or inclined plates), (3) bubble pressure (Jaeger, Sugden, Ferguson), (4) size of drops (weight or volume), (5) shape of drops or bubbles, (6) dynamic methods (ripples, etc.). In almost every case the elementary theory of the method needs correction if accurate results are required. The choice of a method will depend on whether the primary consideration is extreme absolute accuracy, practical convenience, speed of working, small size of sample, or study of time effects. The capillary rise method is the most accurate absolute method, but results accurate to better than 0·5% can be obtained much more conveniently by the drop weight, ring detachment, or bubble pressure methods. As elsewhere in physical chemistry, it is rarely necessary to use a laborious absolute method since most of the difficulties can be circumvented by calibrating the apparatus with a liquid of known surface tension. Some other difficulties can be avoided by employing a *differential* method (see sections 3 and 4 below). Values of surface tension and density data needed for the calibrations will be found in the Appendix.

Scrupulous cleanliness is essential in surface tension work; aqueous solutions, in particular, are highly susceptible to contamination by minute traces of grease, with large reduction in their surface tension. Consequently, from this point of view, the methods in which fresh drops or bubbles are repeatedly formed offer an advantage over static methods such as capillary rise. For the same reason a form of apparatus that can be readily cleaned and protected from contamination is desirable and, in addition, it should be suited to immersion in a thermostat to control the temperature within 0·1°. Four convenient methods are described in this section, and some other methods particularly suitable for the study of surface films and small changes in surface tension are given in chapter 15.

3. CAPILLARY RISE (*Differential form with calibration*)

The elementary technique and theory are well known but inaccurate. With proper corrections and experimental procedure high accuracy (better than 0·05%) has been achieved, but the method is then slow and difficult. A modification which is much simpler and yet capable of yielding results accurate to 0·1% is described below. The chief difficulties met with in the absolute method are as follows: (i) the difficulty of selecting and measuring accurately the bore of a capillary tube of uniform bore: this is overcome in the present modification by calibrating the tube with water or benzene, the surface tensions of which have been accurately established, and by working always with the meniscus at one position in the tube; (ii) the necessity for having a very wide surface of liquid ($>$4 cm) for the lower level if it is to be assumed plane, and the consequent need for much liquid; (iii) the associated experimental difficulty of accurately measuring the level of a wide surface of liquid: these last two difficulties are avoided by using two capillary tubes of different bore.

A differential capillary rise apparatus is shown in Fig. 6D.2. Two capillary tubes *a* and *b* are selected, having internal diameters of about 0·3–0·4 mm and 1·5–2 mm respectively. They are cleaned with warm chromic acid, rinsed with clean water, and the capillary rise with water, h_w, is noted roughly for each. Marks are then scratched

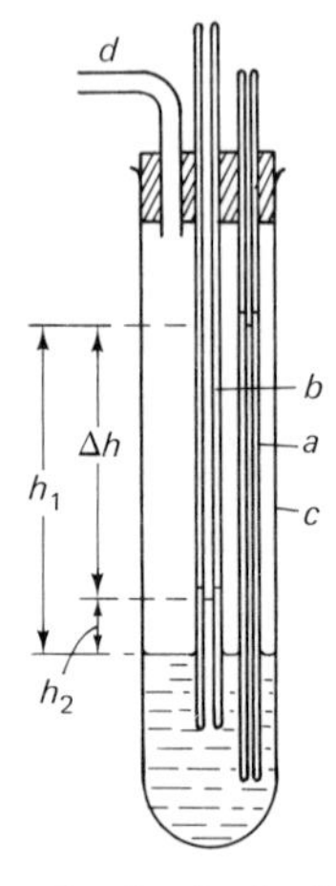

FIG. 6D.2 Determination of surface tension by the differential capillary rise method.

on the tubes at a distance say ($h_w + 1$ cm) above the end of each. This procedure ensures that, whatever the liquid to be tested, the tubes can always be lowered into the liquid until the meniscus stands opposite the mark. The tubes must next be calibrated with water or benzene—preferably the latter, unless aqueous solutions are to be studied.

The capillary tubes are mounted in a boiling tube *c* of good optical quality and supported in a cork as close together as possible so that they can be viewed simultaneously in the telescope of a vernier travelling microscope or cathetometer reading to 0·05 mm or better. The cork should provide a sliding fit so that the tubes can be raised or lowered.

Tubes *a*, *b* and *c* are thoroughly cleaned with chromic acid, washed well under a running tap, rinsed several times with rectified spirit and then twice with pure benzene. Parts of the glassware that will come into contact with the experimental liquid must not now be touched with fingers. The apparatus is assembled as shown, with a quantity of pure benzene in c, and the whole assembly is set up in a glass-sided thermostat or large beaker at $20{\cdot}0 \pm 0{\cdot}1\,^{\circ}\mathrm{C}$. The capillaries are moved up or down until the meniscus in each is at the mark. The liquid is made to rise and fall in the capillaries by gently blowing or sucking through tube d so that the glass walls become wetted with benzene, thus ensuring a contact angle of zero. The liquid levels should return to an accurately reproducible position a minute or two after repetition of this treatment, otherwise a dirty tube is indicated. The final position of the bottom surface of each meniscus is measured on a travelling microscope or cathetometer (set up precisely vertical). The liquid level in c is noted approximately on the same instrument. The two capillary rises, h_1 and h_2, need not be known to better than 0·1 mm but the difference $(h_1 - h_2)$ is measured as accurately as possible. These determinations complete the calibration of the tubes. The apparatus may now be cleaned and the measurements repeated in just the same way with the liquid of unknown surface tension.

Calculation. The *elementary* theory of capillary rise equates the upward pull of surface tension, $2\pi r\gamma$ (r = internal radius, γ = surface tension), with the weight of the column of liquid (density ρ), $\pi r^2 h\rho \boldsymbol{g}$, i.e. $\gamma = \frac{1}{2} rh\rho \boldsymbol{g}$. Hence, for two tubes, $r_1 h_1 = r_2 h_2$, and the difference of level

$$\Delta h = (h_1 - h_2) = \frac{\gamma}{\rho}\left[\frac{2}{\boldsymbol{g}}\left(\frac{1}{r_1} - \frac{1}{r_2}\right)\right].$$

The term in square brackets is a constant for the apparatus, and is therefore obtained from the calibration. Hence, when measurements are made on another liquid of density ρ' and capillary difference $\Delta h'$, its surface tension follows at once from $\gamma'/\gamma = \Delta h'\rho'/\Delta h\rho$. When the differential method is used with this simple theory it gives results accurate to about 1%, which is rather better than would be obtained by the one-tube method. However, only a little further calculation is needed to obtain an accuracy of about 0·1% from the same data, as follows.

For greater accuracy the Poisson–Rayleigh formula can be used to allow for the weight of liquid contained in the meniscus. This correction is equivalent to addition of a small height to the capillary rise, h. The effective height H is given by

$$H = 2\gamma/r\rho\boldsymbol{g} = h + r/3 - 0{\cdot}1288r^2/h + 0{\cdot}1312r^3/h^2$$

For tubes having $r < 0{\cdot}2$ mm the last two terms are quite negligible, and even for tubes up to $r = 1$ mm the last term can be neglected if an accuracy of 0·1% is sufficient.

When these corrections are applied to the differential apparatus, the two capillary rises are given by $H_1 = 2\gamma/r_1\rho g = h_1 + r_1/3$ for tube *a*, and $H_2 = 2\gamma/r_2\rho g = h_2 + r_2/3 - 0{\cdot}129 r_2^2/h_2$ for tube *b*. The differential height Δh is therefore given by

$$\Delta h = (h_1 - h_2) = \frac{2\gamma}{\rho g}\left(\frac{1}{r_1} - \frac{1}{r_2}\right) + \frac{1}{3}(r_2 - r_1) - 0{\cdot}129\frac{r_2^2}{h_2}$$

The approximate magnitudes of the three terms on the right-hand side with the present apparatus are 6·0, +0·03 and −0·0009 for water and 2·7, +0·03 and −0·002 for benzene (all in cm). It is seen that the first correction term, $\frac{1}{3}(r_2 - r_1)(=B)$, is quite important, but r_1 and r_2 need not be known to better than 5% for an accuracy of 0·1% in Δh or γ. Consequently, r_1 and r_2 can be calculated sufficiently accurately from the uncorrected theory, i.e. $r_1 \approx 2\gamma/h_1\rho g$ and $r_2 \approx 2\gamma/h_2\rho g$, taking the known values of γ and ρ for the calibrating liquid. The procedure, therefore, is to calculate first r_1 and r_2 and then the two correction terms, and finally insert the accurate value of Δh in the working equation to obtain an *accurate* value for the constant term $(1/r_1 - 1/r_2)(=A)$. The apparatus is then fully calibrated and can be used in the same manner with a liquid of unknown surface tension (but known density ρ'). The capillary rises h'_1 and h'_2 are measured approximately but their difference $\Delta h'$ accurately. The working formula is, as before,

$$\Delta h' = \frac{2\gamma'}{\rho' g}A + B - C \quad (\text{where } C = 0{\cdot}129 r_2^2/h'_2)$$

Now A is known accurately from the calibration, while B and r_2 are known sufficiently closely. Since the last term C is very small, h'_2 need be noted only roughly.

It is possible, in fact, to dispense with measuring h'_1 and h'_2 separately and measure only $\Delta h'$. The calculation is then made by successive approximations as follows. First an approximate value of γ' (say γ'') is obtained by neglecting the C term and writing $\Delta h' \approx 2\gamma'' A/\rho' g + B$. Now since r_2 is known, the C term is given approximately by $-0{\cdot}129 r_2^2/h_2 = -0{\cdot}129 r_2^3 \rho' g/2\gamma''$. A corrected value of γ' (say γ''') is now calculated from the full equation, and if γ''' is significantly different from γ'' the C term can be recalculated and a new value γ'''' obtained, the process being repeated until no further change in γ' is produced. Probably one recalculation will be enough to reach a constant value of γ' to within 0·1%.

4. SUGDEN'S BUBBLE-PRESSURE METHOD

When a bubble is formed slowly at a jet (radius r) which dips to a depth of h in a liquid (density ρ, surface tension γ), the pressure in the tube increases to a maximum which corresponds with a hemispherical bubble and then decreases as the bubbles break away. On simple theory, the maximum bubble pressure should be given by $(h\rho \boldsymbol{g}+2\gamma/r)$, and this may be measured on an external U-tube manometer. This is the basis of Jaeger's method. The theory, however, needs refinement to allow for the departure of bubble shape from hemispherical form, and the method shares with capillary rise the difficulties associated with having a wide, plane, liquid surface (see above). These complications were overcome by Sugden[1] by using two jets at the same depth and calibrating the apparatus with a standard liquid. An empirical equation accurate to 0·1% was found, namely:

$$\gamma = A(P_1-P_2)\left[1+0{\cdot}69r_2\frac{\rho \boldsymbol{g}}{(P_1-P_2)}\right]$$

where A is a constant of the apparatus, P_1 and P_2 are the pressures (in N m^{-2}) required to form bubbles at the narrow and the wide jet respectively, and r_2 is the radius of the *wider* jet. The method is one of the most convenient for an accuracy of about 0·3%. The apparatus is shown in Fig. 6D.3. The bubbler tube A contains the two jets at

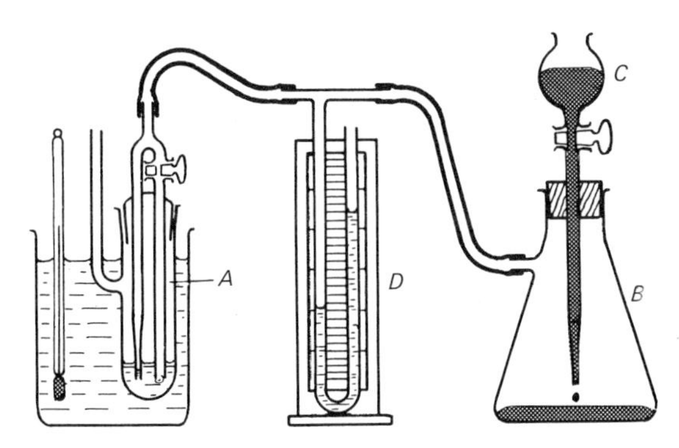

FIG. 6D.3 Determination of surface tension by Sugden's maximum bubble-pressure method.

the same level, of radii r_1 = about 0·1 mm and r_2 = 1–2 mm. The jets are made by pulling down quill tubing, cutting off cleanly, and grinding on a glass plate with fine carborundum powder and water until the tips are smooth, sharp edged, and perpendicular to the sides. The bubbler is connected to the flask *B* and dropping funnel *C* containing water (or mercury) by means of which the pressure can be slowly increased. The pressure inside the tubes is measured on a U-tube manometer *D* containing a non-volatile liquid such as dibutyl phthalate (density = 1 046·6 kg m^{-3} at 20°C, 1 042·6 at 25°C, other values by linear interpolation to the manometer temperature). The apparatus should, of course, be free from leaks.

To calibrate the apparatus, the bubbler and jets are first carefully cleaned and then rinsed with water, alcohol, and finally benzene. Enough pure benzene is poured into *A* to cover the jets to a depth of about 5 mm. Bubbles are formed *very slowly* at each jet in turn by increasing the pressure in the apparatus by running a slow stream of liquid from the dropping funnel *C* (the stem of which should be constricted). The rate of bubbling should not exceed about 1 bubble per second. The *maximum* bubble pressures are read on the manometer *D* by means of a cathetometer, lens and scale, or vernier calipers and converted to N m^{-2}. The radius of the wider jet, r_2, must be measured with a travelling microscope, a mean of radii in several directions being taken. The constant of the apparatus, *A* in Sugden's equation above, can now be calculated.

Having determined the apparatus constant *A* the surface tension of various liquids at any temperature may be determined.

5. THE DROP-SIZE METHOD

When a drop of liquid is formed *very slowly* at a jet (Fig. 6D.4), the quantity of liquid which eventually falls off is a *definite* function of the radius of the jet, *r*, the density of the liquid, *ρ*, and its surface tension *γ*, but it cannot be calculated from any *simple* theory. How-

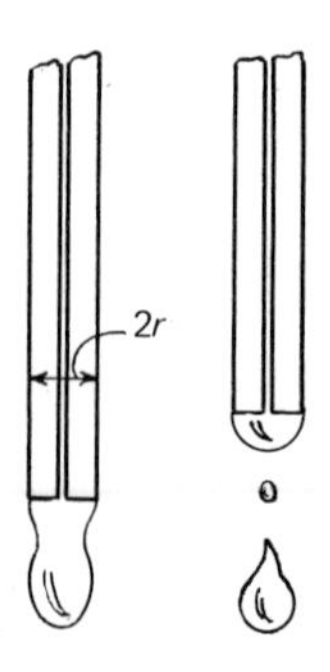

FIG. 6D.4 Stages in the detachment of a drop from a jet.

ever, Harkins and Brown[2] (1919) made a very careful experimental study of drop sizes, and prepared tables which can be used to obtain surface tension with an accuracy of 0·1% from careful measurements of drop size. The surface tension is calculated from the formula $\gamma = F \times mg/r$ where m is the mass of a drop, and F is a factor depending on v/r^3, v being the volume of the drop. Some values of F are given in the following table.[3]

v/r^3	F	v/r^3	F	v/r^3	F
17·7	0·2305	3·433	0·25874	1·4235	0·26544
13·28	0·23522	2·995	0·26065	1·3096	0·26495
10·29	0·23976	2·637	0·26224	1·2109	0·26407
8·190	0·24398	2·3414	0·26350	1·124	0·2632
6·662	0·24786	2·0929	0·26452	1·048	0·2617
5·522	0·25135	1·8839	0·26522	0·980	0·2602
4·653	0·25419	1·7062	0·26562	0·912	0·2585
3·975	0·25661	1·5545	0·26566	0·865	0·2570

Apparatus similar to that of Morgan[5] (Fig. 6D.5) is usually employed for measuring surface tensions by the drop-weight method. Liquid in bottle A is very slowly forced through the capillary tube to

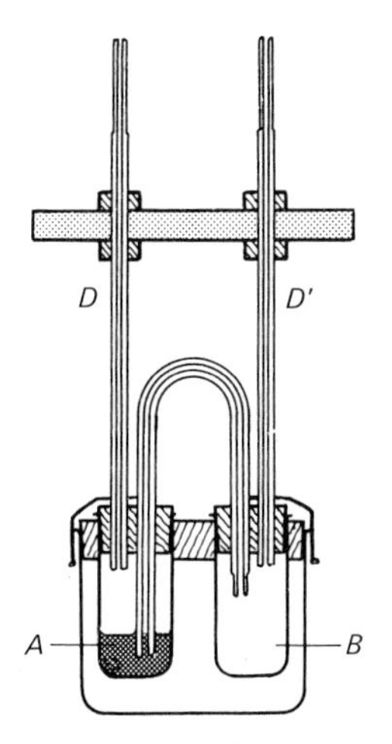

FIG. 6D.5 Apparatus for determination of surface tension by the drop-weight method.

grow a drop on the special jet in bottle B. The drop must fall away by gravity only (not vibration or kinetic energy of flow). Some 30 drops are collected and weighed and the surface tension calculated by use of the above tables. (See references for other details.)

6. INTERFACIAL TENSIONS

The drop-size method is particularly suitable for measurement of *interfacial tensions* between two immiscible liquids. Further, since much larger drops are obtained than with drops in air, it is possible to measure their size volumetrically. A very convenient apparatus for measuring small volumes accurately is the micrometer syringe (Fig. 6D.6(a)), in which a graduated micrometer *A* propels the piston of a hypodermic syringe *B* of known cross-section. This instrument

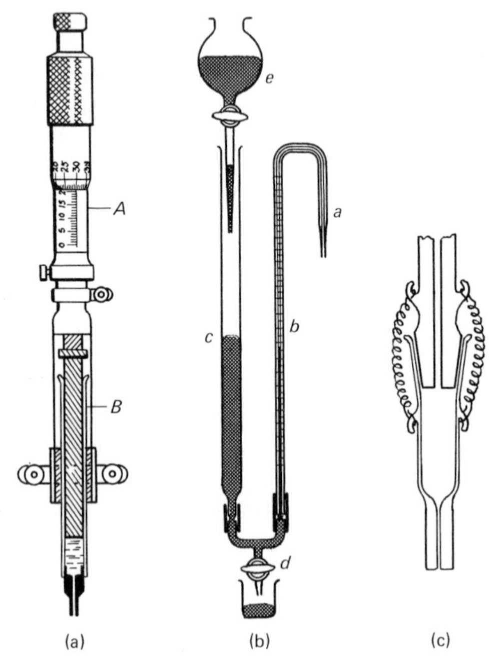

FIG. 6D.6 Determination of drop-volume. (a) Micrometer syringe (1 cm^3). (b) Microburette (1 cm^3). (c) Detachable jet for interfacial-tension measurements.

reads down to 0·0001 cm^3 and consequently the volume of one drop can be determined accurately. If a micrometer syringe is not available, a microburette readable to 0·001 cm^3 can be constructed from a graduated 1 cm^3 pipette (Fig. 6D.6(b)). In this microburette the liquid is displaced by mercury, and the position of the mercury meniscus can be read very precisely. To charge the burette with the liquid, the jet *a* is dipped into a beaker of the liquid and *clean* mercury is run into the tube *c* from the dropping funnel *e* until it overflows through *a*. The tap *d* is then slowly opened, and as the mercury runs out it draws liquid behind it into the graduated limb *b*. The tap is closed when the liquid-mercury interface reaches the bottom of the scale. The liquid is delivered by running mercury into *c* from the tap funnel. As the bore of the tube *c* is much greater than that of *b* it is easy to control the movement of the mercury level, and a very slow rate of rise can readily be produced.

The jet used for interfacial tension determinations must be large if the v/r^3 value is to fall within the accurate range of the table of Harkins and Brown (namely, 10–0·8); for the water–benzene system an outside diameter of 7–10 mm and bore of about 0·4 mm are suitable. The jet should be ground truly cylindrical and accurately perpendicular at the end: the edges must be sharp. The diameter of the jet is measured by a travelling microscope. It is advisable to have the jet detachable from the burette, as shown in Fig. 6D.6(c).

To make a determination of interfacial tension, the jet is immersed in the less dense of the two liquids, and drops of the heavier liquid are slowly formed. The drop can be expanded fairly fast at first, but the delivery of the last 5% should take at least one minute. Clearly, the correct drop-volume (v) is not obtained unless the *previous* drop has also been correctly produced. The other quantity needed for the calculation is the effective density of the heavier liquid when suspended in the lighter, i.e. the difference between their densities $(\rho_1 - \rho_2)$. These should be determined very accurately by means of a pyknometer at the temperature used in the experiment. The interfacial tension ${}_1\gamma_2$ can then be calculated from the expression ${}_1\gamma_2 = \{[v\ (\rho_1 - \rho_2)\boldsymbol{g}]/r\}F$, the value of F being obtained from the v/r^3 function by interpolation of the tables.

EXPERIMENT

Determine the interfacial tensions of the benzene–water, chloroform–water, and medicinal paraffin–water interfaces. Discuss the results with reference to Antonoff's rule, namely,

$$ {}_1\gamma_2 \approx {}_1\gamma_{\text{Air}} - {}_2\gamma_{\text{Air}} $$

7. THE RING-DETACHMENT METHOD (DU NOUY)[4]

For many purposes, e.g. study of biological fluids, colloidal solutions, etc., the surface tension can be determined quickly and with sufficient accuracy by measuring the force required to detach a horizontal ring of platinum wire (radius R) from the surface of the liquid. On the elementary theory, the force P should be given by twice the perimeter

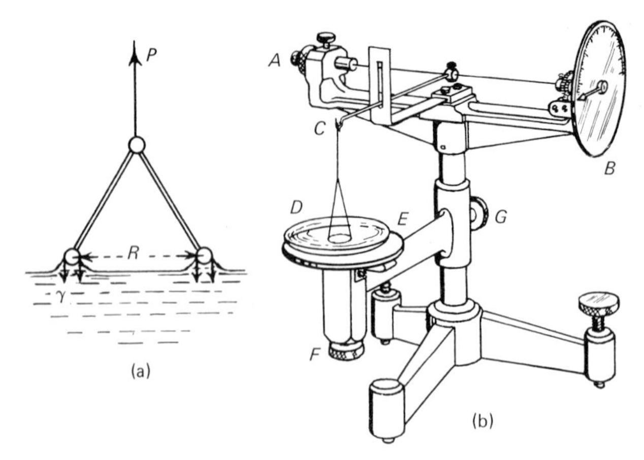

FIG. 6D.7 Determination of surface tension by du Nouy's method. (a) Principle of the method. (b) Torsion balance.

of the ring times the surface tension, i.e. $P = 4\pi R\gamma$ (see Fig. 6D.7(a)). More precisely, the pull must be multiplied by a correction factor F which varies from 0·75 to 1·1, and depends on the dimensions of the ring. The apparatus employed is generally du Nouy's torsion balance (Fig. 6D.7(b)).

BIBLIOGRAPHY 6D: Surface tension

Partington, Vol. II, section 8 (G).
Weissberger, Vol. I, Pt 1, chapter XIV.
Adamson, *Physical Chemistry of Surfaces*, 1960 (Interscience, New York).
Adam, *Physics and Chemistry of Surfaces*, 3rd edn, 1941 (Oxford University Press, Oxford).
Padday in Matijevic (ed.), *Surface and Colloid Sci.*, 1969, **1**, 39–251 (Wiley-Interscience, New York).
[1] Sugden, *J. Chem. Soc.*, 1922, **121**, 858; 1924, **125**, 27.
[2] Harkins and Brown, *J. Amer. Chem. Soc.*, 1919, **41**, 499.
[3] Reproduced by permission of the McGraw-Hill Book Co., Inc., from *International Critical Tables*, Vol. 4, Copyright 1928, McGraw-Hill Book Co., Inc.
[4] Du Nouy, *J. Gen. Physiol.*, 1919, **1**, 521.

7
Thermodynamic properties of dilute solutions

Determination of molecular weights and activity coefficients: chemical equilibria in solution

7A Introduction: the 'colligative' properties

Dilute solutions have played an important part in the history of physical chemistry because many of their properties are found to be closely related—at least as a first approximation—to the *molar* concentration of solute which they contain, and to depend relatively little on the nature of solute. Consequently, measurements of these properties have provided methods of determining *molecular weights* of substances in solution.

When a substance B is dissolved in a solvent A, the vapour pressure of A is reduced from its normal value, p_A°, to a lower value, p_A^s, which is given according to Raoult's law by $p_A^s = p_A^\circ N_A$, where N_A is the *mole fraction* of A in the solution defined by

$$N_A = \frac{\text{no. of moles of } A}{\text{total no. of moles}}$$

Solutions which obey this law at all concentrations are called *ideal*. In practice, Raoult's law is found to hold approximately for solutions in which solute and solvent are of very similar chemical nature. More often, however, Raoult's law holds only for the solvent, A, and then only for solutions that are very dilute in B (i.e. as $N_A \rightarrow 1$). In such solutions the solute, B, tends to follow Henry's law, namely, $p_B = kN_B$, where k is a constant; or $p_B = p_B^\circ N_B f_B$, where f_B is the 'activity coefficient' of B.

If w g of solute of (unknown) molecular weight m are dissolved in W g of solvent of (known) molecular weight M, the partial pressure of A over the solution is given by

$$p_A^s = p_A^\circ \frac{W/M}{(W/M + w/m)}$$

if the solution is ideal. Consequently, a study of the vapour pressure of the *solvent* can give information about the *molar* composition of the solution, and if the weight composition is known, the molecular weight of the *solute* can be calculated. The lowering of the vapour pressure can be measured by the gas saturation method (chapter 6), by a dew-point method (chapter 8) or by the so-called 'isopiestic' method in which the solution is brought to vapour equilibrium with solutions of a substance of known vapour pressure, but although these methods are frequently used for studying the thermodynamic properties of solutions, they are rarely employed for molecular weight determinations. In place of the lowering of vapour pressure it is more convenient to measure the corresponding *elevation of the boiling point* (ΔT_b) or *depression of freezing point* (ΔT_f). The origin of these phenomena can be understood from the vapour pressure–temperature diagram (Fig. 7A.1). The solution, having a lower

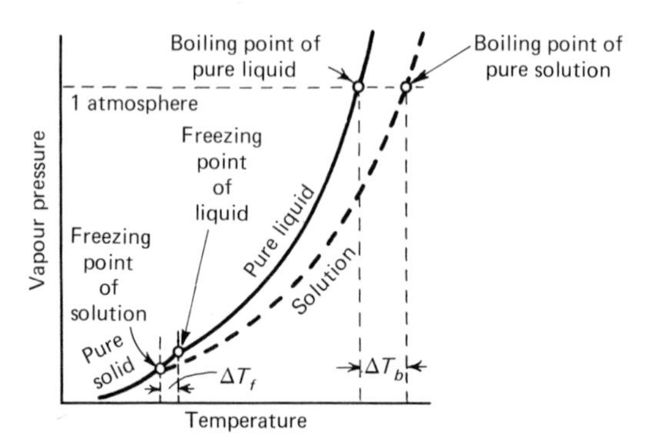

FIG. 7A.1 Diagram showing the effect of a non-volatile solute in reducing the vapour pressure, raising the boiling point and depressing the freezing point of a solution.

vapour pressure than the pure solvent (assuming solute B to be non-volatile), must be raised to a higher temperature to bring its pressure up to 1 atm, i.e. its b.p., and similarly the solution must be cooled to a *lower* temperature for its vapour pressure to coincide with that of the pure solid (i.e. its f.p.). It can be shown by thermodynamics that the elevation of the boiling point (the solute being non-volatile) or depression of the freezing point (assuming that pure solvent crystallizes out on cooling) for a dilute solution in which the solvent obeys Raoult's law are given by the same formula, namely:

$$\Delta T = \frac{RT^2}{L} \cdot N_B$$

where ΔT is the elevation (or depression), L is the molar latent heat of vaporization (or crystallization), and T is the boiling point (or freezing point) on the Kelvin scale. On rearrangement, the equation gives for the molecular weight of the solute $m = Kw/\Delta TW$ kg, where $K(=RT^2M/L)$ is a constant for the solvent, known as the boiling point ('ebullioscopic') constant (K_b) or freezing point ('cryoscopic') constant (K_f) in the respective cases. The table below gives the approximate values of these constants for some common solvents.

Solvent	*f.p.* (°C)	K_f	*b.p.* (°C)	K_b	*Pressure* correction*
Acetic acid	16·6	3·900	—	—	—
Acetone	—		56·2	1·710	0·004
Benzene	5·4	5·120	80·2	2·530	0·007
Chloroform	—	—	61·2	3·630	0·009
Ethyl alcohol	—	—	78·5	1·220	0·003
Nitrobenzene	5·3	7·200	—	—	—
Water	0·0	1·858	100·0	0·505	0·001

* The correction values in the last column are per 10 mm, and should be added to K_b when the barometric pressure is above 760 mm and subtracted when it is below 760 mm (see Hoy and Fink, *J. Phys. Chem.*, 1937, **41**, 453).

The f.p. depression is easier to measure accurately than is the b.p. elevation. It has therefore been used frequently for studying the behaviour of solutes which give abnormal results for 'molecular weights' on account of *association* (e.g. carboxyllic or hydroxy-compounds in benzene), *dissociation* (e.g. salts in water), hydration, or complex formation (e.g. $AgNO_3 + NH_3$, $KBr + HgBr_2$). Ideally, the method provides means of counting the total number of molecules or ions in the solution, since ΔT gives N_B, the mol. fraction, but the theory is limited to solutions which are more or less ideal. Electrolytes, in particular, are grossly non-ideal because the coulombic interaction between the ions has an important influence even in dilute solutions, and consequently the above theory does not hold. Instead, measurements of freezing point depression are used to study empirically the activity coefficients of electrolyte solutions.

Osmotic pressure. Another property which is closely related to those discussed above is the *osmotic pressure*. This is the hydrostatic pressure which must be applied to the solution to bring its vapour pressure up to that of the normal pure solvent. The osmotic pressure of very dilute, ideal solutions is given by $\pi = n\boldsymbol{R}T/V$, where n is the number of moles of solute dissolved in a solution of volume V. Osmotic pressure therefore provides another means of determining molecular weights. However, the technique—especially the preparation of semi-permeable membranes—is outside the scope of this book.[1] The method is used for high polymers and proteins; measurements are made at a series of concentrations and extrapolated to infinite dilution to eliminate the effect of non-ideality of the solutions. This work is now normally done using commercially available instruments which eliminate the lengthy equilibration otherwise required.

In recent years the *scattering of light* by solutions has been shown to be related to their osmotic pressure, and an important new method of determining molecular weights has been developed.[2]

BIBLIOGRAPHY 7A: The colligative properties

Guggenheim, *Thermodynamics*, 4th edn, 1959 (North-Holland Pub. Co., Amsterdam).

Caldin, *An Introduction to Chemical Thermodynamics*, 1958 (Oxford University Press, London).

[1] For technique of osmometry see Weissberger, Vol. I, Pt 1, chapter XV.

[2] A good review of the method is given by Doty and Edsall, *Advances in Protein Chemistry*, 1951, **6**, 37. See also Appendix I.

7B Depression of the freezing point

1. BECKMANN'S (SUPERCOOLING) METHOD

The principle of the classical method developed by Beckmann is to find the arrest that occurs in the cooling curve of the solution, corresponding to the onset of crystallization of solid. The solution is cooled very slowly by a freezing mixture. Crystallization usually does not begin until the temperature has fallen a little below the true freezing point, i.e. the solution supercools, but as soon as nucleation has taken place the latent heat of crystallization tends to raise the temperature. The temperature cannot rise above the freezing point, but if the cooling is too rapid it may well never reach the true melting point. The proper experimental technique must therefore be adhered to closely to obtain reliable results.

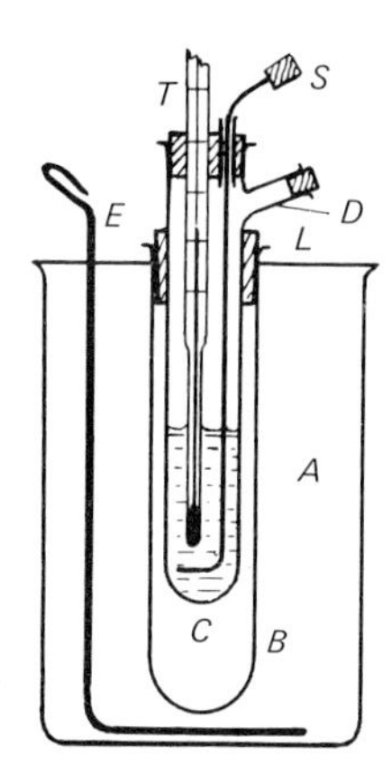

FIG. 7B.1 Beckmann's freezing point depression apparatus.

Apparatus (Fig. 7B.1). The solution is contained in the freezing-point tube *C* which has a side tube *D* for insertion of the solute, and is fitted with a sensitive thermometer *T* (see below) and a stirrer *S*, to the upper end of which a non-conducting handle of cork or wood is attached. The stirrer passes through a short piece of glass tubing inserted in the cork of the freezing-point tube so that it moves freely. The tube *C* is supported by a ring of asbestos inside a wide boiling tube *B* which serves as an air-mantle separating *C* from the freezing bath. This ensures a slower and more uniform rate of cooling of the liquid. The freezing bath is contained in a glass jar *A*, on top of which rests a lid of brass or, better, plastic. The lid has a central hole to take the tube *B* and also holes for a stirrer *E* and thermometer.

2. THE BECKMANN THERMOMETER

In order that the determination of the depression of the freezing point can be made with sufficient accuracy, it is necessary to be able to read the temperatures with an error not exceeding 0·001–0·002°C. A Beckmann thermometer is therefore used. This instrument has a large bulb and a range of only five or six degrees. It is graduated in 0·01°C and can be read to the nearest 0·001°C with the aid of a special lens. It does not indicate absolute temperatures, but merely differences of temperature. The special feature of the Beckmann thermometer is that the amount of mercury in the bulb can be adjusted to make the thermometer read at any desired region of temperature. The regulation is effected by transferring some mercury into a small reservoir at the upper end of the capillary, or vice versa. Clearly, the lower the temperature for which the thermometer is required to read, the greater must be the quantity of mercury in the bulb.

WARNING. *Beckman thermometers are very fragile and expensive and must therefore be handled with special caution. They should always be returned to their case when out of use, and particular care is needed when mounting the thin outer sheath in a clamp or other support.*

Setting the Beckmann thermometer. Before using the Beckmann thermometer, it must be 'set', i.e. the amount of mercury in the bulb must be adjusted so that at the particular temperature of the experiment the end of the mercury thread is on the scale. This is done as follows.

First hang the thermometer in a beaker of water which has been adjusted, with the aid of an ordinary thermometer, to the temperature at which the experiment is to be conducted, and see whether or not the top of the mercury of the Beckmann thermometer stands on the scale. If it does not, then suppose in the first place that it does not rise as far as the scale; that is, suppose there is too little mercury in the bulb. In this case, place the thermometer in a bath the temperature of which is sufficiently high to cause the mercury to pass up to the top and to form a small drop at the end of the capillary. Now invert the thermometer, and tap it gently so as to collect the mercury in the reservoir at the end of the capillary and to join with the mercury there. Return the thermometer carefully, without shaking, to the upright position, and place the bulb again in a bath, regulated for the required temperature. Since the scale of the Beckmann thermometer does not extend upwards to the end of the capillary, the temperature of this bath must be at least 2–3°C *higher* than the highest temperature to be met with in the experiment. The mercury in the bulb will contact and draw in more mercury from the reservoir. After several minutes, when the thermometer will have taken the temperature of the bath, strike the upper end of the thermometer gently against the palm of the hand so as to cause the excess of mercury to break off from the end of the capillary. Make sure, now, that the amount of mercury has been properly regulated, by placing the thermometer in a bath at a temperature which is equal to the highest that will occur in the experiment, and see that the mercury stands on the scale. If it stands above the scale, too much mercury has been introduced, and some of it must be got rid of by driving the mercury once more up into the reservoir and shaking off a little of it from the end of the capillary. Of course, if the mercury is found to stand too low on the scale, then more mercury must be introduced into the bulb in the manner described above, these operations being repeated until the proper amount of mercury has been introduced. *This must always be tested by placing the thermometer in a bath at the temperature of the experiment and making sure that the mercury remains on the scale.*

On account of the so-called 'thermal after-effects' met with in the case of glass, owing to which glass, after being heated, does not immediately acquire its original volume, it is advisable to have at least two Beckmann thermometers, one for use at lower, the other for higher temperatures. Also, it is worth mentioning that it is often possible to get the mercury to pass from the bulb to the reservoir or vice versa simply by inverting the thermometer and tapping it gently. This avoids heating the thermometer.

3. PRECAUTIONS

The following precautions should be observed in carrying out determinations by the supercooling method:

3.1. *The temperature of the cooling bath must not be too low*, otherwise the supply of heat by crystallization of the solvent may not be great enough to counteract the abstraction of heat by the cooling bath. In practice, the temperature of the cooling bath should not be more than 3 or 4°C below the freezing point of the liquid.

3.2. *The amount of supercooling should not exceed 0·3 °C.* If a larger amount of supercooling occurs, so large is the quantity of solid which eventually separates that the concentration of the solution is thereby appreciably increased, with consequent spurious further depression of the freezing point. When the solution has supercooled by about 0·1°C it is stirred vigorously to encourage crystallization to begin. The stirrer must be of glass or stainless steel: copper is attacked by, e.g., benzoic acid. If this does not succeed, the solution must be 'seeded' by adding a small crystal of frozen solvent. Another method is to insert into the solution a fine glass capillary which has been standing in a tube containing solvent, the tube having been cooled until the solvent is partly frozen.

3.3. *The stirring should not be too rapid and should be as uniform as possible* otherwise too much heat is generated. An up-and-down movement of the stirrer at the rate of about once per second is sufficient.

3.4. *The thermometer should always be tapped before a reading is taken.* As the bore of the thermometer is very narrow, the mercury is apt to 'stick', and the object of tapping the thermometer gently with the finger or a small padded 'hammer' is to overcome this hysteresis of the meniscus.

EXPERIMENT

Determine the molecular weight of a substance (e.g. naphthalene or benzoic acid) from its depression of the f.p. of benzene.

Procedure. First set up the apparatus (section 7B.2) completely, and see that the stirrer in the f.p. tube works smoothly without striking against the bulb of the thermometer. Remove the thermometer and stirrer from the f.p. tube, and fit the latter, which must be clean and dry, with an unbored cork. Weigh this tube, and then pour in 15–20 g of *pure* benzene, and weigh again. For this purpose a balance weighing to a centigram should be used. Now set the Beckmann thermometer so that at the temperature of 5·4°C (m.p. of benzene) the mercury stands not lower than the middle of the scale. Dry the thermometer thoroughly and insert it, along with the stirrer, in the f.p. tube, so that the bulb of the thermometer is completely immersed in the benzene. Fill the vessel *A* with water and ice, so that a temperature of about 2–3°C is obtained. This can be regulated by varying the amount of water and ice. The f.p. of the benzene is then determined.

In doing this, make a first approximate determination by placing the f.p. tube directly in the cooling bath, so that the temperature falls comparatively rapidly. When solid begins to separate, quickly dry the tube and place it in the air-mantle in the cooling bath; stir slowly and read the temperature when it becomes constant. Now withdraw the tube from the mantle and melt the solid benzene by means of the hand. If in this operation the temperature of the liquid is raised more than about 1°C above the f.p., place the tube again directly in the cooling bath and allow the temperature to fall to within about half a degree of the f.p. as determined above; quickly dry the tube and place it in the air-mantle and allow the temperature to fall, stirring slowly all the while. When the temperature has fallen to about 0·2–0·5°C below the approximate f.p. found above, stir more vigorously. This will generally cause the crystallization of the benzene to commence, and the temperature will begin to rise. (If not, 'seed' with a crystal of frozen solvent.) Stir slowly again, and with the help of a lens, read the temperature every few seconds, tapping the thermometer firmly with the finger each time before doing so. Note the highest temperature reached. Again melt the solid benzene which has separated out, and re-determine the f.p. in the manner just described. Not fewer than three concordant readings of the f.p. should be made, the mean of these being then taken as the f.p. of the benzene. The deviations of the separate readings from the mean value should not exceed 0·002°C.

The f.p. of the solvent having been determined, a weighed amount of the substance ($\approx$0·1 g), compressed into tablet form, is now introduced into the benzene through the side tube *D* of the apparatus. The amount taken should be sufficient to give a depression of the f.p. of not less than 0·2°C. After the substance has dissolved, the f.p. of the solution is determined in exactly the same manner as described for the pure solvent, first an approximate and then not fewer than five accurate determinations being made. In each case note the degree of supercooling.

Two or three further additions of the substance should be made, and the f.p. of the solution determined after each addition. The total depression of the f.p. should not exceed about 0·5°C. Plot the depression ΔT against the *total* weight of substance added. From the slope of the graph calculate the molecular weight of the solute. The error should not exceed 3–5%.

4. ABNORMAL MOLECULAR WEIGHTS

EXPERIMENT

In the manner previously described, determine the apparent molecular weight of benzoic acid in benzene, and from the numbers obtained calculate the degree of association, assuming that two single molecules combine to form one compound molecule.

Calculation. The degree of association can be calculated in the following manner: If x represents the degree of association, or the fraction of the total number of molecules which combine to form larger molecules, and if n represents the complexity of the new molecules, then of each mole of substance taken there will be $1-x$ mole unassociated, and x/n mole associated. Consequently, instead of there being 1 mole there will be only $(1-x+x/n)$ or $[1-x(1-1/n)]$. In other words, the number of dissolved molecules has decreased in the ratio of $1:1-x(1-1/n)$. The depression of the f.p. is proportional to the number of moles (in a given weight of solvent); hence, if d_t represents the depression calculated on the assumption of no association, and d_0 the depression actually obtained

$$\frac{d_0}{d_t} = \frac{1-x(1-1/n)}{1}, \quad \text{or} \quad x = \frac{d_t - d_0}{d_t(1-1/n)}$$

5. THE EQUILIBRIUM METHOD

Instead of slowly freezing a solution of known concentration one can bring a solution to equilibrium with excess of pure frozen solvent and determine the temperature and concentration of the equilibrium mixture. This method is particularly suitable for aqueous solutions, and is capable of giving very high accuracy.[1] The determinations are best carried out in a Dewar vacuum vessel of 200–250 cm^3 capacity, closed by a rubber stopper through which pass a Beckmann thermometer and a stirrer. A glass tube, wide enough to allow the passage of a pipette, also passes through the rubber stopper. While the experiment is in progress, this tube is plugged with cotton-wool. The Dewar vessel is packed in a wooden box with cotton-wool. The box should also be closed by a lid pierced with holes for the thermometer and stirrer, and for the passage of a pipette. Distilled water, sufficient in amount to cover the bulb of the thermometer, is placed in the Dewar vessel which is then practically filled with broken ice prepared from distilled water. The contents of the vessel are stirred and the equilibrium temperature read as carefully as possible on the Beckmann thermometer. The water is then poured away and about 100 cm^3 of 0·2–0·3 molal solution, previously cooled to 0°C by standing in a jar of ice, are poured into the Dewar vessel. More pure crushed ice is added if necessary. When the temperature has become constant, 20–25 cm^3 of the solution are withdrawn by means of a fine-pointed pipette, and run into a small weighed flask. The weight of the solution is determined and its concentration then ascertained by titration or by gravimetric analysis or by a physical method of analysis such as determination of the refractive index, conductivity or absorption of light.

6. CALCULATION OF OSMOTIC COEFFICIENTS AND ACTIVITY COEFFICIENTS FROM FREEZING POINT DATA FOR DILUTE SOLUTIONS

Solvent activity. In the first instance, a freezing point depression gives a measure of the *activity* of the solvent in the given solution at its freezing point (T).

Provided the solution is so dilute that the heat of dilution is negligible, it can be shown that the activity of the solvent in the solution $(a_1)_T$ relative to that of the pure solvent, is given by

$$\log_e (a_1)_T = -\int_T^{T^\circ} \frac{L_f}{RT^2}\,dT$$

where L_f is the molar heat of fusion of the pure solid solvent, T° is the f.p. of the pure solvent, and T is that of the solution. L_f is itself dependent on temperature, and can be expressed in terms of T by means of the Kirchhoff equation, thus,

$$(L_f)_T = (L_f)_{T^\circ} + (C_p^S - C_p^L)(T^\circ - T)$$

where C_p^S and C_p^L are the molar heat capacities of the pure solid and liquid respectively.

In the case of water, $(L_f)_{T^\circ}$, the heat of fusion at 0°C, is 6 017 J, and $(C_p^S - C_p^L)$ is -38 J K^{-1}. Hence, putting the change of f.p. as $\Delta T = T^\circ - T$, it follows as a first approximation

$$\log_e (a_w)_T = -0{\cdot}009696\,\Delta T - 0{\cdot}0000051\,(\Delta T)^2$$

Measurements of the f.p. depression, ΔT, therefore give at once the activity of the solvent, $(a_1)_T$.

If the solution were ideal, according to Raoult's law a_1 would be equal to the mol fraction of solvent in the solution, N_1.

Osmotic coefficients. For *non-ideal* solutions it is often convenient to discuss the thermodynamic behaviour of the solvent in terms of its *osmotic coefficient* g_1 (introduced by Bjerrum*), which is defined by $a_1 = (N_1)^{g_1}$. If the molecular weight of the solvent is M, and the total *molality* (moles per kg of solvent) of solutes in the solution is m, then, for a dilute solution, it follows to a good approximation

$$\log_e a_1 = g_1 \log_e (N_1) = -g_1 M_1 m$$

* Bjerrum's g = van't Hoff's i/v = Lewis and Randall's $(1-j)$.

Clearly, an ideal solution corresponds to $g=1$. For non-ideal solutions $(1-g_1)$ is a convenient measure of the deviation from ideality.

For the important case of an aqueous solution of an electrolyte of molality m which dissociates into ν ions per molecule, it follows that the osmotic coefficient of the water, g_w, is given by

$$g_w = -55{\cdot}51 \log_e a_w/\nu m$$

Substituting for $\log_e a_w$ in terms of the f.p. depression,

$$g_w = [0{\cdot}5382\Delta T - 0{\cdot}00028(\Delta T)^2]/\nu m \qquad (1)$$

Activity coefficients. The osmotic coefficient of the solvent, g_1, is related to the 'practical' activity coefficient of the solute, γ_2, through the Gibbs–Duhem relation. For a binary solution containing n_1 moles of solvent species (1) and n_2 moles of solute (2)

$$n_1 \, d(\log a_1) + n_2 \, d(\log a_2) = 0$$

Putting $\log a_1 = -g_1 m_2 M_1$ and $a_2 = \gamma_2 m_2$ leads to the equivalent form

$$-d \log \gamma_2 = -dg_1 + (1-g_1) \, d(\log m_2)$$

On integration from infinite dilution (where $\gamma = g = 1$) to molality m, this gives

$$-\log \gamma_2 = (1-g_1) + \int_0^m (1-g_1) \, d(\log m)$$

In principle, therefore, γ for the solute at molality m can be obtained from f.p. depression determinations made on a series of solutions from infinite dilution up to m, the integral being obtained graphically. In practice, the relative error incurred in f.p. measurements on extremely dilute solutions excludes the possibility of obtaining accurate results by this procedure. Consequently, it is usual to assume some suitable form of law to govern g in extremely dilute solutions, and to perform the above integration only over the range of solutions which can be studied experimentally with accuracy. Thus, if the activity coefficient at molality m' is γ' (supposed known) then its value γ'' at another molality m'' is obtained from

$$\log_{10} \frac{\gamma'}{\gamma''} = (g' - g'')/2{\cdot}303 + \int_{m'}^{m''} (1-g) \, d(\log_{10} m) \qquad (2)$$

Electrolytes. The above equations apply equally to a solute which is an electrolyte as to one which is a non-electrolyte; in the former case, γ is the mean ion activity coefficient (cf. section 13A). However, the osmotic behaviour of salts in extremely dilute solutions is entirely different from that of non-electrolytes. Henry's law is the limiting law for non-electrolytes—i.e. $a_2 \propto m_2$, which corresponds to $\gamma_2 = \text{const}$ or $(1-g_1) \propto m_2$. For electrolytes, the limiting law is that of Debye and Hückel (cf. section 13A), according to which $-\log \gamma_\pm$ (or $1-g$) is proportional to $\sqrt{m}$. By assuming appropriate interpolation formulae, the range between infinite dilution and the lowest experimentally accessible concentration can be included in the above integration.[1]

EXPERIMENT

Determine the osmotic coefficients (and activity coefficients) of solutions of an electrolyte (e.g. hydrochloric acid) from about 10 to 1 000 mol m^{-3} *by the equilibrium f.p. method.*

Procedure. Prepare some pure ice, break it into small pieces, and half-fill a Dewar vessel with it. Add enough distilled water, previously cooled to 0°C, to just fill the interstices of the cracked ice. Insert a Beckmann thermometer set so that the ice-point reads near the top of the scale, and stir up the ice and water until a steady reading is recorded.

Now add some pure hydrochloric acid (also previously cooled to 0°C) and obtain a new equilibrium f.p. Note the steady temperature which is reached, and then take a sample of the equilibrium solution by means of a pipette with a fine point. If the solution is very dilute, at least 100 cm^3 should be taken for analysis. Determine the concentration of the solution by titration as accurately as possible. (As the end-point of iodine titrations is sharper than that of acid-alkali titrations in the case of dilute solutions, it is advantageous to determine the acid by running it into an excess of a KI–KIO_3 mixture and titrating the iodine liberated with $Na_2S_2O_3$, using starch at the end-point. If desired, the chloride present could be determined instead by a potentiometric titration (section 13G).)

Now make a second addition of HCl. Obtain a new equilibrium temperature and take a sample of the liquid for analysis. Continue making additions of HCl to obtain further values of the f.p. at increasing concentrations until a depression of 2–3 degrees is reached.

Treatment of results. Calculate the osmotic coefficients of the solutions by use of equation (1) given above, taking $\nu=2$ for HCl. Plot a graph of g against $\sqrt{m}$. Note that the experimental error on g becomes excessive for very dilute solutions. However, it is known that $g \rightarrow 1$ as $\sqrt{m} \rightarrow 0$, and, further, the line for g on the graph becomes asymptotic to the limiting law of Debye and Hückel in extremely dilute solutions, namely, for a 1:1 electrolyte in water at 0°C,

$$1-g = 0{\cdot}374\sqrt{m}$$

Plot this line on the graph, and interpolate the line for g. Use the smoothed curve for the following exercise in the calculation of activity coefficients. (It will be appreciated that f.p. data of extremely high accuracy are called for in this field.)

It may safely be assumed that solutions of HCl conform fairly closely to the Debye-Hückel equation up to concentrations of 0·001 molal. At this concentration the activity coefficient (for 0°C) is given (section 13A) by

$$-\log_{10} \gamma_{\pm} = 0{\cdot}488\sqrt{m}$$

Compute values of $\gamma_{\pm}$ for 0·01, 0·05, 0·1, 0·5 and 1 molal by graphical integration of equation (2) above, plotting $(1-g)$ against $\log_{10} m$ from $m=0{\cdot}001$ upwards.

Other applications. Solutions of H_2SO_4 give much lower values of g than those for HCl. This is partly because of the effect of the divalent SO_4^{--} ions, and partly because the second stage of dissociation, $HSO_4^{-} \rightarrow H^{+} + SO_4^{--}$, is weak.

Colloidal electrolytes, such as dodecyl sulphonic acid, behave as normal electrolytes in extremely dilute solutions, but above a certain concentration (the critical micelle concentration) further increase of concentration lowers the f.p. very little, and g falls to abnormally low values. This is due to aggregation of the molecules into colloidal micelles, with the result that the number of osmotically active particles is much less than that calculated on the assumption of complete dissociation.

F.p. determinations on weak acids such as acetic can be used to deduce dissociation constants provided allowance is made for the activity coefficients of the ionic species (cf. section 13G).

7. RAST'S MICRO-METHOD OF DETERMINING MOLECULAR WEIGHTS

The basis of Rast's method of determining molecular weights is the use of *camphor* as a solvent. The f.p. depression constant K_f of camphor is exceptionally large, namely, about 40°C for 1 mole of solute in 1 kg of solvent, as compared with 5°C for benzene and 1·86°C for water. Consequently, a 10% solution in camphor may have a m.p. many degrees lower than that of pure camphor, and the depression can be measured sufficiently accurately with an ordinary thermometer instead of a Beckmann thermometer. By using a micro m.p. method, the determination of molecular weight can be carried out with as little as 0·2 mg of solute. Such small quantities have, of course, to be weighed on a micro-balance; if an ordinary analytical balance, weighing to ±0·1 mg is used, the minimum quantity of substance required will be 10 mg since this can be weighed to the nearest 1%, which is the limit of accuracy expected of the method.

'Camphor' is not a unique substance. The common variety is either Japan camphor, which is *d*-2-camphanone ($C_{10}H_{16}O$, m.p. 178–179°C, [α] = +44·26°C), or artificial camphor (*dl*-). 'Camphor' from Borneo, Malaya or Sumatra is *d*-borneol ($C_{10}H_{17}OH$, m.p. 208°C, [α] = +37·4°C). Commercial samples vary slightly in m.p. and optical rotation. Camphor is too waxy to be ground to a powder *dry*, but is easily ground when moistened with ether. Owing to the variable character of samples of camphor it is necessary to determine the m.p. and the f.p. depression constant of the camphor used in every experiment. Acetanilide (mol. wt 0·1351 kg) or naphthalene (mol. wt 0·1281 kg) are suitable substances for calibration. In view of the necessity of making this standardizing measurement as well as the determination of m.p. of the pure camphor, there is no need to apply emergent stem corrections if all the temperatures are taken on the same thermometer.

Camphor is a good solvent for many substances. In using it to find molecular weights the implicit assumption is made that dilute solutions in camphor are 'ideal', and, further, that the camphor crystallizes out first on cooling, since the simple f.p. depression theory does not apply when solid solutions, eutectics or compounds separate. The validity of these assumptions has apparently never been proved. In practice, results within 5% are generally obtained. The method cannot be used for substances which are insoluble in camphor, react with it chemically, or decompose when heated to the m.p. of camphor. It has been applied successfully to liquid solutes.

EXPERIMENT

Determine the f.p. depression constant of camphor using acetanilide as solute at three concentrations, about 5, 10 and 20% by weight. Hence calculate the latent heat of fusion of camphor in joules per kg.

Preparation of the solution. Prepare a thin-walled glass tube about the size of an ignition tube (say about 1 × 5–10 cm), one end being pulled off and rounded by blowing into a slight bulb. Dry the tube in the oven, cool it, and weigh it as accurately as possible (±0·1 mg). By means of a 'micro-spatula' (i.e. a stainless steel or platinum wire or thin glass rod flattened at one end) introduce about 10 mg of the substance to be used as solute into the bottom of the tube and weigh again. Now add about 100 mg of camphor and weigh a third time.

Before the contents of the tube are melted to form a uniform mixture, the open end of the tube must be sealed off and pulled out to a long, stout fibre to prevent volatilization of camphor. The tube is now placed in a small bath of high-boiling paraffin oil heated to the melting point of camphor, and when melted the contents are thoroughly mixed by rotating the tube, which is then left to cool in the bath. This precaution is to prevent condensation of camphor on the upper part of the tube, which might otherwise occur.

Determination of a melting point. When a homogeneous solution has been made and cooled, open the tube and introduce some of the contents into a thin-walled m.p. capillary ready for the determination of its m.p. The capillary is made by drawing down a thin glass tube such as an ignition tube, which should be absolutely clean. The final capillary should be about 2 mm wide at the bottom, 4–5 cm long, and sealed neatly at the bottom without a large bead of glass. Particles of solid are introduced by means of a fine glass ramrod. Prepare a similar capillary with pure camphor. Attach the capillary tubes to the sides of a suitable thermometer by means of a thin rubber band (cut from the end of a piece of rubber tubing). The particles of solid should lie alongside the thermometer bulb. Several m.p. tubes may be run at the same time. The procedure is now the usual one commonly employed for finding the m.p. of organic solids. High-boiling paraffin or glycerol may be used as heating bath, and a small beaker is a convenient container.

Clamp the thermometer with its attached capillary tubes so that the bulb and samples are just immersed in the heating bath. Raise the temperature rapidly to the neighbourhood of the m.p., and then warm *very slowly* with thorough stirring. A light glass-rod stirrer and a micro-burner may be used. Watch the samples closely. Near the m.p. the particles take on the appearance of melting ice, but still contain skeleton crystals. The point to be noted is the temperature at which the last crystals just disappear. Record this temperature as accurately as possible. The point when crystals first appear on *slow* cooling may also be determined. The measurements of m.p. and f.p. should be repeated several times until concordant results are obtained before dismantling the apparatus.

BIBLIOGRAPHY 7B: Depression of the freezing point

Weissberger, Vol. I, Pt 1, chapter III.

[1] Guggenheim, *Thermodynamics*, 4th edn, 1959 (North Holland Publ. Co., Amsterdam); Harned and Owen, *Physical Chemistry of Electrolytic Solutions*, 3rd edn, 1958 (Reinhold, New York).

7C Elevation of the boiling point

The true b.p. of a liquid is the temperature at which its vapour and liquid are in equilibrium under a pressure of 1 atm. This temperature cannot be determined accurately by simply inserting a thermometer into the boiling liquid, as the temperature in the liquid is variable and too high on account of local superheating and the effect of hydrostatic pressure. In the case of a *pure* liquid the *condensation* temperature can be determined instead by placing a thermometer in the condensing vapour, but in the case of solutions this method is not applicable since the condensation temperature then differs from the b.p. owing to fractionation.

All modern forms of apparatus make use of a principle introduced by Cottrell (1919); the thermometer is hung *in the vapour* above the boiling solution and a stream of boiling liquid is sprayed by a simple 'pump' on to the thermometer *above the bulb*. The liquid is slightly superheated (e.g. by about 0·05°C) when it emerges from the pump, but in flowing down the thermometer it very quickly comes to equilibrium with the surrounding vapour, cooling slightly by evaporation. The liquid dripping from the bulb is therefore at its true b.p.

Two forms of b.p. apparatus ('ebulliometers') are shown in Fig. 7C.1: (a) is the apparatus of Washburn and Read, and (b) is the

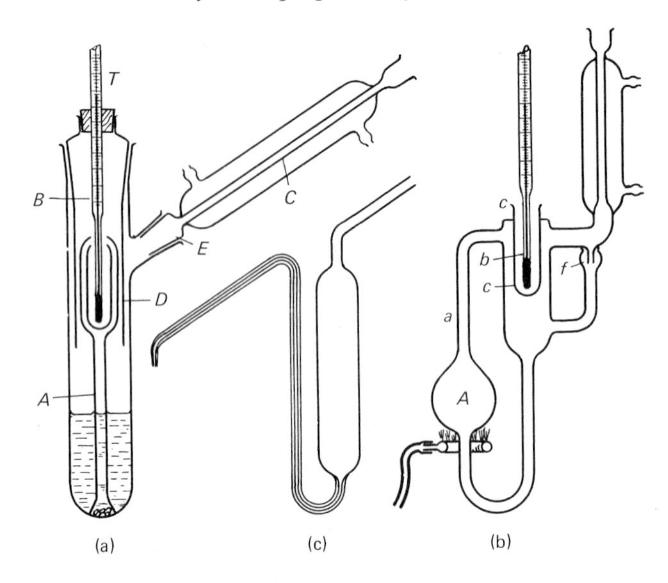

basic form of ebulliometer favoured by Swietoslawski. In both types of apparatus the boiling liquid is carried upwards by the stream of vapour bubbles through a narrow tube A, *a* (the 'pump' or 'lift') so that liquid squirts continually on to the thermometer stem *B* or inserted well *c* into which the thermometer is dipped. In the former apparatus the lift divides into two or three branches symmetrically arranged round the thermometer. The tube *D* is to act as a radiation screen round the thermometer and to keep off cold liquid which may run down the condenser *C*. Swietoslawski's ebulliometer incorporates a drop-counter *f* which indicates the rate at which the liquid is distilling. Although originally heated by gas, ebulliometers are now heated by an electric mantle.

EXPERIMENT

Determine the molecular weight of a substance of low volatility by the ebullioscopic method, e.g. camphor, anthracene or ethyl benzoate in benzene, or azobenzene in chloroform.

Procedure. A suitable amount of pure solvent is placed in the clean, dry boiling vessel, the quantity being determined by weighing. A weighed delivery pipette similar to that shown in Fig. 7C.1(c) may conveniently be used. The apparatus is assembled and wrapped with asbestos paper. If gas heating is used, a small burner capable of good regulation should be employed and protected from draughts by asbestos boards. The Beckmann thermometer is set to the temperature at which the solvent boils and the apparatus is then assembled, and the solvent is brought to the boil.

It is essential to adjust the rate of boiling to the optimum. If the ebulliometer is functioning correctly there ought to be an appreciable range over which the heating can be altered without any change in thermometer reading, and the middle of this range should be used. It is therefore necessary to make some preliminary readings of temperature at different heating rates, the latter being noted, for example, by the setting of the gas-heating control or by the current passing (if electric heating is used) or by the rate of fall of drops from the condenser. For example, it may be found that the thermometer reading remains unchanged over the range 10–20 drops per minute; 15 drops per minute would therefore be chosen. Once the optimum rate of heating has been found, it should be closely adhered to throughout the subsequent work.

After the b.p. of the pure solvent has been found, a weighed pellet of solute (prepared in a tablet press) is introduced by way of the condenser, and the new b.p. is determined exactly as before. Further pellets may be added subsequently if several concentrations are to be studied.

The above readings can be used to calculate the molecular weight of the solute, assuming the theoretical value of the b.p. elevation constant. The result, however, is subject to two experimental errors; firstly, the solution is slightly more concentrated than when it was made up because a small, unknown amount of solvent is present in the condenser, and secondly, the graduations on the Beckmann thermometer may not be precisely equal to centigrade degrees.

A rough correction for the first error is to subtract 0·2 g from the weight of solvent in making the calculations. A better method is to calibrate the ebulliometer with a solute of known molecular weight and use the empirical elevation constant so obtained in subsequently studying the elevation produced by the solute of unknown molecular weight. If the solutes are of similar chemical type, this comparative procedure will also tend to eliminate errors due to deviations from Raoult's law. The same rate of boiling must obviously be used throughout.

Pressure correction. One difficulty that militates against more widespread application of ebulliometry is the sensitivity of the b.p. to changes of barometric pressure; changes are liable to occur while a series of measurements is in progress. The pressure-dependence of the b.p. can be calculated from the Clapeyron–Clausius equation in the form $dP/dT = PL/RT^2$. P now becomes the barometric pressure B and T is the b.p. Hence, $dT/dB = RT^2/LB = K_b/MB$, where K_b is the b.p. elevation constant (section 7A) and M the molecular weight of the solvent. Hence, a change of barometric pressure equal to dB produces a change of b.p., $dT = K_b\,dB/MB$. The following list gives the increase of the normal b.p. ($B = 760$ mm) of some common solvents produced by an increase of barometric pressure of 1 mm: acetone 0·039°C, benzene 0·043°C, carbon tetrachloride 0·044°C, chloroform 0·042°C, ethyl alcohol 0·033°C, methyl alcohol 0·033°C, water 0·037°C. It is seen therefore that if the thermometer reads to 0·001°C, a significant change will be noticed if the barometric height varies by as little as 0·025 mm.

For ebullioscopic measurements of high accuracy it is usual to avoid the necessity of making the pressure corrections by having a second apparatus in which the pure solvent is kept boiling, or by using Swietoslawski's differential ebulliometer. The latter contains two thermometer wells, one bathed with boiling liquid from the Cottrell lift and the other in the condensing vapour. The Beckmann thermometer can be quickly transferred from one well to the other to obtain the difference between the b.p. of the solution and that of the pure solvent.

Modern high-precision ebulliometers employ a differential, multiple-junction thermocouple in conjunction with a sensitive galvanometer to measure the b.p. elevation.[1] The apparatus is calibrated with a solute of known molecular weight.

BIBLIOGRAPHY 7C: Elevation of the boiling point

Weissberger, Vol. I, Pt 1, chapter VIII.
Swietoslawski, *Ebulliometric Measurements*, 1945 (Reinhold, New York).
[1] Ray, *Trans. Faraday Soc.*, 1952, **48**, 809.

7D Distribution of a solute between immiscible solvents

When a substance is 'extracted' from aqueous solution by means of ether—a common operation in organic chemistry—the extraction is not complete, but the solute distributes itself between the two solvents according to its 'solubility' in each. Such solutions cannot both obey Raoult's law, but as a first approximation they generally follow Henry's law, according to which the partial pressure (or activity) of the solute is proportional to its concentration (provided the solutions are very dilute). The proportionality constants are different, however, for the two solvents. It follows, therefore, that at a given temperature the solute distributes itself in a constant ratio between the two liquids. This ratio, called the distribution (or partition) coefficient, is generally approximately independent of the total concentration (Nernst's Distribution Law).

Deviations from the law occur when the solute undergoes chemical changes in one of the phases and not in the other. The law is then applicable to the concentration of any particular chemical species in the solutions, although not to the total concentration of the substance. Measurements of the distribution can thus afford a method of studying such processes as association, dissociation, solvation, hydrolysis, complex formation and 'salting-out'. Approximate values of the corresponding equilibrium constants may be obtained, but the method is limited quantitatively by the assumption that Henry's law holds for the species concerned. The following examples will serve to illustrate the method. Since no special experimental technique is required, further examples will not be given here in detail.

(a) *Simple partition: iodine between water and carbon tetrachloride.*

Here the molecular condition of the solute is the same (I_2) in both solvents and the distribution coefficient k is practically independent of concentration.

(b) *Complex formation: iodine between carbon tetrachloride and solutions of potassium iodide.*

The distribution ratio varies with concentration because in the aqueous phase the reaction $I_2 + I^- \rightleftarrows I_3^-$ occurs, with equilibrium constant $K_c = [I_3^-]/[I^-][I_2]$.

K_c is also called the *stability constant* of the tri-iodide ion. Since KI and KI_3 are ionic and hence insoluble in the non-polar solvent CCl_4, the only species undergoing distribution is molecular I_2, whose concentration is governed by its characteristic distribution coefficient k. Hence, measurement of the concentration of I_2 in the CCl_4 layer, together with a knowledge of k, gives the *equilibrium concentration* of I_2 in the aqueous phase. This concentration is less than the total, titratable iodine present, as the rest is bound as KI_3. The equilibrium constant can therefore be calculated.

EXPERIMENT

Determine the equilibrium constant of the reaction $KI + I_2 = KI_3$ *by the distribution method.*

Practical details. One must first determine the distribution coefficient of iodine between water and a suitable non-miscible solvent, such as carbon tetrachloride or carbon disulphide. To do this, prepare a saturated solution of iodine in, say, carbon tetrachloride at the ordinary temperature, and shake up 20 cm^3 of this solution with 200 cm^3 of water in a stoppered bottle immersed in the water of a thermostat at 25°C. After equilibrium has been attained, allow the bottle to stand in the thermostat for 20–30 min so as to secure complete separation of the two liquid layers. The concentration of iodine in the carbon tetrachloride and in the water is then determined by pipetting out a given volume of the solutions (say, 5 cm^3 of the carbon tetrachloride solution and 50 or 100 cm^3 of the aqueous solution), and titrating with 50 mol m^{-3} thiosulphate solution. (A small quantity of a concentrated solution of potassium iodide is added to the carbon tetrachloride solution in order to ensure complete extraction of the iodine, and the two layers should be well shaken during the titration.) The determination is repeated, using (a) 10 cm^3 of saturated iodine solution, 10 cm^3 of carbon tetrachloride, and 200 cm^3 of water; (b) 15 cm^3 of iodine solution, 5 cm^3 of carbon tetrachloride, and 200 cm^3 of water. The distribution coefficient is then

$$k = \frac{\text{concentration in } CCl_4}{\text{concentration in } H_2O}$$

Similar experiments are then carried out using a solution of potassium iodide of known concentration (say 100 mol m^{-3}). The iodine in the two layers is accurately titrated by means of thiosulphate solution. The concentration of iodine in the two solvents is thereby known. From the concentration of iodine in carbon tetrachloride, the concentration of free iodine in the aqueous solution can be calculated from the distribution coefficient. The titration value of the iodide solution gives the total iodine, and the difference between this and the amount of free iodine gives the iodine combined with potassium iodide to form KI_3. The amount of iodide which has thus combined with iodine can be calculated, and if this amount of iodide is subtracted from the original amount of iodide present, the amount of uncombined iodide is obtained. In this way, the amounts of KI, I_2 and KI_3 in a given volume of solution can be obtained, and the concentrations calculated. The equilibrium constant is then

$$K_c = \frac{[KI_3]}{[KI][I_2]} = \frac{[I_3^-]}{[I^-][I_2]}$$

It is instructive to work out the error in K_c which results from a 1% error in each of the titrations.

Other systems suitable for study by the distribution method:

Simple partition: succinic acid between ether and water.

Dimer formation of carboxylic acids: benzoic acid between water and benzene.

Electrolytic dissociation: monochloracetic acid between water and ether.

Hydrolysis of salts: aniline hydrochloride between water and benzene.

Complex formation: ammonia between chloroform and copper sulphate solutions.

'*Salting-out*' *and* '*salting-in*': benzoic acid between benzene and aqueous solutions of various electrolytes; tetraethyl-ammonium iodide salts 'in'.[1]

Measurement of activity coefficients: hydrochloric acid between water and benzene, and between salt solutions and benzene.

BIBLIOGRAPHY 7D: Distribution of a solute between immiscible solvents

See standard textbooks of Physical Chemistry.
[1] Bockris, Bowler-Reed and Kitchener, *Trans. Faraday Soc.*, 1951, **47**, 184.

7E Homogeneous equilibria in solution

1. THEORY

The reversible reaction $KI(aq)+I_2(aq)=KI_3(aq)$ discussed in the previous section is an example of a large class of reactions which exhibit homogeneous equilibria in solution. The state of equilibrium is governed rigorously by a thermodynamic equilibrium constant: for example,

$$K_{Th} = a_{KI_3}/a_{KI} \times a_{I_2}$$

where the a terms are thermodynamic *activities*.

In the classical treatment of equilibria by Guldberg and Waage, it was assumed that the 'active mass' of a substance in solution could be taken as equal to its *concentration*. This is now known to be only approximately correct. Nevertheless, in many cases an equilibrium constant formulated in terms of concentrations is found to be approximately constant for a series of solutions of different concentration. Strictly, the concentrations should be multiplied by *activity coefficients* (cf. section 13A) but these may be nearly constant, or may partially cancel, as in the KI_3 equilibrium.

Many methods have been employed for arriving at the composition of equilibrium solutions, and hence determining equilibrium constants. A number of examples will be found in different places in the present book, and only one other need be given here, namely, a case in which the amount of product formed in solution can be determined by optical measurements of the intensity of colour of the solution.

2. COMPLEX-FORMATION BETWEEN SALICYLIC ACID AND FERRIC SALTS

Many phenolic compounds give strongly coloured products when treated with solutions of ferric salts. The coloured substance appears to be a weak 'complex' which is in equilibrium with the reactants.

Foley and Anderson[1] studied the stoichiometry of such complexes by measuring the intensity of colour produced with different proportions of the reactants. A theory due to Job[2] was employed which indicates that when *equimolar* solutions of two reactants are mixed in various proportions, the maximum amount of equilibrium product is formed when the proportions of reactants employed correspond to the empirical formula of the product. For example, if the stoichiometry of the reaction can be represented by $a\mathrm{A}+b\mathrm{B} \rightleftarrows \mathrm{A}_a\mathrm{B}_b$, then the greatest amount of $\mathrm{A}_a\mathrm{B}_b$ is present if a parts of a solution of A are mixed with b parts of an (equimolar) solution of B.

EXPERIMENT

Determine the empirical formula and approximate standard free energy of formation of the coloured complex between salicylic acid and ferric ions.

$$\mathrm{Fe}^{+++} + x\,(\mathrm{sal}) \rightleftarrows \mathrm{Fe}^{+++}(\mathrm{sal})_x$$

Procedure. The optimum pH for the complex is about 2·6–2·8. Here the phenolic –OH group and the carboxylic –COOH group are both undissociated, and much of the hydrolysis of ferric salts is suppressed.

This pH may be obtained sufficiently nearly by working throughout with very dilute solutions of the substances in approx. 2 mol m^{-3} HCl; prepare 2 dm^3 of this acid.

Prepare 0·5 dm^3 of a 1·00 mol m^{-3} solution of salicylic acid in 2 mol m^{-3} HCl, and 0·5 dm^3 of a solution of ferric ammonium alum, $Fe_2(SO_4)_3.(NH_4)_2SO_4.24H_2O$, in 2 mol m^{-3} HCl, the latter to contain 1·0 kg *ions* of Fe^{+++} per m^3. This solution hydrolyses on standing and must therefore be used at once for preparing the equilibrium mixtures.

By means of two burettes, prepare a series of mixtures of the salicylic acid and ferric salt in different proportions, ranging from 1:10 to 10:1.

Determine the intensity of the colour (% absorption) of the mixtures by means of a photo-electric colorimeter or a spectrophotometer (section 9D). Use a filter (or wavelength setting) of complementary colour to that of the solution—i.e. that which shows the smallest percent transmission for the solution.

Take 10 cm^3 each of 1·00, 0·75, 0·50, 0·25 and 0·1 mol m^{-3} ferric solution and add *excess* solid salicylic acid in order to convert the ferric salt wholly into the complex form. Determine the percent absorption of these solutions.

Treatment of results. Plot a graph of percent absorption against molar composition of the mixtures made from equimolar solutions. The maximum in the curve gives the empirical formula of the complex.

Calculate the *optical density* (section 9C) of the five solutions used in the second part of the experiment, and plot the resulting values against concentration of ferric ion. If Beer's law is obeyed, the graph should be a straight line through the origin, and the slope of the line gives the extinction coefficient of the complex. This graph can now be used to calculate the concentration of the complex present in the different reaction mixtures. Since the formula of the complex has been found, it is now possible to calculate also, by difference, the concentration of *uncombined* salicylic acid and Fe^{+++} present in the various mixtures, and hence to calculate the classical equilibrium constant K_c for the reaction. The error in K_c due to the uncertainty in the determinations should be calculated.

The free energy of formation of the complex is given by

$$\Delta G = -\boldsymbol{R}T \log_e K_c = -2{\cdot}303\boldsymbol{R}T \log_{10} K_c$$

If the units of concentration are mol m^{-3}, neglecting the difference between concentration and activity, ΔG is the free energy change accompanying the formation of 1 mole of complex in a 1 mol m^{-3} solution from 1 mol m^{-3} solutions of salicylic acid and Fe^{+++} salt. Since the standard state for a solution corresponds to an activity of 1 mol (kg solvent)$^{-1}$ or approximately 1 mol dm^{-3} the corresponding *standard* free energy change ΔG° is given by the same expression but with K_c now calculated using concentrations expressed in mol dm^{-3}.

In considering the probable electronic structure of the coloured complex, one must take the following points into consideration. (1) The profound change of colour shows that the electronic condition of the Fe^{+++} ion has been changed; the colour of the ferric ion is probably connected with an internal electronic transition involving the incomplete 3*d* shell. (2) At pH 2·6–2·8 both the phenolic –OH group and the carboxylic –COOH will be undissociated. (3) No similar coloured complex is formed by ferric ions with either *meta*- or *para*-hydroxybenzoic acid.

BIBLIOGRAPHY 7E: Homogeneous equilibria in solution

Chemical Society Special Publications: Nos 6 (1957) and 7 (1958).

[1] Foley and Anderson, *J. Amer. Chem. Soc.*, 1948, **70**, 1195; *ibid.*, 1949, **71**, 909; Foley and Turner, *ibid.*, 1949, **71**, 912; Meek and Banks, *ibid.*, 1951, **73**, 4108.

[2] Job, *Ann. de Chimie*, 1928, **9**, 113.

8
Phase equilibria

The condition for equilibrium in heterogeneous systems is given by the phase rule: $F = C + 2 - P$, where F is the number of degrees of freedom, C the number of components, and P the number of phases present. For example, when $P = C + 2$, the system has no degrees of freedom; the temperature and pressure under which it can exist are perfectly definite. If one of the variables, temperature or pressure, is fixed, then equilibrium will be definitely fixed when $P = C + 1$. Thus, if the pressure is fixed (say, atmospheric) then equilibrium between a solid and a liquid phase of a single substance will be defined by a definite temperature, namely, the m.p.

Owing to the importance of heterogeneous equilibria, a number of different cases will be studied. The equilibrium between a pure liquid and its vapour, and the f.p. and the b.p. of dilute solutions have already been considered. (Chapters 6, 7.)

8A Vapour pressure of salt hydrates

A salt hydrate is a two-component system. When dehydration occurs a second solid phase, anhydrous salt or lower hydrate, is formed, and the system then consists of three phases—two solid phases and one vapour phase. The system therefore has only one degree of freedom, and the vapour pressure will vary only with temperature, and not with composition. If the temperature is kept constant the vapour pressure will be constant, and may be measured, for example, by the dew-point method.[1]

EXPERIMENT
Determine the vapour pressure of sodium sulphate decahydrate.

The apparatus (Fig. 8A.1) consists of a wide-mouthed bottle *B* which is closed by a rubber stopper. Through this stopper pass a

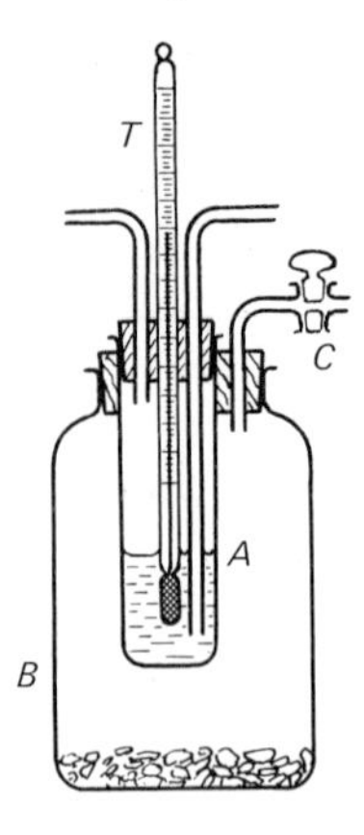

FIG. 8A.1 Determination of aqueous vapour pressure by the dew-point method.

closed tube of bright silver, *A*, and the glass tube *C*, which is furnished with a stopcock. The mouth of the silver tube also is closed by a rubber stopper, through which pass a thermometer *T* graduated in tenths of a degree, and two narrow glass tubes. One of these reaches nearly to the bottom of the silver tube, the other just passes through the stopper. Before being used for an experiment, the silver tube, which must be brightly polished, is dipped into boiling distilled water. In this way, a boundary line is produced which becomes visible when dew is deposited on the tube and renders the deposit more easily detectable.

A quantity (10–15 g) of roughly powdered sodium sulphate decahydrate is placed at the bottom of the dry bottle *B* and the rubber stopper, carrying its silver and glass tubes, is placed firmly in position. A quantity of ether is poured into the silver tube which is then stoppered, and the bottle is supported in a beaker of water, the temperature of which is maintained constant at, say, 20 or 25°C. Tube *C* is connected with a water pump and the bottle is partially exhausted. The stopcock on *C* is then closed.

When the bottle and its contents have taken the temperature of the bath, air is blown through the ether by means of a rubber bulb tube. The ether is thereby caused to evaporate rapidly, and the temperature of the silver tube falls. Stop evaporating the ether, allow the temperature of the silver tube to rise, and note the point at which the film of dew disappears. Several readings of the appearance and disappearance of the dew should be made, and the mean taken. This is the dew-point. From the tables of vapour pressure of water, the pressure corresponding to the dew-point is obtained, and this gives the vapour pressure of the system $Na_2SO_4.10H_2O$–Na_2SO_4–H_2O (*vapour*) at the temperature of the bath.

In the same way, one may determine the dissociation pressures of $CuSO_4.5H_2O$ and of $MgSO_4.7H_2O$ at, say, 20–30°C. The method may also be used to determine the vapour pressure of aqueous solutions. The technique has been refined so that the dew-point can be determined to $\pm 0{\cdot}01$°C.

The vapour pressure of a hydrate may also be determined by finding the concentration of a solution of sulphuric acid (or potassium chloride, etc.) with which the hydrate is in equilibrium. The vapour pressure of the solution can be obtained from tables. This is the principle of the isopiestic method for measuring the v.p. of solutions.

Salts in contact with their saturated solutions provide convenient means of controlling the *relative humidity* in a closed vessel. A list of suitable salts is given in the Appendix, Table A7.

BIBLIOGRAPHY 8A: Vapour pressure of salt hydrates

Weissberger, Vol. I, Pt 1, chapter IX.
Findlay (revised Campbell and Smith), *The Phase Rule and its Applications*, 9th edn, 1951 (Dover, New York).

[1] Cumming, *J. Chem. Soc.*, 1909, **95**, 1772; Hepburn, *ibid.*, 1932, 550.

8B Determination of solubility

Gases are miscible in all proportions, but other phases more often show only partial miscibility. One often needs to know the solubility of a gas in a liquid or solid, of a liquid in another liquid, or of a solid in a liquid. Solutions of solids in solids are important among metals.

1. SOLUBILITY OF A GAS IN A LIQUID

In the equilibrium between a gas and a liquid, only two phases are present—the system, therefore, is bivariant. In order that the condition of equilibrium may be defined, it is necessary to fix two of the independent variables, temperature and pressure.

The solubility of a gas in a liquid can readily be determined by means of the apparatus shown in Fig. 8B.1, except in those cases where the gas is very soluble (ammonia, hydrogen chloride, etc.).[1] The apparatus consists of a gas-measuring burette *A* connected with a levelling tube *B*. The burette is furnished with a 3-way tap *a* which connects on the one side with the gas supply, and on the other, with a

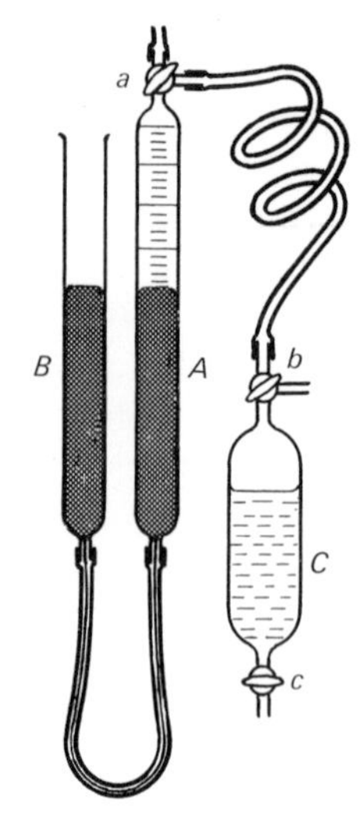

FIG. 8B.1 Determination of the solubility of a gas in a liquid.

tube leading to the 'absorption pipette' *C*, also furnished with the 3-way tap *b*, and an ordinary tap *c*. The gas burette and absorption pipette are connected by a length of thick-walled rubber tubing of narrow bore.

Before carrying out a determination, the volume of the absorption pipette between the two taps must be determined by means of water or mercury. For the solubility of gases such as carbon dioxide, the pipette may suitably have a volume of about 100 cm^3. For less soluble gases the volume should be greater.

The solvent to be used for the absorption of the gas must first be freed from air; this is best done by boiling the liquid for some time in a flask with reflux condenser attached. The flask is best fitted up as shown in Fig. 8B.2. The side tube *a* of the round-bottomed distilling flask is connected with a condenser by means of a piece of rubber tubing having a screw clip *c*. Through a rubber stopper in the neck of the flask passes a glass tube *b* which reaches nearly to the bottom of the flask. The other end of the tube is closed by rubber tubing and a clip.

While the liquid is being boiled, the clip on the tube *b* is closed while there is free connection with the condenser. After the liquid has been boiled for 10–15 min, the clip *c* is closed and the flame at the same time removed. After the liquid has cooled down to the ordinary temperature the flask is inverted and connected, by means

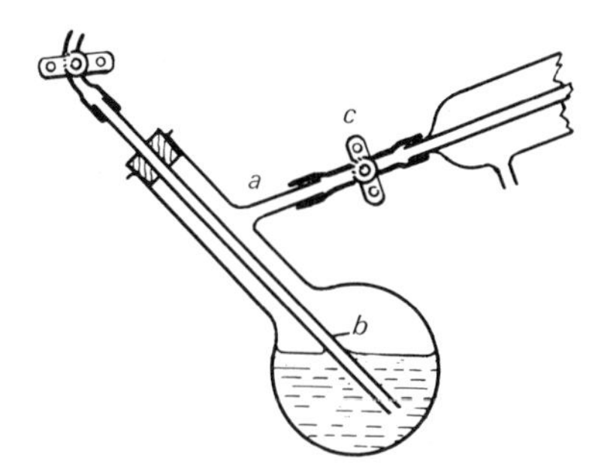

FIG. 8B.2 Preparation of air-free liquid.

of the side tube, with the lower outlet tube of the absorption pipette. The absorption pipette is exhausted by means of a pump attached to the tube *b* of the absorption pipette and the taps then closed. The clip on the tube *b* of the distillation flask is now opened in order to bring the interior of the flask to atmospheric pressure, and then, *as soon as possible*, the clip *c* is opened and also the lower tap of the absorption pipette. The solvent is then drawn into the pipette from the bottom of the liquid in the boiling flask, to which the air admitted into the flask will not have had time to diffuse.

EXPERIMENT
Determine the solubility of carbon dioxide in water at 25°C.

Procedure. The apparatus having been fitted together, the absorption pipette, filled with air-free water, is placed in a thermostat at 25°C. The mercury in the measuring burette is raised until it completely fills the burette, and a current of moist carbon dioxide from a Kipp apparatus is allowed to pass through the connecting tube and to escape into the air through the 3-way tap on the absorption pipette. When all the air has been swept out, the tap *a* is turned and the burette filled with carbon dioxide. After adjusting the levels of the mercury, the volume is read off. A slight increase of pressure is now established in the burette and the tap *a* turned so as to make connection between the burette and the pipette. The tap *c* of the pipette is opened and then the tap *b*, and 20–30 cm^3 of water allowed to run out into a flask. The weight of water run out is then determined. The volume of the water in the pipette, and also the air space, can thus be calculated, since the total volume of the pipette is known.

The pipette is replaced in the thermostat and is shaken carefully from time to time, the gas in the burette being always in communication with the pipette. As absorption of the gas proceeds, the levelling tube of the burette is raised so as always to maintain the gas at atmospheric pressure. When the absorption of gas ceases, the volume left in the burette is read off.

Calculation. If the solubility S of a gas in a liquid is defined by the ratio of the volume of gas absorbed to that of the absorbing liquid, one obtains

$$S = \frac{1}{V_2}\left[v_1\left(\frac{P_1-p_1}{P_2-p_2}\cdot\frac{T_3}{T_1}\right)-v_2\left(\frac{P_2-p_1}{P_2-p_2}\cdot\frac{T_3}{T_2}\right)-V_1\right]$$

where v_1 is the initial and v_2 the final volume of gas in the burette; P_1 is the initial and P_2 the final barometric pressure; p_1 is the partial pressure of the water vapour at the initial absolute temperature of the burette and p_2 the partial pressure at the temperature of the thermostat; T_1 is the initial and T_2 the final absolute temperature of the burette, and T_3 the absolute temperature of the thermostat; V_1 is the volume of the gas space and V_2 the volume of the liquid in the pipette.

Instead of filling the burette with moist gas, it is better to use dry gas. This may be obtained directly by allowing dry ice to evaporate in a sealed vessel fitted with a tube through which the CO_2 gas escapes. Alternatively gas from a Kipp apparatus can be dried by inserting a short tube of Anhydrone between the connecting tube and the absorption pipette; also, the 3-way tap of the absorption pipette should be kept closed except when it is necessary to allow gas to pass from the burette to the pipette. Under these conditions, the solubility is given by the expression

$$S = \frac{1}{V_2}\left\{\left[(v_1-v_2)\frac{P}{P-p_2}\cdot\frac{T_3}{T_1}\right]-V_1\right\}$$

The barometric pressure P is assumed not to change during the experiment.

The solubility S of carbon dioxide in water at 25°C is equal to 0·82.

2. SOLUBILITY OF A LIQUID IN A LIQUID

When ether is shaken with water a definite amount of the ether dissolves in the water, and similarly a definite amount of water dissolves in the ether. One thus obtains two liquid solutions the composition of which depends on the temperature. At each temperature, therefore, there will be two solubility values, one representing the solubility of ether in water, and the other the solubility of water in ether.

When the liquids are such that the amount of one of them can conveniently be determined by analysis, the mutual solubility curve is easily determined. The two liquids are shaken together in a stoppered bottle immersed in a thermostat, and the bottle is then allowed to remain undisturbed until the two liquid layers have separated. A quantity of each layer is then pipetted out, weighed, and the amount of one of the components determined by analysis. Such a method, for example, can be employed in the case of aniline and water, the aniline being titrated by means of a solution of potassium bromate and bromide of known concentration. Similarly, also, with phenol and water. Physical methods of analysis can often be used.

In many cases, however, the analytical method is not convenient and it is therefore necessary to employ the synthetic method. Weighed amounts of the two components are placed in a small glass tube, and the end of the tube then drawn out and sealed off. So long as two liquid layers are present, a turbid liquid is formed on shaking the tube, but at the temperature at which one of the layers just disappears, this turbidity also disappears, and a single homogeneous solution is now obtained which represents a saturated solution of one of the liquids in the other. Since the composition of the solution and the temperature at which the turbidity disappears are known, a point is obtained on the mutual solubility curve of the two liquids. By varying the initial amounts of the two liquids, the complete solubility curve can be obtained. The maximum temperature on a plot of temperature against composition is the critical solution temperature. It is greatly influenced by the presence of impurities, so its value can be used as a test of purity.

The system *n*-butyl cellosolve (ethylene glycol monobutyl ether) + water is unusual in that these liquids are miscible in all proportions at room temperature, but separate into two phases in the region 50–130°C. Above 130°C they are once more completely miscible. The miscibility gap is therefore a closed loop. The phenomenon of 'salting-out'[2] is readily demonstrated with this system.

EXPERIMENT

Determine the mutual solubility curve of n-butyl cellosolve and water.

A tube about 10 cm long and 1 cm wide is used with a constriction near the open end or a stopper held in place by two springs.

Two sets of experiments are done, each starting with 6 cm^3 of water and 4 cm^3 of butyl cellosolve in the tube initially, delivered from burettes. In the first set consecutive additions of 2 cm^3 water are made; in the second the additions are of butyl cellosolve. In each case the miscibility temperature is found. The tube is placed in a holder formed of stout copper wire, and immersed in turn in a large beaker full of water, in which also is a thermometer graduated in tenths of a degree. The water is warmed and, during this time, the tube with the two liquids is shaken at frequent intervals. At first the temperature may be allowed to rise rapidly until the turbidity shows signs of appearing, after which the temperature must be raised only slowly, and the tube must be shaken frequently. At the moment when the turbidity appears on shaking, the temperature is read. The temperature of the bath is now allowed to fall very slowly, and the point is noted at which the turbidity just begins to disappear once more. The experiment is repeated once or twice, and the mean of the temperatures taken. The results are then plotted with percentage amounts of one component as abscissae against temperatures at which homogeneity occurs as ordinates, and a smooth curve is drawn through these points. The experiment may be repeated using 1 000 mol m^{-3} KCl solution instead of water to demonstrate the 'salting-out' effect.

Additional experiments may be carried out with the following pairs of liquids: phenol and water, iso-butyric acid and water, hexane and methyl alcohol, carbon disulphide and methyl alcohol, acetylacetone and water.

3. THREE-COMPONENT LIQUID SYSTEMS

The mutual solubility of a pair of partially miscible liquids may be markedly altered by the addition of a third component. In general, when the third component is soluble in only one of the other two components, the mutual solubility of the two liquids is diminished; but when the third component dissolves readily in each of the other two components the mutual solubility of the latter is increased. This behaviour is illustrated by the system chloroform–water–acetic acid. When acetic acid is added to a heterogeneous mixture of chloroform and water at a definite temperature, the mutual solubility of these two liquids is increased, and a point is reached at which the mixture becomes homogeneous. The amount of acetic acid which must be added to bring about homogeneity at the given temperature will depend on the relative proportions of chloroform and water in the original mixture. Similarly, when water is added to a homogeneous mixture of chloroform and acetic acid (two completely miscible liquids), a heterogeneous mixture (two liquid solutions) is formed when a certain amount of water, depending on the initial composition of the mixture, has been added.

EXPERIMENT
Determine the limit of homogeneous phase in the system chloroform–acetic acid–water at 18 °C.

Mixtures of chloroform and acetic acid of varying composition are made up in a set of five stoppered bottles of 70–80 cm^3 capacity. These mixtures may suitably contain 10, 20, 40, 60 and 80% of chloroform by weight respectively. Since the relative densities of chloroform and acetic acid are 1·50 and 1·05 respectively, the mixtures may be prepared by running into the bottles from burettes the following volumes of chloroform and acetic acid:

chloroform (cm^3)	1·67	3·33	6·66	9·99	13·32
acetic acid (cm^3)	21·43	19·05	14·30	9·53	4·77

The bottles containing the solutions are placed in a thermostat at 18°C. After they have taken the temperature of the bath, water is run from a burette, in small quantities at a time, into each of the bottles in turn; and after each addition the bottle is well shaken. Addition of water is continued until a turbidity appears on shaking. The final additions should be made drop by drop, care being taken not to add too much water, especially in the case of the solutions richest in chloroform. Where quantities of less than 2 cm^3 are to be measured, a microburette is used. The bottles should be returned to the thermostat every few minutes, and the temperature of the mixture checked with a thermometer.

From the amount of the water added and the initial amounts of chloroform and acetic acid, the percentage composition of the mixture when turbidity makes its appearance is calculated, and the results are plotted in a triangular diagram. The points obtained are joined by a smooth curve, which may be completed by means of the following data: A saturated solution of water in chloroform contains 99% of chloroform, while the conjugate saturated solution of chloroform in water contains 0·8% of chloroform.

4. THE TRIANGULAR DIAGRAM

A triangular diagram usually consists of an equilateral triangle. The length of the side is taken as unity or 100, and represents, therefore, the sum of the fractional or percentage amounts of the three components; each corner of the triangle represents 100% of one of the components (Fig. 8B.3). In plotting the composition of a ternary

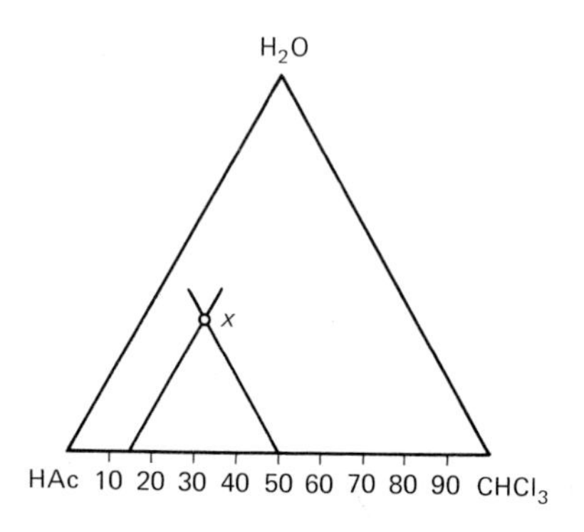

FIG. 8B.3 Representation of the composition of a ternary mixture by means of triangular co-ordinates.

mixture, two points are marked on one side of the triangle, representing the percentage amounts of two components, and from these points lines are drawn parallel to the other two sides of the triangle. The point of intersection gives the composition of the ternary mixture. Printed triangular co-ordinate graph paper is obtainable.

Thus, to represent the composition, 15·1% of chloroform, 50·2% of acetic acid and 34·7% of water, the side of the triangle for chloroform and acetic acid (Fig. 8B.3) is marked at the two points representing 15·1% of chloroform and 50·2% of acetic acid, and lines are drawn from these points parallel to the sides of the triangle. The point of intersection, *x*, represents the composition of the above ternary mixture.

Results. The curve obtained by joining the experimentally determined points forms the boundary between homogeneous and heterogeneous mixtures. A mixture of chloroform, acetic acid and water represented by any point *outside* the curve, towards the acetic acid corner of the triangle, will form only one homogeneous solution, while any mixture represented by a point *within* the curve will separate into two layers.

If it is required to draw *tie-lines* across the miscibility gap to indicate *conjugate solutions*, pairs of mutually saturated solutions must be analysed. Since the binodal curve has been determined in the above experiment it will suffice to analyse for acetic acid only.

EXPERIMENT
Explore the range of miscibility of water, butyl cellosolve and benzene at 25 °C.

Both water and benzene are completely miscible with butyl cellosolve (ethylene glycol mono *n*-butyl ether) at room temperature, but are not miscible with each other. Thus by adding butyl cellosolve to a mixture of benzene and water there will be a change from a two phase system (immiscibility) to a one phase system (miscibility).

Calculated quantities of benzene and water are run from burettes into a stoppered bottle or boiling tube, which is then placed in a thermostat at 25°C. Then, at the required temperature, the tube is removed from the thermostat and butyl cellosolve is added from a burette until the two phases become miscible. (See section 8B.3 for details.)

The following compositions of mixtures of benzene and water are convenient: 10, 20, 30, 40, 50, 60, 70, 80, 90 and 95 mole % benzene. The total quantity of benzene–water mixture required is 0·2 mole for the water-rich mixtures, increasing to 0·4 mole for the benzene-rich mixtures.

Plot the results on triangular graph paper (section 8B.4). The method may be checked by reversing the above procedure for one or two mixtures, i.e. preparing a suitable solution of (a) benzene and butyl cellosolve and adding water until the solution becomes turbid, and (b) water and butyl cellosolve and adding benzene.

Additional experiments. The system benzene–acetic acid–water may be studied at 20°C in the manner described above. Further, the boundary curve for chloroform, acetic acid and water may be determined at a series of temperatures, and the curves plotted on the same triangular diagram. The construction of a three-dimensional model would also be instructive.

It will also be of interest to study, between 50 and 100°C, the system water–phenol–aniline and the system water–ethyl–succinic nitrile at 60°C.

5. SOLUBILITY OF SOLIDS IN LIQUIDS

When a solid is brought into contact with a liquid in which it can dissolve, a certain amount of it passes into solution, until the solution is *saturated.* In all determinations of the solubility, it is necessary, not only to determine the amount of dissolved substance in the solution, but also to ascertain the nature of the solid phase which is in equilibrium with the solution.

The amount of substance dissolved depends also on the temperature; the solubility of a substance, or the number of grams of substance dissolved by a given weight of the solvent, may either increase or decrease with rise of temperature, according to whether the dissolution is endothermic or exothermic, respectively. In all cases, however, the solubility curve of any substance is continuous so long as the solid substance in contact with the solution remains unchanged. If, however, a change in the solid phase occurs, the solubility curve will show a 'break' or discontinuous change in direction.

It takes an appreciable time to saturate a liquid with a solid. The length of time required not only varies with the state of subdivision of the solid and the efficiency of the shaking or stirring, but is also dependent on the nature of the substance. In all cases, therefore, care must be taken that sufficient time is allowed for equilibrium to be established, more especially when changes in the solid may occur.

In cases where the solubility increases with rise of temperature, the time required for the attainment of equilibrium can generally be shortened by preparing a saturated solution at a higher temperature and allowing it to cool to the desired temperature in contact with the solid phase. Particular care, however, must be exercised in cases where the solid phase itself undergoes change with the temperature.

6. DETERMINATION OF THE SOLUBILITY

The production of a saturated solution can be carried out very simply using a stoppered flask in which the solid and solvent are shaken by a mechanical agitator whilst the flask is immersed in a thermostat. Even more convenient is the submersible magnetic stirring head (Rank Bros, Bottisham, Cambridge) shown in Fig. 8B.4. A magnet sealed in inert plastic is placed in the flask which is supported in the thermostat tank over the stirring head. This contains a rotating magnet which in turn rotates the stirring magnet in the solution.

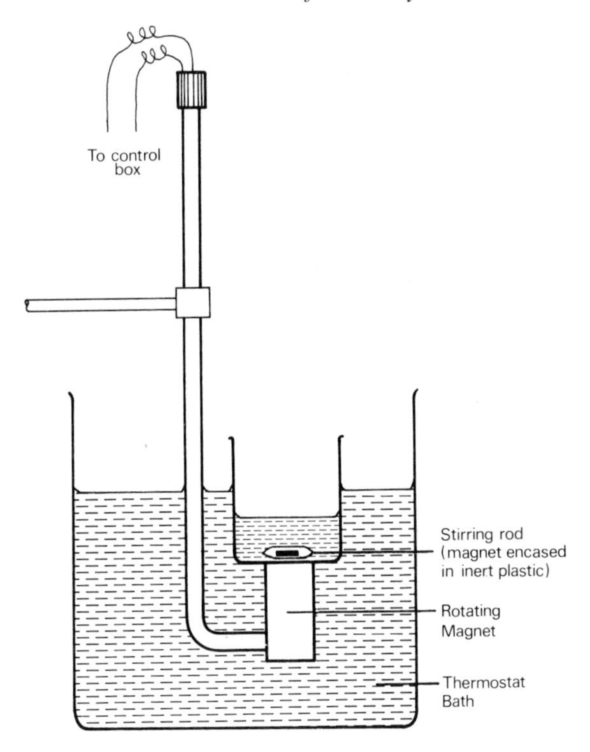

FIG. 8B.4 Submersible magnetic stirrer. (Courtesy Rank Brothers, Cambridge.)

The progress of solution towards saturation can be tested by withdrawing some of the solution from time to time, and determining the amount of dissolved substance, e.g. by means of refractive index measurement (section 9E). This requires to be done once only, the solution in other experiments being then well stirred for a period somewhat longer than that required for complete saturation.

When saturation has been effected, the solution must be analysed. The stirrer is removed from the tube, and the latter closed with an unbored cork, the solubility tube being kept meanwhile in the thermostat. After the solid has subsided, a known volume of the solution is transferred to a tared weighing bottle by means of a pipette to the end of which is attached by rubber tubing a short glass tube filled with cotton-wool to act as a filter. The solution is then weighed. The density of the solution can in this way be determined. The amount of solid in solution is determined in an appropriate manner, most simply (if allowable) by evaporation on the water-bath, and drying, if necessary, at a slightly higher temperature, or by gravimetric or volumetric analysis.

When the temperature of experiment is fairly high, it may be necessary to warm the pipette first before withdrawing the solution, otherwise solid may separate out in the pipette.

EXPERIMENT

Determine the solubility of potassium chloride from 10 to 50 °C.

Fit up a thermostat and regulate the temperature so as to be at about 5–8°C, the variations of temperature being not greater than 0·1°C.

In a tightly stoppered flask place a quantity of *finely powdered* potassium chloride and water. Arrange the agitator so the flask is held in the thermostat bath. The solid and solution should now be shaken for 2–3 hours, and a quantity of the solution removed as described above, weighed, and evaporated to dryness, or analysed by titration. To the solution in the tube add a further quantity of *finely powdered* potassium chloride, and allow the shaking to continue for another period of 2–3 h. Again determine the composition of the solution. If this agrees with the former determination, it shows that saturation was complete in the first period of 2–3 h; but if the amount of dissolved solid is greater in the second case, the stirring must be continued for some time longer, with addition, if necessary, of more potassium chloride until the concentration of the solution becomes constant. This gives the solubility at the particular temperature of the experiment, and should be checked by a second, independent determination. Express the solubility as grams of salt to 1 kg of water.

Having determined the solubility at a temperature between 5 and 8°C, raise the temperature of the thermostat by 5 or 10°C and make another determination of the solubility at this higher temperature. Make further determinations at intervals of about 10°C up to 50–55°C. Instead of making two determinations at each temperature, as mentioned above, the first series of determinations can be checked by approaching saturation from the side of super-saturation, i.e. by allowing the solution to cool down from a higher temperature *while in contact with the solid.* The solid phase must be present, and the solutions should be stirred.

Results. The results are plotted in rectangular co-ordinates, the temperatures being plotted as abscissae, and the solubility (grams of salt to kg of water) as ordinates. Draw a smooth curve through the points so obtained, and from the curve read off the solubility at every five degrees.

When the van't Hoff reaction isochore is applied to solubilities, one obtains the expression

$$\frac{\mathrm{d}\log_e S}{\mathrm{d}T} = \frac{Q}{RT^2}$$

where S is the solubility in any units and Q the heat of solution ($\mathrm{J\ mol^{-1}}$). Assuming (what is generally not quite true) that Q is independent of the temperature, this expression yields, on integration,

$$\log_{10} S_2 - \log_{10} S_1 = \frac{Q(T_2 - T_1)}{2{\cdot}303 \times 8{\cdot}184 \times T_2 \times T_1}$$

The value of Q calculated from the solubility may also be compared with the (different) heat of solution, determined calorimetrically (section 9A).

Additional exercise. The solubility of benzoic acid may also be determined at temperatures between 20 and 60°C. The amount of benzoic acid in a definite volume and weight of saturated solution can be determined by titration with 250 mol $\mathrm{m^{-3}}$ NaOH, using phenolphthalein as indicator, the solution of benzoic acid being first diluted as may be necessary. (Solubility at 20°C = 2·90, and at 60°C, 11·55 g per kg of water.)

In the case of potassium chloride and benzoic acid the solid phase remains unchanged throughout the whole range of temperature. A case may now be studied where the solid phase undergoes change.

EXPERIMENT

Determine the solubility of sodium sulphate from 10 to 50°C.

Details. The determinations of the solubility are carried out exactly as described above. Between 28 and 35°C, determinations should be made at every two degrees. The results are then plotted graphically as before, the solubility being calculated in grams of *anhydrous* salt to 1 kg of water.

At the temperature of 30°C, and also at the temperature of about 35°C, the excess of solid in contact with the solution should be rapidly separated by filtration with the aid of a water-pump, using for the filtration merely a loose plug of cotton-wool in the stem of the funnel, or a sintered glass filter may be used. The solid is then rapidly pressed between filter-paper, and the amount of water of crystallization determined in the ordinary way. The solubility curves obtained from the above determinations should be produced so as to cut each other. The point of intersection gives the *transition point* of the phase reaction

$$Na_2SO_4.10H_2O \rightleftarrows Na_2SO_4 + 10H_2O$$

7. SOLUBILITY AND IONIC STRENGTH

The solubility of a salt (e.g. barium iodate $Ba(IO_3)_2$) is changed by the presence of other electrolytes in three ways:[3]

(a) it can be decreased by adding a salt with a common ion (e.g. $BaCl_2$),
(b) it can be increased by adding a substance that reacts chemically with one of the ions (e.g. EDTA), and
(c) it can be changed by raising the ionic strength of the solution by means of an 'inert' electrolyte such as KCl.

The experiment described below is concerned with the last case; however the ionic strength effect plays a role even in cases (a) and (b).

Qualitatively, case (c) arises because cations and anions are coulombically attracted to each other. Thermal forces alone would tend to create a purely random arrangement of the ions in solution but, under the simultaneous influence of electrical forces, a loose kind of structure develops. Each cation then has, on the average, more anions than cations in its neighbourhood and vice versa, and this surrounding 'cloud' of ions is called an 'ionic atmosphere'. As a result the ions cannot move about entirely independently of each other and thus the effective or thermodynamic concentration—the activity—is smaller in dilute solutions than the analytical or stoichiometric concentration. Activity (a) and molality (m) of any species i are related by the activity coefficient f:

$$a_i = m_i f_i$$

The ionic atmosphere effect is the more pronounced the more ions there are in the solution: f depends upon the ionic strength I, which is defined by

$$I = \tfrac{1}{2} \sum_i m_i z_i^2$$

with the summation taken over *all* the ions in the given solution. z_i is the charge number of ion i, e.g. 2 for Ba^{++}. The quantitative relation between f and I, derived in 1923 by Debye and Hückel, is discussed in section 13A.5.

Let us now consider the solubility equilibrium

$$Ba(IO_3)_2(s) \rightleftarrows Ba^{++} + 2IO_3^-$$

The thermodynamic equilibrium constant is the solubility product K_s expressed in activities. Since $a_i = m_i f_i$ and the activity of a pure solid is unity,

$$K_s = a_{Ba^{++}} a^2_{IO_3^-} = m_{Ba^{++}} m^2_{IO_3^-} f_{Ba^{++}} f^2_{IO_3^-} = s(2s)^2 f_{Ba^{++}} f^2_{IO_3^-}$$

where s, the solubility, is the molality of the salt in the saturated solution. The activity coefficients of the Ba^{++} and IO_3^- ions always occur as the product and it is convenient to introduce the mean activity coefficient

$$f_{\pm} = (f_{Ba^{++}} f^2_{IO_3^-})^{\frac{1}{3}}$$

so that $K_s = 4s^3 f^3_{\pm}$ or

$$sf_{\pm} = \left(\frac{K_s}{4}\right)^{\frac{1}{3}} = \text{constant} \tag{1}$$

The relationship $sf_{\pm}$ = constant is quite generally applicable to the solubility of a salt of *any* valence type in the presence of other 'inert' electrolytes.

According to the extended Debye and Hückel theory (section 13A.5)

$$\log_{10} f_{\pm} = \frac{-A|z_+ z_-|\sqrt{I}}{1 + Ba\sqrt{I}} + bI \tag{2}$$

where A and B are known constants, a is the distance of closest approach of the cation and anion, and b is an unknown parameter. In water at 25°C, $A = 0{\cdot}511$. The distance a is often around 0·3 nm (3×10^{-10} m) which makes $Ba \approx 1$, and the resulting simplified version was suggested by Guggenheim:

$$\log_{10} f_{\pm} = \frac{-A|z_+ z_-|\sqrt{I}}{1 + \sqrt{I}} + bI \tag{3}$$

If the bI term is negligible we obtain Güntelberg's equation, which contains neither of the empirical parameters a and b:

$$\log_{10} f_{\pm} = \frac{-A|z_+ z_-|\sqrt{I}}{1 + \sqrt{I}} \tag{4}$$

In very dilute solutions this can be further simplified to the limiting Debye–Hückel equation

$$\log_{10} f_{\pm} = -A|z_+z_-|\sqrt{I} \tag{5}$$

These equations give the variation of the solubility of barium iodate with ionic strength. Equation (1) together with the limiting Debye–Hückel equation (5) gives

$$\log_{10} s - 2A\sqrt{I} = \tfrac{1}{3}\log_{10}\left(\frac{K_s}{4}\right) \tag{6}$$

and with the Guggenheim version (3):

$$\log_{10} s - \left(\frac{2A\sqrt{I}}{1+\sqrt{I}}\right) = \tfrac{1}{3}\log_{10}\left(\frac{K_s}{4}\right) - bI \tag{7}$$

Equation (6) can be tested by plotting $\log_{10} s$ against $\sqrt{I}$; Eq. (7) by plotting the left-hand side of Eq. (7) against I; the slopes of the lines obtained can be compared with those predicted by the various versions of the Debye–Hückel theory.

EXPERIMENT

Investigate the variation of the solubility of barium iodate with ionic strength.

0·5 g $Ba(IO_3)_2$, finely powdered, is placed in each of six long necked 100 cm^3 flasks. 50 cm^3 of water is added to the first flask; 50 cm^3 of 10, 20, 50, 100 and 200 mol m^{-3} KCl to the others. The flasks are stoppered and shaken in a 25°C thermostat, preferably overnight. The thermostat temperature should be measured. 10 cm^3 portions from the flasks are titrated against thiosulphate solution. The flasks are allowed to stand without shaking for 5 min to allow the solid to settle, and the sample drawn into a pipette through a glass wool filter (in a polythene tube attached to the pipette). The portion is run into an equal volume of water; 0·1 g KI and 1 cm^3 2 000 mol m^{-3} H_2SO_4 are added and the titration performed with 10 mol m^{-3} thiosulphate using starch indicator: the thiosulphate can be standardized against KIO_3.

Calculation. Calculate the solubility s and ionic strength I for each solution, remembering to include contributions to I from the Ba^{++} and IO_3^- as well as from the K^+Cl^- ions. The solubilities and ionic strength should strictly be in units of molality, i.e.

mol (kg $H_2O)^{-1}$; the approximation $m \approx 0{\cdot}001\ c$, where the concentration c is in mol m^{-3}, can be used for this experiment. Test equations (6) and (7) of section 8B(7) as suggested there; find the value of K_s from the intercept of the plot which best fits the results.

BIBLIOGRAPHY 8B: Solubility

Weissberger, Vol. I, Pt 1, chapter XI.
Seidell, *Solubilities of Inorganic and Organic Compounds*, 4th edn, 1958 (Van Nostrand, New York).
[1] Findlay and Creighton, *J. Chem. Soc.*, 1910, **97**, 536; van Slyke, *J. Biol. Chem.*, 1939, **130**, 545.
[2] Gross, *Chem. Revs.*, 1933, **13**, 91.
[3] Davies, *Electrochemistry*, chapters 5 and 8, 1967 (Newnes, London).

8C Transition points

1. It is well known that there are many substances which are capable of existing in more than one crystalline form, e.g. sulphur. In general, these different polymorphous forms are not equally stable at a given temperature. Thus, at the ordinary temperatures, rhombic sulphur is the more stable form, and monoclinic sulphur, if kept sufficiently long, will change spontaneously into rhombic. If, however, the temperature is raised to, say, 100–110°C, it is found that the monoclinic crystals can be kept indefinitely without undergoing change, while the rhombic crystals pass into monoclinic. At this temperature, therefore, the monoclinic is the more stable form.

At a temperature of about 96°C, it is found that the two forms are equally stable, and that neither form changes into the other on keeping. This temperature is known as the *transition temperature*, or transition point. Graphs of the molar *free energies* of the two forms cross at this temperature, the upper curve being that of the metastable form at other temperatures.

Not only are transition points found with polymorphous substances, but they are found, in general, also with salt hydrates. When a salt combines with water to form one or more different hydrates, it is found that under given conditions of temperature, etc., only one of the hydrates, or it may be the anhydrous salt, is stable. Thus, when sodium sulphate decahydrate is heated above 33°C, it decomposes into anhydrous sodium sulphate and a saturated solution of this salt. On the other hand, on allowing a saturated solution of sodium sulphate to cool down in presence of anhydrous sodium sulphate, it is found that when the solution is cooled below about 33°C the anhydrous salt takes up water and forms crystals of the decahydrate. The temperature of (approximately) 33°C therefore constitutes a transition point or inversion point for the change $Na_2SO_4 . 10H_2O \rightleftarrows Na_2SO_4 + 10H_2O$. Similar relationships are, in general, found with other salt hydrates, e.g. sodium carbonate hydrates.

2. DETERMINATION OF THE TRANSITION POINT

Various methods have been employed for the determination of the transition point of a polymorphous solid or of a salt hydrate. The different methods, however, are not equally suitable in every case; nor are the values obtained by the different methods always identical. It is well, therefore, to determine the transition point by different methods. The more important of these are: solubility, thermometric, dilatometric and tensimetric methods.

1. *Solubility method.* This has already been studied in the preceding section.

2. *Thermometric method.* The thermometric method depends on the fact that change from one form to another on passing through the transition point is accompanied by a heat effect—absorption or evolution of heat. Thus, when $Na_2SO_4 . 10H_2O$ breaks up into Na_2SO_4 plus solution, heat is absorbed, while the reverse change is accompanied by evolution of heat.

EXPERIMENT

Determine the transition point of sodium sulphate by the thermometric method.

A quantity (30–50 g) of recrystallized sodium sulphate decahydrate is placed in a thin glass tube so as entirely to surround the bulb of a thermometer, graduated in tenths of a degree. The tube is placed in a large beaker of water, the temperature of which can be very slowly raised by means of a small flame, and be kept uniform by means of a stirrer. The temperature of the bath is raised to about 32°C, at which it may be kept for several minutes, and then very gradually raised until the Glauber's salt becomes partially liquid. The temperature of the bath is then kept constant. The partially liquefied mass in the tube is now well stirred by means of a ring stirrer of glass or platinum, as in carrying out a f.p. determination with the Beckmann apparatus (sections 7B(1–3)), and the temperature of the mass read from time to time. Meanwhile the temperature of the bath may be allowed to rise very slowly (1°C in 5 min), and the temperature of the partially liquefied mass should be read every minute. After the temperature of the bath and of the partially liquefied mass has risen to about 34°C, allow the temperature to fall slowly. Meanwhile stir the sodium sulphate and solution well, and read the temperature every minute.

The temperature readings for the mixture in the tube are plotted against the time, and in this way two curves will be obtained, one for rising and the other for falling temperature, each showing an approximately horizontal portion. Owing to suspended transformation, these two horizontal portions may not coincide.

Repeat the experiment, but allow the temperature of the bath to rise more slowly between 32 and 33°C. Again read the temperature on the thermometer in the tube every minute, and plot the results as before.

Similar determinations may also be carried out with the salts given in Table A2 of the Appendix.

Dilatometric method. Since, in the majority of cases, transformation at the transition point is accompanied by an appreciable change of volume, it is only necessary to ascertain the temperature at which this change occurs in order to determine the transition point. For this purpose the dilatometer is employed, an apparatus which consists of a bulb with capillary tube attached, and which constitutes a sort of large thermometer (Fig. 8C.1). Some of the substance to be examined is passed into the bulb A through the tube B, which is then

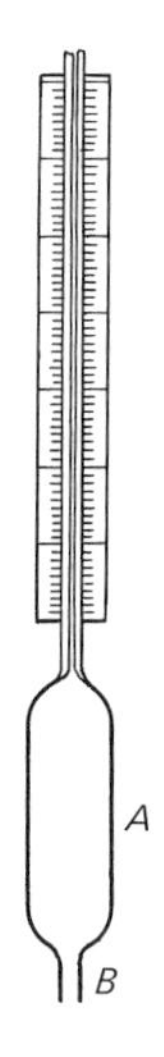

FIG. 8C.1 Dilatometer.

sealed off or closed by a ground glass stopper. The rest of the bulb and a small portion of the capillary tube are then filled with some liquid which is without chemical action on the substance under investigation. A liquid, however, may be employed which dissolves the substance slightly.

In using the dilatometer, two methods of procedure may be followed. According to the first method, the dilatometer containing the form stable at lower temperatures is placed in a thermostat and maintained at a constant temperature until it has taken the temperature of the bath. The height of the meniscus is then read on a millimetre scale attached to the capillary. The temperature of the thermostat is slowly raised, and the height of the meniscus at each degree of temperature noted. If no change takes place in the solid, the expansion will be practically uniform, or the rise in the level of the meniscus per degree of temperature will be practically the same at the different temperatures, as represented by the line *AB* in Fig. 8C.2.

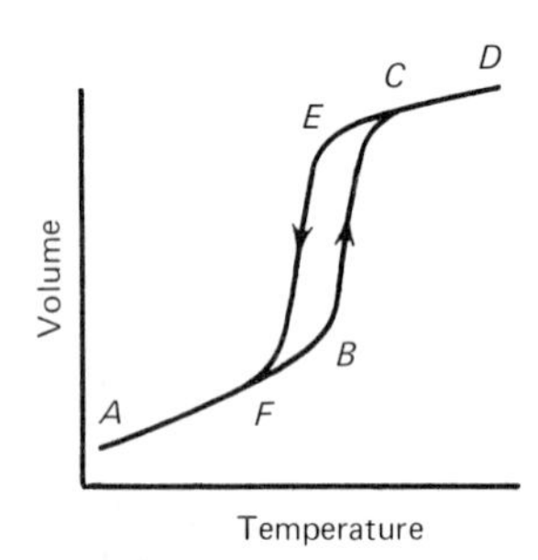

FIG. 8C.2 Typical dilatometric records for heating and cooling a solid through its transition temperature.

On passing through the transition point, however, there will be a more or less sudden increase in the rise of the meniscus per degree of temperature (line *BC*), if the change in the system is accompanied by increase of volume. Thereafter, the expansion will again become uniform (*CD*). Similarly, on cooling, contraction will at first be uniform, and then at the transition point there will be a large diminution of volume (*DE*, *EF*).

If the transformation occurred immediately the transition point was reached, the sudden expansion and contraction would take place at the same temperature. There is, however, always a certain time lag, so that, with rising temperature, the relatively large expansion does not take place until a temperature somewhat higher than the transition point, and with falling temperature, the contraction occurs at a temperature somewhat below the transition point (e.g. *BC* and *EF*). The amount of lag will vary from case to case and will depend on the rapidity with which the temperature is raised and the velocity with which the system changes. After the transition point has been ascertained approximately in this way, the determination is made with greater care by allowing the temperature in the neighbourhood of the transition point to alter more slowly. In this way the amount of lag is diminished.

EXPERIMENT

Determine the transition point for Glauber's salt and anhydrous sodium sulphate.

Procedure. Invert the dilatometer (Fig. 8C.1), and drop into the bulb a small glass bead with stalk, so as to close the end of the capillary tube, and so prevent it being blocked by solid material. Then introduce a quantity of powdered Glauber's salt until the bulb is half or three-quarters full, and seal the end of the tube *B*. The dilatometer must now be filled with some measuring liquid, e.g. petroleum or xylene. This is best done by attaching an adapter to the end of the capillary tube by means of a rubber stopper. A quantity of petroleum is introduced into the wider portion of this tube, and the dilatometer then exhausted by means of a water-pump. On now allowing air to enter, petroleum is driven down into the bulb. The operation is repeated until all the air is withdrawn from the dilatometer and replaced by petroleum. Tap the tube so as to displace any adhering air-bubbles. The excess of petroleum is then removed from the capillary by means of a long, finely drawn capillary tube, so that when the dilatometer is placed in the thermostat, the petroleum meniscus remains on the scale.

Immerse the bulb of the dilatometer completely in the water of a thermostat, the initial temperature of which may be 25–26°C. After about 5 or 10 min, read the level of the petroleum, and then slowly raise the temperature, 1°C in 5 to 10 min, and at each degree again read the level of the meniscus in the capillary. At 32 to 33°C it should be found that the rise of the meniscus per degree of temperature is relatively very large; and that as the temperature is raised above 33°C the rise per degree becomes less again and nearly uniform. This shows that the transition point is between 32 and 33°C. Carry out the same series of observations in the reverse order, allowing the temperature to fall from about 35 or 36° to about 28°C. Then make a more accurate determination by allowing the temperature to alter very slowly from 31 to 34°C.

Vapour pressure methods. When the systems undergoing change at the transition point possess a measurable vapour pressure, e.g. salt hydrates, measurements of the latter may be used to determine the transition point. This depends on the fact that at the transition point the vapour pressure of the two systems becomes equal. A differential manometer is generally used.

BIBLIOGRAPHY 8C: Transition points

Findlay (revised Campbell and Smith), *The Phase Rule and its Applications*, 9th edn, 1951 (Dover, New York).

8D Freezing points of binary mixtures

1. When a homogeneous binary liquid mixture is cooled, a solid eventually separates, but its composition depends entirely on the nature of the substances concerned. If the two components are chemically dissimilar but form no compound, then each lowers the f.p. of the other; they crystallize out separately, and a *eutectic* system is obtained, having a minimum f.p. (Fig. 8D.1). If, on the other hand, a stable compound (*congruently melting*) is formed, then the f.p. curve shows a maximum together with two eutectic minima. In some systems more than one compound is formed, while others exhibit weak compound formation (*incongruently melting compound*).

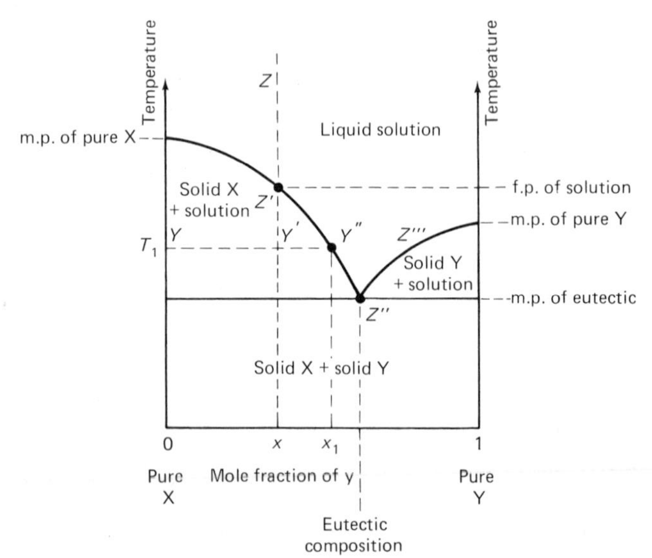

FIG. 8D.1 Phase diagram for a typical binary mixture with simple eutectic formation.

Other types of solid–liquid equilibria are found with two components of similar chemical type; if they are isomorphous they may form a *continuous series of solid solutions* (with or without a maximum or minimum in the f.p. curve). If the miscibility of the components in the solid state is only partial (the pure components not being isomorphous), then a general eutectic or a peritectic f.p. diagram is obtained.

Solid–liquid equilibria are generally investigated by *thermal analysis*—that is, a study of the temperature–time curves obtained when the liquid is slowly cooled. Suppose, for example, the components X and Y form a simple eutectic system (Fig. 8D.1). Imagine a certain mixture prepared and melted and then placed in a shielded container and allowed to cool. At first its temperature would fall regularly along a curve *AB* (Fig. 8D.2) until the f.p. point of the solution

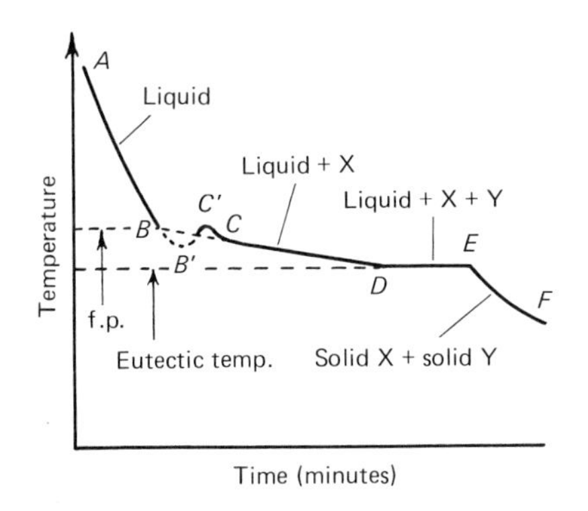

FIG. 8D.2 Typical cooling curve for a binary liquid mixture forming a eutectic.

were reached, at which the liquid became saturated with X; this cooling corresponds to *ZZ′* on the phase diagram, Fig. 8D.1. With crystallization of X there would be a reduced rate of fall of temperature owing to evolution of latent heat, but, since the solution would be constantly growing more concentrated in Y, the f.p. would not remain constant but would fall along a curve such as *CD*. In practice, the transition from *AB* to *CD* is rarely ideal; more often a solution *supercools* below its true f.p. *B* to some temperature *B′* and then, when crystallization of X does commence, there is a sudden evolution of heat which temporarily sends the temperature up to *C′*.

CD on the cooling curve corresponds to *Z′Z″* on the phase diagram: at any temperature between these points the liquid composition is given by the corresponding abscissa value; for example at

T_1° the solution composition is x_1 (Fig. 8D.1). By the '*lever rule*' the proportions of pure X and solution of composition x_1 formed at T_1 from the original liquid (composition x) are in the ratio $Y'Y'':YY'$.

Eventually the solution reaches saturation with respect to Y as well as X. When this occurs, solid X and solid Y continue to crystallize out side by side in the same proportions as they are present in the solution; the temperature therefore remains constant—the eutectic point. (Z'' on the phase diagram.) Finally, when the last drop of solution has solidified and the system consists entirely of solid X and solid Y, the solid mixture continues to cool along the iine EF.

EXPERIMENT

Determine the f.p. curve of mixtures of ortho-nitrophenol and para-toluidine.

The f.p. of mixtures of *o*-nitrophenol and *p*-toluidine can be determined in essentially the same manner as the f.p. of aqueous solutions (section 7B). A test-tube closed by a cork through which pass a thermometer (graduated in tenths) and a glass rod, the end of which is bent into a loop to serve as a stirrer, is used as the f.p. tube. It is supported by means of a cork ring in a wider tube which acts as an air-mantle.

After weighing the f.p. tube, a quantity of one of the components is placed at the bottom of the tube, and its weight ascertained. Sufficient of the substance must be taken to ensure that, when molten, it completely covers the bulb of the thermometer. The cork with thermometer is now inserted and the tube, surrounded by its air-mantle, supported in a beaker of water. The temperature of the bath is raised until the substance melts in the f.p. tube. The bath temperature is then allowed to fall very slowly, and the temperature of the molten substance read off at intervals of half-a-minute or a minute. Stir the molten mass slowly all the time. At a certain point it will be found that crystals begin to separate from the molten mass and the temperature rapidly rises and remains constant. This constant temperature is the f.p. of the one component. Repeat the determination once or twice.

Now add, from a weighed tube, a quantity of the second component, raise the temperature until all the solid has melted and then allow it to fall slowly and as uniformly as possible. Plot the cooling curve (temperature against time) and note the temperature at which there is a 'break' in the cooling curve. Allow the temperature to fall still further until another arrest is shown on the cooling curve and the temperature remains constant until complete solidification takes place. This is the eutectic point. For each composition of the mixture, note the eutectic temperature reached and also the time during which the temperature remains constant at the eutectic point.

Make further additions of the second component, and determine the cooling curve, f.p. and eutectic temperature for each mixture. Mixtures varying in composition by about 10% of the added component should be made.

Results. Plot the values of the f.p. (as ordinates) against composition in mole fractions as abscissae.

With *o*-nitrophenol + *p*-toluidine a diagram showing two curves meeting at the eutectic point is obtained. Other organic systems suitable for study in the manner described above are naphthalene–*p*-nitrotoluene; naphthalene–α-naphthylamine; phenol–urea.

2. Phase diagrams are particularly important in the chemistry of processes at higher temperatures, e.g. in connection with alloys, bricks, refractories, etc., and consequently thermal analysis is most often carried out with thermocouples. Quite small thermal changes can be detected by *differential thermal analysis*. In this technique the two junctions of a thermocouple are embedded in similar masses of material which are heated together in a furnace, but one material is inert to heating while the other undergoes a thermal process, e.g. evolution of moisture, recrystallization, etc. The thermocouple then shows a temperature difference between the materials when such a process occurs.

The following experiment illustrates the use of a thermocouple for simple thermal analysis at moderate temperatures (300–500°C).

EXPERIMENT

Determine the phase diagram of the system $PbCl_2$–KCl.

The apparatus is shown diagrammatically in Fig. 8D.3(a). *A* is a small electric furnace constructed with Nichrome wire, and controlled by a 'Variac' or other variable transformer. The melts are prepared in 'Pyrex' test tubes. Temperatures are measured by a chromel–alumel thermocouple encased in a 'Pyrex' sheath *C*. The cold junction is kept in an ice–water mixture in a small Dewar vessel. The temperature is best measured on a potentiometric recorder (section 4C(4)).

Procedure. The temperatures to be measured lie between 400 and 500°C, i.e. 15–22 mV. It is convenient to 'back-off' about 13 mV,

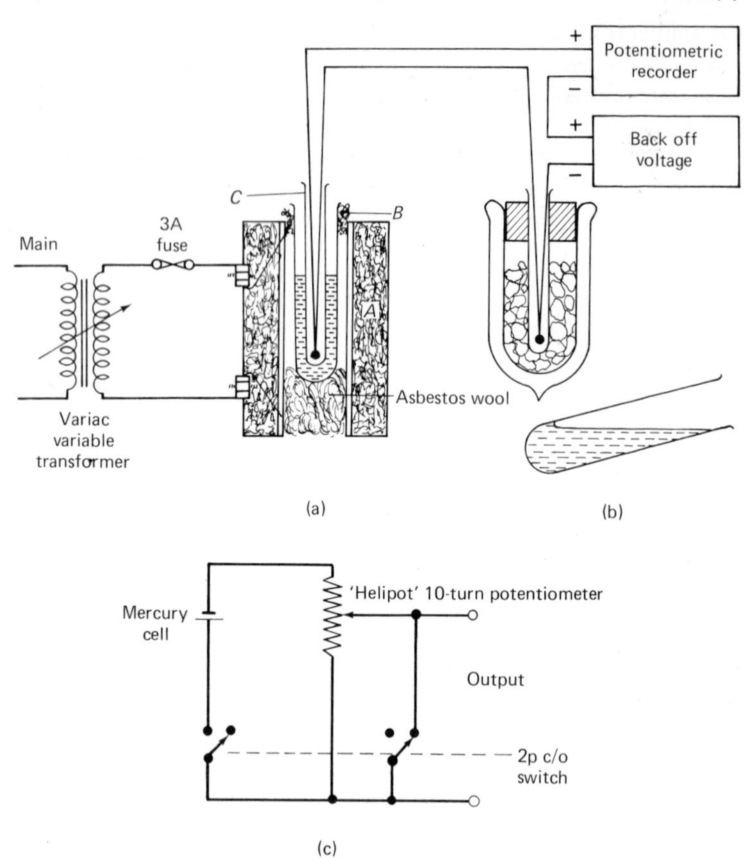

FIG. 8D.3 Determination of cooling curves by means of a thermocouple.

so the voltage to be measured becomes 2–9 mV, and can be displayed on the 10 mV range of the recorder.

The 'back-off' potentiometer voltage source (Fig. 8D.3(c)) should be set to 13 mV using the recorder as a voltmeter, and inserted in the circuit. When its switch is 'off' the terminals are shorted so that the recorder measures the full thermocouple voltage. The thermocouple–back-off–recorder combination is calibrated experimentally by determining the cooling curves for pure zinc and $PbCl_2$ (m.p. 419·5°C and 501°C resp.). If the recorder deflection is V_1 mV at 419·5°C and V_2 mV at 501°C, the temperature T for any intermediate deflection V mV is given by

$$T = 419{\cdot}5 + 81{\cdot}5\,(V - V_1)/(V_2 - V_1)\ ^\circ\mathrm{C}$$

This assumes that the thermocouple voltage is a linear function of temperature over this range. Before beginning ensure that the end of the thermocouple is at the end of its glass sheath. Each sample is melted in turn in the furnace, using enough to give a melt 2–3 cm in depth (about 10 g sample). The variac is set to zero volts and increased gradually to achieve the desired temperature. When melting the sample the 'back-off' device is switched off and the rise in temperature followed with the recorder on the 50 mV range. Do not heat above a temperature given by 22 mV. The current is then reduced so the melt cools: it should be well stirred with the thermocouple sheath.

The 'back-off' device is switched on and the recorder set to the 10 mV range to record the cooling curve, using a chart speed of about 600 mm per hour. Eight mixtures of $PbCl_2 + KCl$ covering the range 0–2 g of KCl in 10 g $PbCl_2$ are made up. Each mixture is melted in turn and its cooling curve determined, ensuring that the samples are constantly stirred. Immediately the sample has solidified, the furnace is heated up to melt the sample. The thermocouple sheath is removed and the test tube allowed to cool out of the furnace, clamped in a tilted position as shown in Fig. 8D.3(b). If this procedure is followed the melt may be re-used if necessary but otherwise it is liable to crack the tube on solidification. The temperatures of the two arrests are measured from the cooling curve (Fig. 8D.2). They may appear only as kinks in the cooling curve. Calculate the mole fraction of KCl in each mixture and plot the portion of the phase diagram covered. The point Z'' (Fig. 8D.1) is best found by extrapolating the curves ZZ'' and $Z'''Z''$ since the sharp change in gradient is sometimes not observed, because of supercooling.

BIBLIOGRAPHY 8D: Freezing point of binary mixtures

Findlay (revised Campbell and Smith), *The Phase Rule and its Applications*, 9th edn, 1951 (Dover, New York).
Weissberger, Vol. I, Pt 1, chapter VIII.

8E Equilibria between liquids and vapours

1. BOILING-POINT CURVES OF TWO MISCIBLE LIQUIDS

If the external pressure is maintained constant, the b.p. of a mixture of two completely miscible liquids will vary with the composition.

When the curve of total v.p. shows neither a maximum nor a minimum, the b.p. curve will also show neither a maximum nor a minimum, and the b.p. of all mixtures will be intermediate between the b.p. of the pure components, Fig. 8E.1(a). The line, however, will show a curvature which will be all the greater the more widely the b.p. of the components are separated. When a solution which gives such a b.p. curve is distilled, the vapour (distillate) will be richer than

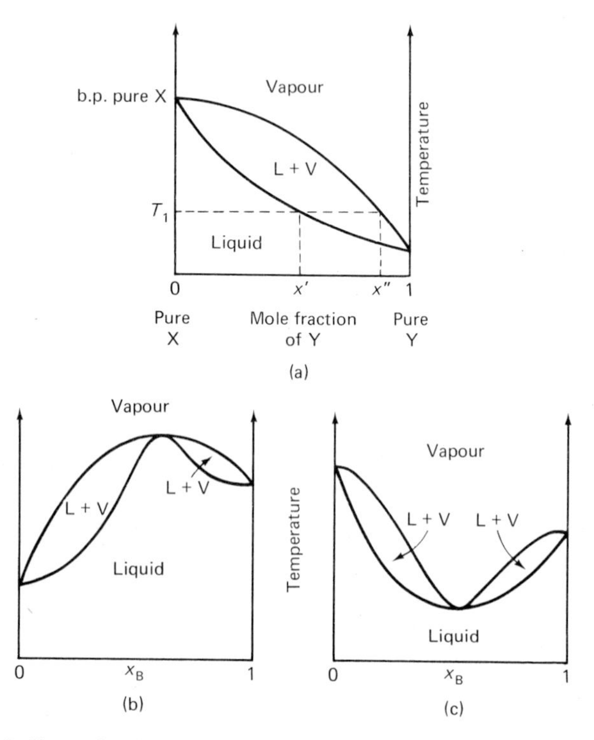

FIG. 8E.1 Boiling point diagrams for zeotropic mixtures (a), and for azeotropic mixtures with maximum (b) and minimum (c) boiling points.

the liquid in the component of higher v.p., and the b.p. will rise as the distillation proceeds. By repeated fractional distillation of such a solution, or by distillation through an efficient fractionating column, a complete separation of the two components may be effected.

When the v.p. curve of the liquid solutions passes through a maximum the b.p. curve will pass through a minimum, Fig. 8E.1(b), and when the v.p. curve passes through a minimum the b.p. curve will pass through a maximum, Fig. 8E.1(c). In such cases, a separation of the mixture into its pure components cannot be effected by distillation, but only a separation into one or other of the components and a mixture of constant b.p. (*azeotropic mixture*), the composition of which corresponds to the maximum or minimum point on the b.p. curve.

EXPERIMENT
Determine the equilibrium liquid–vapour curves for binary mixtures of completely miscible liquids.

The b.p. curves should be determined for one or more of the following binary systems, which are representative of the three classes referred to above:

(a) *Zeotropic mixtures*, with b.p. intermediate between those of the pure components: (i) benzene (80·2°C)–toluene (110·6°C): (ii) benzene–hexane (69·0°C) (Fig. 8E.1(a)).

(b) *Azeotropic mixtures with maximum b.p.*: (i) chloroform (61·2°C)–acetone (56·4°C), (azeotropic mixture 64·5°C, 65·5 mol % of $CHCl_3$): (ii) water (100°C)–formic acid (99·9°C), (azeotropic mixture 107·1°C, 43·4 mol % of water) (Fig. 8E.1(b)).

(c) *Azeotropic mixtures with minimum b.p.*: (i) *iso*-propyl alcohol (82·5°C)–benzene (azeotropic mixture 71·9°C, 39·3 mol % of alcohol). (ii) carbon tetrachloride (76·8°C)–methyl alcohol (64·7°C) (azeotropic mixture 55·7°C, 44·5 mol % of CCl_4): (iii) methyl alcohol–benzene (azeotropic mixture 58·3°C, 61·4 mol % of alcohol) (Fig. 8E.1(c)).

Procedure. Prepare binary mixtures containing 10, 20, 40, 60, 80, 90 mol % of one of the components, and determine the refractive indices of each mixture at a definite temperature (15–20°C), using a Pulfrich or Abbe refractometer, and also the refractive index of the pure components (see section 9E). Draw, on a moderately large scale, a graph showing the relation between composition and refractive index.

First determine the b.p. of the pure components so that the thermometer readings may be checked and any necessary correction introduced. Then determine the b.p. of each of the mixtures and the equilibrium composition of the vapour and of the liquid. For this purpose a quantity of the mixture (70–80 cm^3) is placed in a boiling-tube as shown in Fig. 8E.2(a). The boiling-tube is fitted with a standard ground glass joint and carries a small condenser. Samples of distillate are obtained by tilting as indicated. The alternative

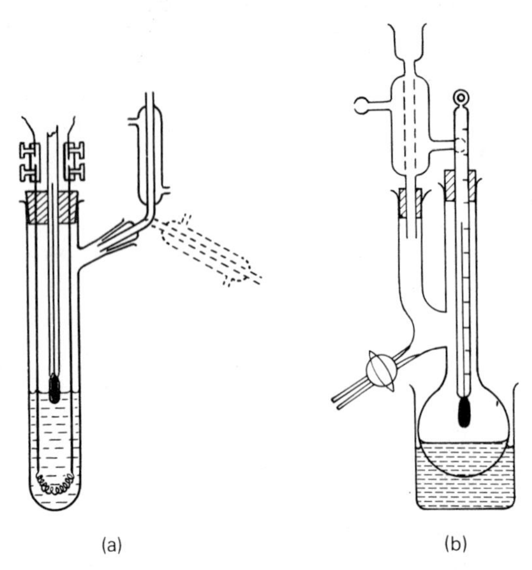

FIG. 8E.2 Simple apparatus for determination of vapour–liquid equilibrium diagrams.

arrangement shown in Fig. 8E.2(b) avoids moving the condenser: samples are obtained via the tap.

The tube is placed in a Dewar vacuum vessel. The apparatus is fitted up so that the liquid can be heated electrically, as shown in the figure. Alternatively the tube can be heated by immersion in a water-bath. A thermometer, graduated in tenths of a degree, passes through the cork fitted in the mouth of the boiling-tube, so that the bulb is partially immersed in the liquid. (For the best accuracy one must use the Cottrell apparatus to avoid errors due to superheating (section 7C).) The liquid is now caused to boil steadily until vapour is freely forming liquid in the condenser. When the temperature of the liquid has become constant, it is noted and the condenser is then rotated so that the condensate runs out into a small, cooled weighing bottle. After about 0·5–1 cm^3 of liquid has been collected (do not collect too much), the heating current is cut off, the cork is removed from the boiling-tube and a small quantity of liquid is removed and run into a small weighing bottle, which is then stoppered and cooled. The refractive indices of the distillate and of the liquid in the boiling-tube are determined under the same conditions as before, and the composition of the boiling liquid and of the vapour in equilibrium with it (the distillate) is read from the composition-refractive index curve.

Proceed in the manner described with each binary mixture, and plot the molar percentage composition of liquid and vapour at each boiling temperature in a temperature–composition diagram.

2. SEPARATION OF MIXTURES BY DISTILLATION

The vapour in equilibrium with a solution of two volatile liquids is richer than the liquid in the component of higher v.p. (lower b.p.), and, consequently, when the liquid mixture is fractionally distilled, each fraction will be richer in the more volatile component than the liquid from which it is distilled. By subjecting these 'fractions' to further fractionation, distillates can be obtained which are progressively richer and richer in the more volatile component, until, in the case of zeotropic mixtures, a complete separation of the components is effected.

When distillation is carried out with a fractionating column, repeated distillation of the fractions is avoided. When distillation is in progress, a temperature gradient is established along the column and, on passing upwards, the vapour which condenses is increasingly rich in the more volatile component. In other words, a process of repeated distillation takes place in the column, and, under theoretically perfect conditions, the pure component of lower b.p. collects in the condenser, and the component of higher b.p. is left in the still.

In the case of liquids which form binary azeotropic mixtures, separation into the two pure components cannot be effected by distillation, but only into an azeotropic mixture and one or other of the components, depending on the initial composition of the mixture. In the case of liquids which give a b.p. curve showing a minimum, the azeotropic mixture will be obtained as the distillate, and the residue will be one or other of the components. When the b.p. curve shows a maximum, the distillate will be one or other of the components, and the azeotropic mixture will be left in the still.

After having obtained the b.p. curves for a series of binary mixtures, the separation of these mixtures by distillation should be studied, and the efficiency of different fractionating columns investigated. The constant boiling mixtures, also, should be prepared by distillation, and their composition be thereby more accurately determined.

Owing to the great importance of distillation, the theory and technique of the process have received great attention. Rather elaborate apparatus is nowadays used in determinations of vapour–liquid equilibria. The efficiency of fractionating columns is tested by means of standard zeotropic mixtures, e.g. CCl_4– C_6H_6, and is expressed as the 'h.e.t.p. number', that is, '*height equivalent to a theoretical plate*'. This means the length of the column required to produce in practice as much separation as one ideal vapour–liquid *equilibrium* would effect. The practical separation obtained depends on the *reflux ratio*, i.e. the proportion of vapour which is condensed and permitted to return down the column; the greater the reflux, the better the separation. For the details of the theory and technique of fractional distillation, one of the monographs cited below should be consulted.

BIBLIOGRAPHY 8E: Equilibria between liquids and vapours

Weissberger, Vol. IV.
Coulson and Herington, *Laboratory Distillation Practices*, 1958 (Newnes, London).

9
Optical measurements in chemistry

9A Light sources

1. CONTINUOUS SPECTRUM SOURCES

The tungsten filament lamp is a convenient source of light for microscopy and absorptiometry in the visible region of the spectrum. It gives a *continuous* spectrum, having maximum intensity in the near infra-red, and falling off to zero in the near ultra-violet. A short-filament or headlamp bulb used at a voltage slightly in excess of the normal rating gives a higher relative intensity at the violet end of the spectrum extending also into the ultra-violet. In the quartz–iodine lamp a tungsten filament is heated in a quartz envelope containing a trace of iodine; this reduces evaporation of the filament enabling its temperature to be increased; this in turn increases both the total light emitted and the relative intensity at short wavelengths. This source is particularly suitable for *photomicroscopy* when ordinary (not panchromatic) plates are used.

The *hydrogen discharge lamp* is the source employed for absorption spectrometry in the ultra-violet; an electric discharge through hydrogen at a few mm pressure gives a light of a pale mauve colour which is found to consist of a number of discrete lines* in the visible spectrum (Balmer series), leading to a continuum in the blue and ultra-violet. The lamps consist of a small, oxide-coated tungsten filament surrounded by a cylinder of tungsten. The discharge passes through a rectangular slit in the screen under a potential of about 100 V, giving an intense, localized light source.

An arc or condensed high-voltage spark between tungsten-steel rods is sometimes used as a quasi-continuous source for visible and ultra-violet absorption spectrometry. Its spectrum contains a very large number of sharp lines.

* A short table of prominent lines is given in the Appendix, Table A8.

2. LINE SPECTRUM SOURCES

Small *gas discharge tubes* containing a low pressure of hydrogen, helium, argon, neon, etc., provide sharp line spectra suitable for calibrating spectrographs in the visible region of the spectrum. These require a 250 V d.c. supply.

Metallic vapour lamps provide the most convenient sources of line spectra of high intensity. The 'Osram' laboratory lamps, containing sodium, cadmium, mercury or mercury+cadmium, made by the General Electric Co. Ltd (London), are particularly valuable for spectrograph calibration, refractometry, polarimetry, etc. The spectra of these metals give a number of intense, well-separated lines of known wavelength, which can be isolated by use of filters or a spectrometer to give monochromatic light of various wavelengths. The 'Osram' lamps (Fig. 9A.1) contain a low pressure of inert gas and a quantity of the volatile metal. The electrodes consist of two

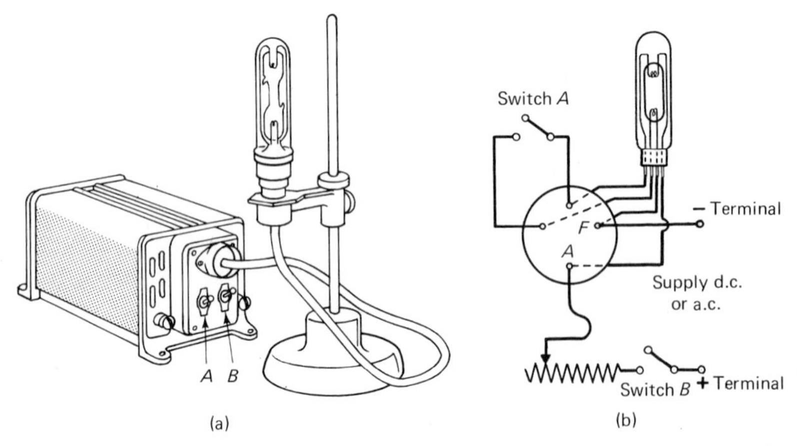

FIG. 9A.1 Laboratory metal vapour discharge lamp. (a) 'Osram' lamp and rheostat. (b) Wiring diagram.

tungsten filaments which are initially heated for about 1 min by passage of a current. This induces thermionic emission from the filaments so that when the heating current is switched off and the full mains voltage applied across the tube, a discharge strikes between the electrodes, being carried initially by the inert gas. As the heat of the discharge vaporizes the metal, the metallic spectrum grows in intensity and supersedes that of the inert gas.

The *sodium lamp* is the usual source for refractometry and polarimetry, as it provides almost monochromatic light—about 99% of the visible radiation emitted by it lies in the 589·0–589·6 nm doublet ('D lines'). If necessary, the faint red, green and blue lines can be almost completely absorbed by passing the light through a few cm of 7% aqueous potassium dichromate solution or equivalent glass or gelatine filter. The 45 W sodium lamps used for street lighting are also very suitable for polarimetry.

Mercury vapour lamps are particularly useful when light of several wavelengths in the visible or ultra-violet is needed as in determinations of optical dispersion, in photochemistry, and ultra-violet photomicroscopy. Many types of lamp are available. The 'Osram' mercury lamp is convenient for refractometry in the visible region. Small street-lighting mercury lamps give larger sources of visible and near ultra-violet light. The small quartz mercury discharge tubes employed in commercial 'black' lamps (for exciting fluorescence) are very intense in the visible and ultra-violet. A small source of still greater localized intensity, particularly suitable for producing a focused beam, is provided by lamps in which a heavy discharge is passed between stout tungsten rods enclosed in mercury vapour in a silica bulb.

Several types of mercury discharge lamp are suitable for irradiating solutions, paints, etc. The chief distinction between them is the pressure at which they operate; low pressure lamps, operating at about 5 000 V, emit a high intensity of the mercury resonance line at 253·7 nm. Higher pressure lamps (running hot) show reversal of this line owing to self-absorption, and give greatest intensity in the 365·0 nm line. Dark goggles should always be worn when working with ultra-violet lamps. The harmful rays are largely cut off by ordinary glass.

3. FILTERS

When a more or less monochromatic source of light is needed, as in 'colorimetry' (absorptiometry) and photochemistry, some form of filter may be applicable. Liquid solutions, coloured glasses and dyed gelatine films are often used. It should be realized that the absorption bands of all filters are relatively broad and gradual in onset. Consequently, the ideal complete cut-off of some regions of the spectrum and complete transmission of a narrow region are rarely obtained. A green filter, for example, would transmit a little yellow and blue light when used with a continuous spectrum source. However, the principal well-separated lines of the mercury spectrum can be isolated in fairly pure form by use of combinations of liquid, glass or gelatine filters. Glass filters are also available for absorbing infra-red (heat rays) and for transmitting the near ultra-violet while absorbing the visible ('black glass').

'Interference' filters are commercially available (e.g. from Messrs Barr & Stroud, London); they consist of metallic interference films on glass deposited by vaporizing the metal, and have the advantage of a narrow transmission band.

When light of very high purity is needed, a single or double *monochromator* is used. This instrument is essentially a spectrometer of large aperture, normally using a diffraction grating. Gas lasers produce a well collimated beam of coherent light at a single wavelength, e.g. 634 nm for the helium–neon laser. As the wavelength cannot be varied, this source is used mainly for Raman spectroscopy rather than for the spectra discussed in this chapter. It is also useful for other measurements, e.g. for the determination of the thickness of thin films by interference fringes.

BIBLIOGRAPHY 9A: Light sources

See texts quoted in bibliography 9B and 9D.
Weissberger, Vol. I, Pt 3, chapter XXVIII.
Brode, *Chemical Spectroscopy*, 1943 (John Wiley, New York).

[1] E.g. from Adam Hilger Ltd, London.

9B Emission spectra*

When a substance in the gaseous state is *excited* by high temperatures or electric discharge, it may emit light, and, owing to the quantization of the permitted energy levels of atoms and molecules, the light will consist of quanta of definite size (or sizes) and hence will give a *line spectrum*. It is well known that the wavelengths λ (or frequency ν, or wave-number $\tilde{\nu}$, where $\tilde{\nu}=1/\lambda$ and $\nu \times \lambda = c$, the velocity of light) of lines of emission spectra are highly characteristic of each substance. Spectrum analysis has therefore long been used for detecting the presence of various elements, and is nowadays extensively employed in routine quantitative analysis (e.g. of alloy steels). Quartz ultra-violet spectrographs of high dispersion are generally used for spectrographic analysis, but the principles of spectrum analysis can be readily illustrated by reference to the glass wavelength spectrometer described below.

* The brief introduction to spectroscopy contained in sections *B* and *C* is confined to visible and near ultra-violet spectra. Although the same general principles apply to other spectral regions, the techniques employed in infra-red and for ultra-violet spectroscopy are not considered here.[1]

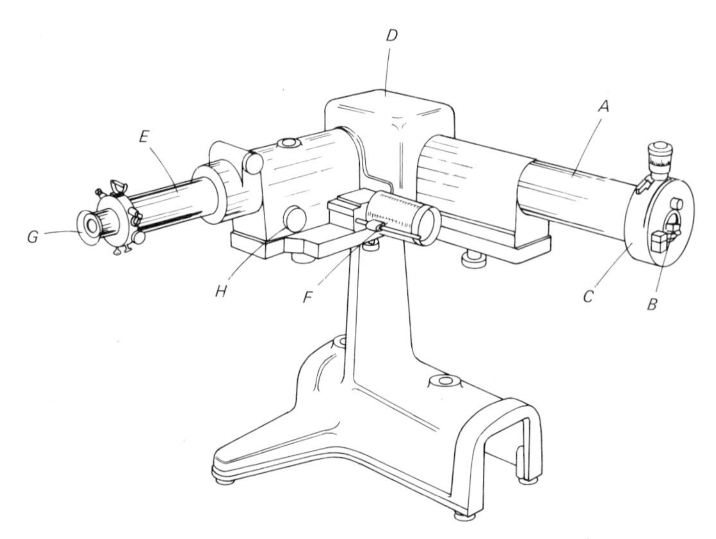

FIG. 9B.1 Constant deviation spectrometer: general view. (Courtesy Messrs Adam Hilger Ltd, London.)

2. THE CONSTANT DEVIATION SPECTROMETER

The determination of wavelengths of spectral lines is rendered very simple by using a spectrometer graduated so that wavelengths can be read directly on a scale. This is the case with the constant deviation spectrometer: the telescope and collimator are fixed at right angles, while the special prism can be rotated by a calibrated micrometer screw. Figure 9B.1 shows the form of this instrument as manufactured by Adam Hilger Ltd (London). The collimator tube *A* carries an adjustable slit *B* and simple shutter *C*. The constant deviation prism *D* has the form shown in Fig. 9B.2(a), from which it can be seen that the ray which travels in the prism exactly parallel with the diagonal *BD* will emerge exactly at right angles to its original direction whereas rays of other wavelength, being differently refracted, emerge in different directions. However, for each wavelength there is one position of the prism which would permit the ray to enter the telescope *E*. The setting of the prism therefore determines the wavelength of light which will be observed at the centre of the eyepiece *G*. The rotation of the turntable carrying the prism is effected by a micrometer screw, the hardened point of which presses against a polished steel plug in the projecting arm attached to the table. The screw carries a drum *F* (see Fig. 9B.2(b)) on which a predetermined wavelength scale has been engraved. This scale is dependent on the

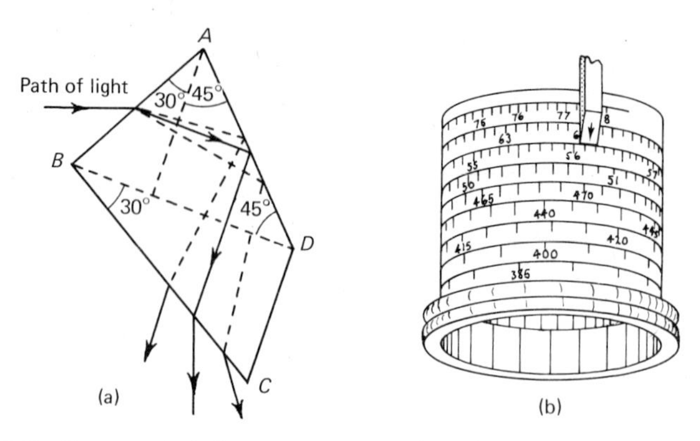

FIG. 9B.2 Constant deviation spectrometer. (a) Path of rays through the prism. (b) Wavelength drum.

dispersion of the prism, but with a given instrument the wavelength can be read off directly as indicated by the index. Focusing of the telescope is effected by means of a milled knob H. The telescope eyepiece G contains a pointer which can be adjusted laterally; it is brought into focus by turning the eyepiece.

Once the spectrometer has been set up the prism should not be touched, but the following adjustments and checks can be made: (1) slit width, (2) verticality of slit, so that the image of a line is parallel with the slit in the eyepiece, (3) focus of eyepiece on the pointer, (4) focus of telescope for sharpness of spectra—separation of the Na-D lines is a good test. The accuracy of setting of the instrument can then be checked by reading the wavelengths of a number of well-defined spectral lines such as those of sodium and mercury (Appendix Table A8). Once the scale has been checked (or calibrated) the instrument can be used to determine the wavelength of spectra.

EXPERIMENTS

A. Examine the light emitted when the chlorides of sodium, potassium, lithium and thallium are vaporized in a Bunsen flame.

B. Measure the wavelengths of as many as possible of the lines of the hydrogen spectrum, and compare the values obtained with those calculated from the expression $\lambda = 3\,645{\cdot}6[m^2/(m^2-4)]$, *where m is successively given the values 3, 4, 5, etc. (Balmer Series.)*

Procedure: A. The substances may be volatilized from a clean platinum wire or fibrous asbestos wick, but a more persistent flame is obtained with an aspirator of the type used in atomic absorption spectroscopy to spray a solution of the salt of the metal into a flame. The lines observed in the flame spectra should be classified roughly according to intensity, e.g. very strong, strong, medium, weak, very weak.

Results. The lists of wavelength recorded should be compared with standard tables of spectral lines. The presence of impurities should be looked for, e.g. Na in K salts and vice versa, Rb in K, etc.

3. SPECTROGRAPHY

The scope of spectrum analysis is greatly extended if the spectra are recorded photographically. In the ultra-violet this method is, of course, indispensable unless a photo-electric spectrophotometer is used.

The eyepiece of the telescope of the Hilger wavelength spectrometer (Fig. 9B.1) can be removed and replaced by a quarter-plate size camera bearing a rack and pinion by which the plate may be moved vertically, so that a number of spectra can be photographed on the same plate. Before exposures are made, it is necessary to make the following adjustments: (1) slit width, (2) slit height (for width of spectrum), (3) focus of telescope for sharpness of image on the plate, (4) angle of tilt of the plate holder. These adjustments can be made with a ground-glass screen in the plate holder, the image of a suitable line source being examined usually with the aid of a hand lens, but the final setting is best found from a number of trial photographs.

The proper alignment of the source on the optical axis of the collimator requires careful attention. A good method of locating the correct position for a small source such as an arc or spark is to place a lighted flashlight bulb in the position that will be occupied by the centre of the spectrum, and to trace back the image of this source where it emerges from the collimator; a darkened room is, of course, necessary for this operation.

Since a small source such as an arc or spark will not illuminate the whole of the slit uniformly (if placed close) or sufficiently intensely (if placed far from it), an auxiliary lens is usually employed to cast an enlarged image of the source on to the slit (Fig. 9B.1). The source should be situated at 4–5 times the focal length of the lens from the slit.

When a spectrum is to be photographed, it is necessary to record on the same plate a spectrum of a known source to provide a wavelength scale. The Hilger wavelength spectrometer bears a small movable prism on the collimator so that light, for example, from a mercury lamp placed at the side, can be reflected into the slit to provide this reference spectrum. The spectrum should be recorded on the top and the bottom of the plate.

The type of photographic plate to be used will depend on the part of the spectrum concerned. 'Ordinary' (Process) plates are sensitive only to the blue, violet and ultra-violet: 'orthochromatic' are also sensitive to green and yellow, but can be processed safely in dim ruby-red light: 'panchromatic' are sensitive to all parts of the visible spectrum, and must therefore be loaded and developed in total darkness. Special 'infra-red' plates extend the photographic range to about 1 200 nm ($\approx 1\ \mu$). A plate giving good 'contrast' is desirable for spectrography, e.g. Ilford 'Xenith' (ortho, to 570 nm) and Ilford 'Rapid Process Panchromatic' plates. Any normal developing and fixing solutions can be used.

EXPERIMENT

Photograph a copper arc spectrum and use it to construct a wavelength calibration chart for the spectrograph.

Procedure. Two rods of high purity copper should be mounted in an insulated, adjustable stand. The arc gap must be aligned with the optical axis of the collimator (see above). The arc is best run from a 200 V d.c. supply with an adjustable rheostat of adequate current carrying capacity and ammeter in series. A current of 5 A is suitable. The arc is easily struck by drawing a carbon rod (on an insulated handle) across the gap between the electrodes after the current has been switched on. The optimum size of gap and value for the series resistance to produce a steady arc is found by trial.

Next, the auxiliary condenser can be set, and a convenient slit width and height can be selected. The wavelength drum should, of course, be set to a suitable position and subsequently not altered. A trial plate can then be exposed, using a range of different exposure times, in order to find a suitable exposure. The best focus and tilt of the spectrograph can next be determined by taking another plate with a number of different settings (all near the visual optimum). This plate should be dried and then examined carefully with a lens to ascertain the settings which give the sharpest lines over the spectral range to be investigated.

Finally, when all the adjustments have been made so that sharp, uniform spectra are attained, the calibrating plate can be taken. The spectrum of a mercury lamp should also be recorded on top and bottom of the same plate. A full record of all the details of the exposures should be made. After processing, the plate can be labelled by writing on the edge with ink.

Construction of calibration chart. When the plate is dry, the positions of as many as possible of the lines of the mercury and copper arcs are to be measured. It is convenient to adopt one particularly prominent and easily identified Hg-line as origin. A straight edge of metal is laid across the plate (the gelatine being uppermost) and carefully aligned with the chosen Hg-line on the two sides of the plate: a fine line is then drawn in the gelatine with a razor blade right across the plate and passing precisely through the centre of the chosen line in the two Hg-spectra.

The distances, positive or negative, of other lines of the Hg- or Cu-spectra can now be measured, taking the mark as origin. The measurements should preferably be made with a travelling microscope.

Next, it is necessary to identify the measured lines by reference to tables of wavelengths. The lines of the mercury arc are fairly easily recognized by inspection of published spectrograms. The general form of the dispersion curve, i.e. wavelength against distance along the plate, can then be drawn, and points for the Cu-spectrum can then be added by reading approximate wavelengths from the provisional chart and then identifying the lines in list of characteristic wavelengths.

Finally, an accurate graph on a large scale can be drawn, plotting wave-number or frequency of the lines against distance from the reference line. This chart will then serve to characterize lines of other spectra which may subsequently be photographed with the same specified setting of the spectrograph.

4. SPECTROSCOPIC ANALYSIS

The *qualitative identification* of elements requires that the substance be volatilized. The flame excitation method previously described is effective for only a few elements, and it is therefore more usual to employ the high temperatures of the carbon arc or of a high-tension spark to volatilize and excite the spectrum of a mixture which is to be examined. Solids can be excited by embedding them in a cavity in the centre of the electrodes of a carbon arc.

For *quantitative* spectroscopic analysis means must also be provided for measuring the *relative intensities* of selected spectral lines. Standard mixtures of known analysis must be used for comparison since there is no simple way of arriving at the percentage of an element in a mixture from a measured intensity. Further, the absolute intensity of an arc or spark is not reproducible from one sample to another. Consequently, a comparison is made between the intensities of two lines, one from, say, the substrate and the other from the substance (e.g. impurity) to be determined.

This method is now being superseded by atomic absorption analysis.[2] The sample is volatilized by spraying a solution into a flame with an aspirator. A hollow cathode lamp for the element to be analysed generates emission in narrow spectral lines corresponding to transitions from higher atomic energy levels to the ground or lowest electronic state. These lines are preferentially absorbed by atoms of the same element in the sample undergoing the converse transition: the absorption is therefore effectively specific to this element. The intensities are measured photo-electrically. This method is extremely sensitive, and can be used to detect trace amounts of many metals.

In the preparation of standard mixtures for spectroscopic analysis one needs chemicals of exceptional purity. Such chemicals are available under the title of 'spectroscopically pure', but they are, of course, very expensive.

EXPERIMENT

Make a spectroscopic examination of a sample of strontium carbonate for the presence of barium.

Procedure. Use the spectroscope which has been calibrated as described in the preceding experiment. Set up a carbon arc with high-purity ('spectroscopic') carbon electrodes, making the lower electrode positive. The arc requires about 60 V and 5 A to maintain it; it can therefore be run from a 110 d.c. supply with a 10 Ω, 10 A resistance. Focus an image of the arc on to the slit of the spectrograph, and select only the centre of the image by means of the wedge or Hartmann diaphragm.

To excite the spectrum of a powder, a small quantity of it is packed into a cavity drilled in the lower carbon electrode.

Photograph (a) the carbon arc alone, (b) the arc with strontium carbonate, (c) the arc with strontium carbonate to which a small amount of barium carbonate (e.g. 0·1%) has been deliberately added.

Identify the principal Sr and Ba lines as far as possible. The persistent Ba lines at 577·77 nm and 553·55 nm are suitable for detecting traces of barium. A rough estimate of the amount of Ba present can be made by comparing the intensity of the Ba line in the sample with that obtained with spectroscopically pure $SrCO_3$ to which known amounts of $BaCO_3$ have been added (with thorough mixing).

Note that the carbon arc gives bands in the near ultra-violet; these are due to CN (formed as a transient entity in the arc). Other band spectra such as those of CaO, CaF, etc., may be encountered in spectrographic analysis.

5. RAMAN SPECTRA

When a beam of monochromatic light is passed through a medium which does not absorb it, a very small proportion of the light is *scattered* in all directions. Only with a colloidal solution is the scattered light perceptible to the eye (*Faraday–Tyndall beam*), but even with a pure, dust-free liquid it is possible to detect scattered light by a suitably long photographic exposure taken, of course, at right angles to the incident beam. (Scattered light must not be confused with fluorescence, which is due to absorption and re-emission; fluorescent light is generally at a longer wavelength than that absorbed.)

The light molecularly scattered by a non-absorbing medium is mostly of the same wavelength as that of the incident light, and a study of the intensity of this scattering forms the basis of an important modern method of determining the molecular size and shape of colloidal particles and macro-molecules.[4]

In addition, however, part of the scattered light is found to have undergone definite displacements of wavelength. The spectrum of the scattered light therefore consists of a strong impression of the spectral line from the source together with a number of other sharp but much weaker lines, some at longer and some at shorter wavelengths than the exciting line. These additional lines are called the *Raman spectrum*.

The differences between the frequencies of the Raman lines and the exciting line correspond to certain of the *frequencies of vibration* within the molecule. Raman spectra, can therefore give structural information similar in character and often complementary to that supplied by infra-red spectra.[5]

The source used for Raman spectroscopy is the intense monochromatic radiation from a laser, normally He–Ne. The liquid sample is normally placed in a glass tube, and the scattered light emerges through the end. It passes through a grating monochrometer, and is detected by a photomultiplier (section 4A(4)).

Raman spectroscopy is also used to measure the concentration of certain molecules in solution for which other methods are inconvenient, e.g. the NO_2^+ ion in nitrating mixtures ($H_2SO_4 + HNO_3$).

BIBLIOGRAPHY 9B: Emission spectra

Candler, *Practical Spectroscopy*, 1949 (Hilger and Watts, London).
Brode, *Chemical Spectroscopy*, 2nd edn, 1943 (John Wiley, New York).
Dingle, *Practical Application of Spectrum Analysis*, 1949 (Chapman and Hall, London).
Nachtrieb, *Principles and Practice of Spectrochemical Analysis*, 1950 (McGraw-Hill, New York).
Smith, *Visual Lines for Spectroscopic Analysis*, 2nd edn, 1952 (Hilger and Watts, London).
Harrison, Lord and Loofbourow, *Practical Spectroscopy*, 1948 (Blackie and Sons, London).

[1] See, for example, Appendix 1, Weissberger, Vol. I, Pt 3, chapters XXVIII and XXIX; Sawyer, *Experimental Spectroscopy*, 2nd edn, 1951 (Chapman and Hall, London); Barnes *et al.*, *Infra-red Spectroscopy*, 1944 (Reinhold, New York).
[2] Dean and Rains (ed.), *Flame Emission and Atomic Absorption Spectrometry*, 1969 (Dekker, New York).
[3] See, for example, Brode, *op. cit.*
[4] Appendix 1; Weissberger, Vol. I, Pt 3, chapter XXXII.
[5] Szymanski, *Raman Spectroscopy*, 1965 (Plenum Press, New York); Gilson and Hendra, *Laser Raman Spectroscopy*, 1971 (Wiley-Interscience, New York).

9C Molecular spectroscopy of diatomic molecules

1. The energy of vibration of the atoms of a molecule is quantized: only certain values permit acceptable solutions of the Schrödinger wave equation. The values of these energy levels can be calculated using quantum mechanics if the variation of the potential energy of the molecule with internuclear distance is known.

2. THE SIMPLE HARMONIC OSCILLATOR

The simplest model for a diatomic molecule is to assume that Hooke's law applies for the force between the atoms, so that for any internuclear separation r, the restoring force is

$$F = -f(r - r_e) \qquad (1)$$

where f is the *force constant* in $\mathrm{N\,m^{-1}}$ and r_e is the mean internuclear distance. The system then behaves as a simple harmonic oscillator (s.h.o.) of frequency $\nu\ \mathrm{s^{-1}}$ given by

$$\nu = \frac{1}{2\pi}\sqrt{\frac{f}{\mu}} \tag{2}$$

where μ = reduced mass $= m_A m_B/(m_A + m_B)$, m_A and m_B being the masses of the atoms (in kg).

The potential energy as a function of the separation of the atoms is

$$V(r) = -\int_{r_e}^{r} F\,\mathrm{d}r = \tfrac{1}{2}f(r-r_e)^2 \tag{3}$$

The corresponding vibrational energy levels are then found from the Schrödinger equation to be

$$E_v = G + (v + \tfrac{1}{2})\boldsymbol{h}\nu \tag{4}$$

where $\boldsymbol{h}$ is Planck's constant, G is the absolute potential energy of the molecule at $r = r_e$ and v takes the values 0, 1, 2, ... Note that the molecule has a minimum vibrational energy in the lowest level of $(E_0 - G) = \frac{1}{2}\boldsymbol{h}\nu$. This is the *zero point energy*, i.e. the vibrational energy of the molecules at absolute zero when all the molecules are in their lowest vibrational level.

The energy levels are evenly spaced, because $E_{v+1} - E_v = \boldsymbol{h}\nu$, independent of v (see Fig. 9C.1). However, experimental spectra show

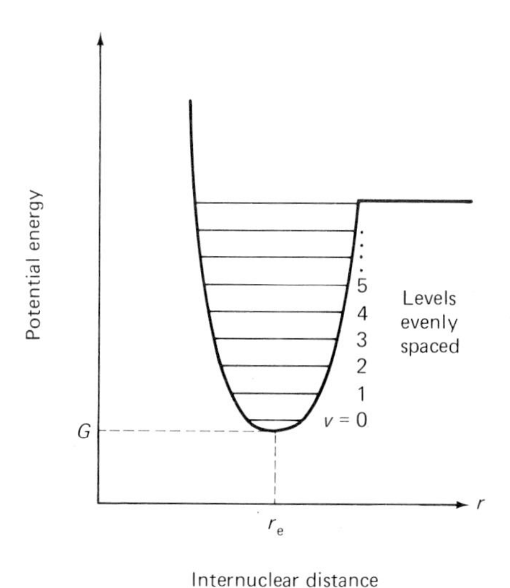

FIG. 9C.1 Potential energy curve for simple harmonic oscillation model.

that vibrational energy levels in fact converge with increasing energy, indicating that the s.h.o. is not a good model for the potential energy of a real diatomic molecule.

3. THE MORSE EQUATION

A better description of the potential energy (for a particular electronic state) is given by the empirical equation suggested by P. M. Morse:

$$V(r) = D[1-e^{-a(r-r_e)}]^2 \tag{5}$$

where a is a constant and D is the dissociation energy (including the zero point energy): $D = E_D - G$ (Fig. 9C.2).

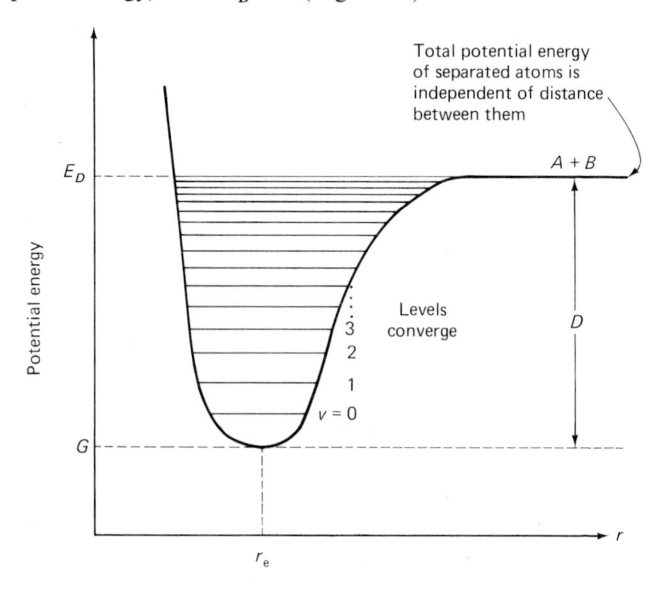

FIG. 9C.2 Morse potential energy curve.

The energy levels given by the Schrödinger equation for this potential energy function are

$$E_v = G + h\nu(v+\tfrac{1}{2}) - xh\nu(v+\tfrac{1}{2})^2$$

where

$$\nu = \frac{1}{2\pi}\sqrt{\frac{2Da^2}{\mu}} \tag{7}$$

and

$$x = \frac{h\nu}{4D} \tag{8}$$

This is now of the same form as the result for the s.h.o. (Eq. (4)) but with the addition of a term in $(v+\frac{1}{2})^2$. ν can be regarded as the frequency of vibration for small oscillations, for which the associated force constant is given by $f=2Da^2=4\pi^2\nu^2\mu$ by Eqs (2) and (7). If $x=0$, Eq. (6) reduces to that for the harmonic oscillator model (4) and therefore x is known as the *anharmonicity constant.*

The energy levels corresponding to the Morse function converge with increasing energy for

$$E_{v+1}-E_v = h\nu-2x(v+1)h\nu = h\nu[1-2x(v+1)] \tag{9}$$

At the *convergence limit* the spacing becomes zero. If this occurs at $v=v_L$

$$E_{v+1}-E_v = 0 = h\nu[1-2x(v_L+1)],$$

from which

$$v_L = \frac{1}{2x}-1 \tag{10}$$

Applying Eqs (6), (10) and (8) we have,

$$E_{\text{limit}}-E_0 = v_L h\nu - xh\nu(v_L^2+v_L)$$

$$= \frac{h\nu}{4x}(1-2x)$$

$$\approx \frac{h\nu}{4x} = D \tag{11}$$

The convergence limit corresponds to the energy E_D required for dissociation.

The zero point energy for the Morse oscillator is obtained by putting $v=0$ in Eq. (6) giving

$$\varepsilon_0 = E_0-G = \tfrac{1}{2}h\nu-\tfrac{1}{4}xh = \tfrac{1}{2}h\nu(1-\tfrac{1}{2}x) \tag{12}$$

EXPERIMENT

The absorption spectrum of iodine.

Theory. Figure 9C.3 shows the potential energy curves for iodine. The vapour at the temperature of this experiment consists mainly of molecules of iodine in the lowest ($v=0$) vibrational level of the lowest (or ground) electronic state, which dissociates to give two normal iodine atoms I (spectroscopic notation $^2P_{\frac{3}{2}}$). When a quantum of visible light is absorbed by a molecule, it is promoted to a higher excited electronic state which is represented by the upper curve in Fig. 9C.3: this would dissociate to give one normal iodine atom and

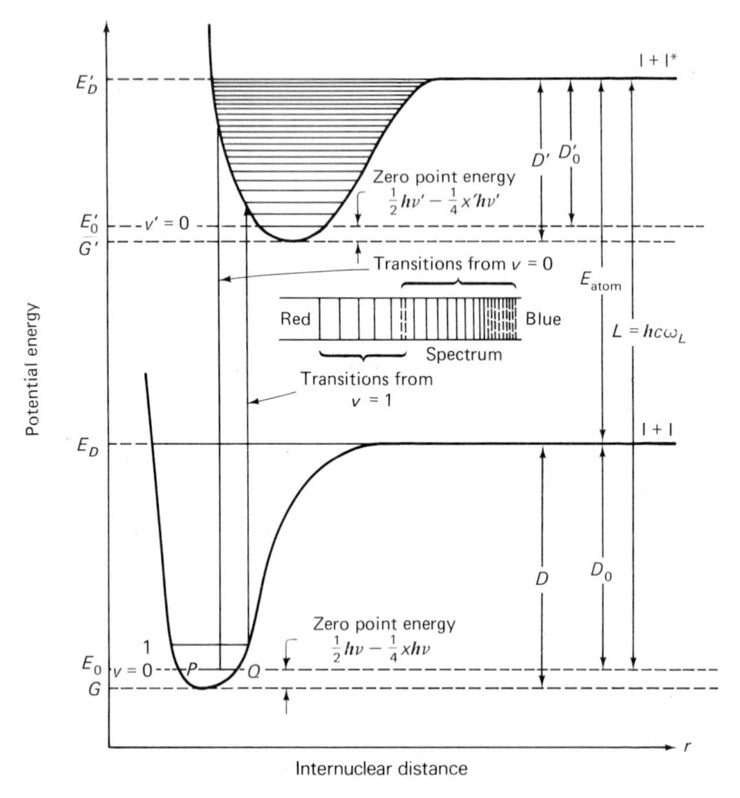

FIG. 9C.3 The adsorption spectrum of iodine from its potential energy curves.

one electronically excited atom I ($^2P_{\frac{1}{2}}$). The two states have different Morse parameters and energies: those for the upper state are indicated by a prime.

The Franck–Condon principle states that the internuclear separation of a molecule is unchanged during the very short time required for the transition from one electronic state to the other. Possible transitions from the $v = 0$ level of the ground state can therefore be represented by vertical lines from any point on PQ, i.e. within the range of possible internuclear distances for molecules in this level. The lowest vibrational level of the excited electronic state which can be reached corresponds to absorption in the yellow part of the spectrum. The set of bands seen for transitions from $v=0$ starts in the yellow and stretches towards the blue, converging to continuous absorption at the dissociation limit. These are the lines to be measured in this experiment. The other set seen in the red correspond to transitions from the $v=1$ level of the ground state ('hot bands'). This set fades out in the yellow unless the gas is so hot that the $v=1$ level is well populated.

The absorption of light to be measured occurs at wavelengths whose energies correspond to transitions between the $v=0$ level of the ground state where

$$E = G+\tfrac{1}{2}\boldsymbol{h}\nu-\tfrac{1}{4}x\boldsymbol{h}\nu = E_0 \qquad (13)$$

and successive levels of the upper state,

$$E' = G'+\boldsymbol{h}\nu'(v'+\tfrac{1}{2})-x'\boldsymbol{h}\nu'(v'+\tfrac{1}{2})^2 \qquad (14)$$

The lines observed therefore have *wave numbers* ω given by

$$E'-E = \boldsymbol{h}c\omega = (G'-E_0)+\boldsymbol{h}\nu'(v'+\tfrac{1}{2})-x'\boldsymbol{h}\nu'(v'+\tfrac{1}{2})^2 \qquad (15)$$

(The wave number ω is the reciprocal of the wavelength, $1/\lambda$. Hence ν, the frequency, is $c\omega$ where c is the velocity of light, $2{\cdot}9979\times10^{-8}$ m s^{-1}. The energy of the corresponding quantum is $\boldsymbol{h}\nu=\boldsymbol{h}c\omega$.)

The energy required to excite an iodine atom, E_{atom}, is known from atomic spectra to be $1{\cdot}508\times10^{-19}$ J. If we can determine ω_L, the value of ω corresponding to dissociation of the upper state, we can obtain the energy

$$L = \boldsymbol{h}c\omega_L = D_0+E_{\text{atom}} \qquad (16)$$

(see Fig. 9C.3) and hence the dissociation energy (D_0) of an iodine molecule in its ground state. We cannot observe ω_L directly because the convergence of the levels (and hence of the absorption lines) with increasing n makes the spectrum appear continuous before the convergence limit is actually reached. We can however evaluate ω_L indirectly from the spacing between successive absorption lines which may be obtained either from Eq. (15) or from Eq. (9); it is given by

$$\Delta\omega = \frac{\nu'}{c}\{1-2x'(v'+1)\} \qquad (17)$$

Thus if we plot $\Delta\omega$ against v', or against any arbitrary serial integer p (Fig. 9C.4), we obtain a straight line of slope

$$\frac{\mathrm{d}\Delta\omega}{\mathrm{d}p} = -\frac{2\nu'x'}{c} \qquad (18)$$

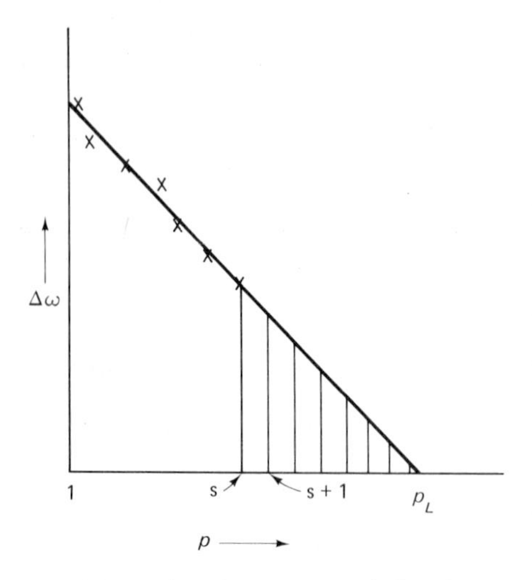

FIG. 9C.4 Birge–Sponer extrapolation plot.

Then

$$\omega_L = \omega_s + \sum_{p=s}^{p=p_L} \Delta\omega \tag{19}$$

where the summation is over the serial integer p from a value $p=s$ at which ω can still be directly observed, to the limit $p=p_L$ which is defined by $\Delta\omega=0$. The sum in Eq. (19) can then be evaluated by adding the ordinates of $\Delta\omega$ from $p=s$ to $p=p_L$ in Fig. 9C.4.

Procedure. The spectrum is observed using the constant deviation spectrometer, whose operation and calibration have been described in section 9B. The iodine is contained in an evacuated glass cylinder about 15 cm in length and 5 cm in diameter with windows at each end. This is heated gently by an electric winding: the heat is adjusted so that enough iodine is vaporized for the spectrum to be observed, but not so strongly that the 'hot bands' (see above) become too intense.

Place the sealed bulb containing iodine in line with the collimator and allow 15 min for the iodine to vaporize after switching on the heating. An electric lamp giving a continuous emission spectrum is placed behind the iodine bulb and the dark bands of the iodine absorption spectrum are observed through the spectrometer. Starting from the one nearest to 543·2 nm, measure the wavelength of every band as far as possible into the blue.

Calculations

(1) Convert the wavelength readings to wave numbers ω expressed in m^{-1}.

(2) Plot $\Delta\omega$ against successive integers $p=1, 2, 3, \ldots$, taking p for the first line (at 543·2 nm) to be unity. Draw the best straight line and extrapolate to the convergence limit at which $\Delta\omega=0$.

(3) Choose a value of p $(=s)$, as near as possible to the limit, for which $\Delta\omega$ lies on or near to the straight line; add to ω_s the sum of the successive values of $\Delta\omega$ for higher values of p, up to the convergence limit, to get ω_L (see Eq. (19) and Fig. 9C.4).

(4) Hence calculate D_0, from Eq. (16) in units of (a) J $molecule^{-1}$, (b) J mol^{-1} and (c) cm^{-1} (the traditional spectroscopic unit).

(5) A complete analysis of the spectrum has shown that the line at 543·2 nm corresponds to $v'=27$; hence calculate, for the excited electronic state of the iodine molecule:

(i) v' at the limit, v'_L.

(ii) The anharmonicity constant x', from Eq. (10).

(iii) The vibration frequency of the upper state ν', from Eq. (18), and the force constant f', from Eq. (2).

(iv) The dissociation energy D' for the upper state, from Eq. (8).

(v) The zero point energy for the excited state, ε'_0 from Eq. (12) and hence D'_0, the dissociation energy corrected for the z.p.e. contribution (see Fig. 9C.3).

(vi) The wavenumber ω_{00} for the lowest transition $v=0$ to $v'=0$ (see Fig. 9C.3).

BIBLIOGRAPHY 9C: Molecular spectroscopy of diatomic molecules

See standard textbooks of physical chemistry, or

Barrow, *Molecular Spectroscopy*, 1962 (McGraw-Hill, New York).

Banwell, *Fundamentals of Molecular Spectroscopy*, 1966 (McGraw-Hill, London).

9D Absorption spectra

1. When a beam of monochromatic light of intensity I_0 passes into a medium which absorbs that wavelength, the intensity is gradually reduced as the beam progresses through the medium, according to Lambert's law, namely, $\log_{10}(I_0/I) \propto d$: $\log_{10}(I_0/I)$ $(=D)$ is called the optical density and I is the intensity of the beam after passing through a thickness d of the medium. If the medium is a solution containing one absorbing substance, then Beer's law generally applies also; this states that the extinction is proportional to the concentration c of the absorbing species. The combined laws therefore give for absorption by a solution:

$$D = \log_{10} \frac{I_0}{I} = kcd$$

The constant k is known as the *molecular extinction coefficient* of the substance since it expresses the extinction resulting from 1 m of unit concentration of the substance. A graph showing how k varies with wavelength of the light is called the *absorption curve* (or spectrum) of the substance.

Molecular band spectra are much less sharp when the substance is examined in solution instead of in the gaseous state, and in most cases the overlap between adjacent lines and bands is so extensive that a broad region of continuous absorption results; the individual lines can no longer be distinguished and only maxima and minima appear on the absorption curve. The 'absorption spectrum' of the substance is then characterized by the wavelengths and intensities of the maxima and minima. The absorption curve depends considerably on the solvent. Much empirical information is available regarding the relation between the absorption spectrum (in the visible and ultraviolet) and chemical structure, particularly of organic compounds.[1]

Absorption spectra have many important applications in analysis, in structure determination and photochemistry, as well as miscellaneous uses such as assessment of colour value of dyes and filters.[2]

2. ABSORPTION SPECTROPHOTOMETRY

An absorption curve is normally obtained using a spectrophotometer, i.e. a monochromator combined with a photometer. The continuous light sources used are a tungsten bulb for the visible and a hydrogen (or deuterium) lamp for the ultra-violet. Light at a particular wavelength is selected by a grating monochromator, passes through a cell containing the sample, and is detected by a photocell or photomultiplier (section 4A(4)), whose photocurrent is displayed on a meter.

In single beam instruments the ratio I_0/I is measured successively at each wavelength for the sample and for a blank, i.e. a cell containing pure solvent. The ratio of $(I_0/I)_{\text{sample}}$ to $(I_0/I)_{\text{blank}}$ gives the optical density due to absorption by the sample.

In double beam instruments the beam is split and passes simultaneously through sample and blank. The intensities are measured alternately by a single photocell or photomultiplier using a rotating mirror, or simultaneously using two photocells, and the optical density displayed.

In modern instruments the spectral region is scanned using an electric motor to change the wavelength scale of the monochromator; a potentiometric recorder (section 4C(4)) is used to record the optical density output, hence drawing the spectrum automatically (see Fig. 9D.1).

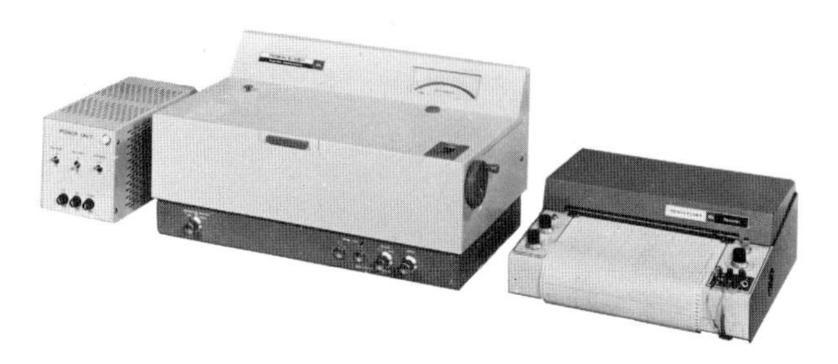

FIG. 9D.1 Visible-u.v. spectrophotometer with recorder. (Courtesy Perkin-Elmer Ltd, Beaconsfield, Buckinghamshire.)

Similar instruments are available for the infra-red. Here a glowing bar (Nernst filament) is used as a source, and a Golay cell as detector. The latter converts the radiation into heat and hence into a minute pressure change in a gas which can be detected.

EXPERIMENT

Determine the free energy of the reaction $N_2O_4 = 2NO_2$.

Introduction.[1] At room temperature nitrogen tetroxide N_2O_4 is partly dissociated into nitrogen dioxide NO_2,

$$N_2O_4\ (g) \rightleftharpoons 2NO_2\ (g) \qquad (1)$$

NO_2 is a brown gas but N_2O_4 is colourless. The position of equilibrium is determined by measuring the light absorbed by the NO_2.

Equilibrium theory. We can construct the following table in which α is the degree of dissociation of N_2O_4, n the number of moles of N_2O_4 which would be present if there were no dissociation ($\alpha = 0$), and p the total pressure of the mixture.

	N_2O_4	NO_2	*Total in mixture*
Number of moles	$(1-\alpha)n$	$2\alpha n$	$(1+\alpha)n$
Mole fraction	$\dfrac{1-\alpha}{1+\alpha}$	$\dfrac{2\alpha}{1+\alpha}$	1
Partial pressure	$\left(\dfrac{1-\alpha}{1+\alpha}\right)p$	$\left(\dfrac{2\alpha}{1+\alpha}\right)p$	p

Hence the equilibrium constant K_p is given by

$$K_p = \frac{p_{NO_2}^2}{p_{N_2O_4}} = \frac{4\alpha^2 p}{1-\alpha^2} \tag{2}$$

The equation of state for the gaseous mixture is

$$PV = \text{total number of moles} \times \boldsymbol{R}T = (1+\alpha)n\ \boldsymbol{R}T \tag{3}$$

where V is the volume of the vessel in m^3, $\boldsymbol{R}$ the gas constant and T the absolute temperature. Thus the equilibrium constant can also be written

$$K_p = \frac{4\alpha^2\ n\ \boldsymbol{R}T}{(1-\alpha)V} \tag{4}$$

The variation of K_p with temperature is governed by the van't Hoff equation

$$\frac{\mathrm{d}\log_e K_p}{\mathrm{d}T} = \frac{\Delta H^\circ}{\boldsymbol{R}T^2} \tag{5}$$

where ΔH° is the standard enthalpy change for the dissociation reaction (1).

As $\mathrm{d}T = T^2\ \mathrm{d}(1/T)$, it follows from Eq. (5) that

$$\frac{\mathrm{d}\log_{10} K_p}{d(1/T)} = -\frac{\Delta H^\circ}{2{\cdot}303\boldsymbol{R}} \tag{6}$$

and therefore that a plot of $\log_{10} K_p$ against T^{-1} is linear and of slope $-\Delta H^\circ/2{\cdot}303\boldsymbol{R}$. (Over a wide temperature range the graph will be a curve owing to the variation of ΔH° with temperature as expressed by the Kirchhoff equation.)

The other two main thermodynamic properties of reaction (1), the change in the standard Gibbs free energy ΔG° and the change in standard entropy ΔS°, are given by the equations

$$\Delta G^\circ = -RT \log_e K_p \tag{7}$$

$$\Delta G^\circ = \Delta H^\circ - T\,\Delta S^\circ \tag{8}$$

In Eqs (2) and (4), if $\boldsymbol{R}$ is taken as 8·314 J mol^{-1} K^{-1} and V is in m^3, K_p will be in N m^{-2}. However, since the standard state of a gas is 1 atm., K_p in Eq. (7) must be expressed in atm. By definition, 1 atm. = 101·325 kN m^{-2}.

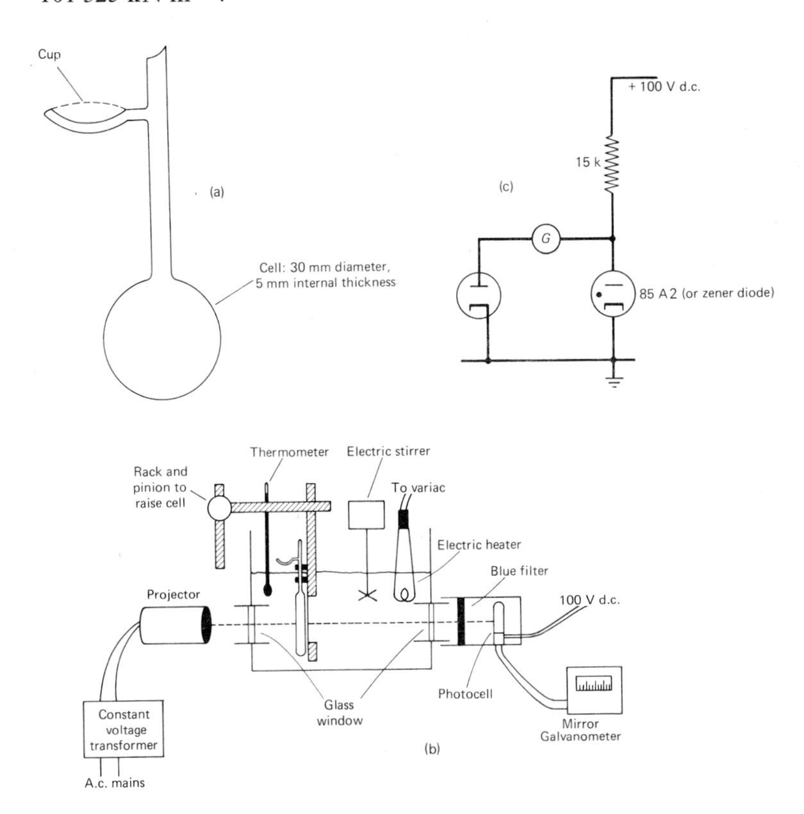

FIG. 9D.2 NO_2–N_2O_4 equilibrium photometer.

Apparatus. The cell used is shown in Fig. 9D.2(a). Before filling it has a constriction at A, and is joined to a vacuum line by a tap. NO_2 from a cylinder (Cambrian Chemicals, London) is purified by freezing and pumping. The cell is gently flamed and pumped to remove water from the walls. When cold it is filled with NO_2 ($\approx\frac{1}{2}$ atm.), and the tap closed. The NO_2 is frozen onto the interior wall of the cup by placing dry ice in the cup. The cell is sealed and removed by heating the constriction. It is left for one hour, and can be tested for vacuum by placing dry ice in the cup. If the brown NO_2 disappears *promptly* the cell is free from air which otherwise hinders passage of the NO_2 into the cup. Once made, the cell can be used repeatedly.

The apparatus is shown in Fig. 9D.2(b). The cell, clamped with spring clips to a support, is placed in a water bath with heater, thermometer and stirrer.

The beam of light from a galvanometer lamp projector or other collimated source, preferably run from a constant voltage transformer, passes through the cell, through a deep blue filter (type not critical) and falls on a vacuum photocell (type not critical). The photocell current is measured on a mirror galvanometer or Scalamp (section 4B). The photocell voltage may be stabilized (Fig. 9D.2(c)) by a neon or zener diode. A rigid rack and pinion mechanism raises and lowers the cell into the light path.

Optical theory. The absorption of (blue) light by the brown NO_2 obeys Beer's law

$$D = \log_{10}\left(\frac{I_0}{I}\right) = kdc_{NO_2} = kd\left(\frac{2\alpha n}{V}\right) \qquad (9)$$

where D is the optical density or absorbance, I_0 and I the light intensities before and after absorption by the gas, k the (decadic) extinction coefficient, d the light path through the gas in m and c the concentration in mol m^{-3}. If k does not vary with temperature, a plot of D against increasing temperature will rise until dissociation to NO_2 is complete ($\alpha = 1$) and then remains constant at a value of

$$D' = kd\frac{2n}{V} \qquad (10)$$

Clearly, at any given temperature,

$$\alpha = \frac{D}{D'} \qquad (11)$$

Combination of (4), (10) and (11) gives

$$K_p = \frac{2RTD^2}{kd(D' - D)} \qquad (12)$$

so enabling the equilibrium constant to be calculated from optical measurements.

Some further comments are necessary regarding Beer's law. The light from the source will be absorbed not only by the gas in the cell, but also by the filter, glass and water, etc., in the light path. Let the optical density due to these when the cell is in the light path be D_A and when it is removed, D_B. If the intensity of the source is I_S and if the intensities seen by the photocell with the cell removed and with it in the light path are I_0 and I respectively, then

$$\log_{10}(I_S/I) = D + D_A$$

$$\log_{10}(I_S/I_0) = D_B$$

So
$$\log_{10}(I_0/I) = D + D_A - D_B$$

If we now measure I_0' and I' without any gas in the light path (the gas having been frozen out inside the cell), then $D=0$ and

$$\log_{10}(I_0'/I') = D_A - D_B$$

Thus
$$D = \log_{10}(I_0/I) - \log_{10}(I_0'/I') \quad (13)$$

As the photocurrent of the cell is proportional to the intensity falling on it, we can use the ratio of the photocurrents for the ratio of the intensities. The source intensity I_S will drift slowly with time; this will not affect the results provided the readings of a given pair (I_0 and I) are taken as soon as possible after each other.

Procedure. It is essential to prevent solid material from obstructing the light path so the heater in the bath must be clean and fresh *distilled* water must be put into the tank before taking readings. Allow the photocell to warm up for 5 min before taking any readings. The positions of the photocell and of the bath should not be altered during the experiment. The readings are sensitive to vibration.

With the water in the tank cold, the level of the water is adjusted so that the light path and cell are under water but the cup is uncovered. Freeze out the NO_2 by drying the cup with filter paper and putting in it a lump of dry ice, with a few drops of acetone to give good thermal contact. (*Caution:* the acetone will bubble violently as the CO_2 comes off!) Measure I' (with the cell in the light path) and I_0' (with the cell out of the light path). Wait a few minutes and repeat to check that all the gas has been frozen out.

With the whole of the cell immersed, measure pairs of values of I and I_0 at about ten temperatures between 20 and 40°C. Switch off the heater prior to taking a reading. To avoid temperature gradients, keep the bath well stirred until a reading is about to be taken.

Heat the water to 100°C and measure I and I_0 at 5° intervals as the water cools to 70°C.

Calculation. Plot the high temperature values of D against T and hence find the limiting value D'.

Given that $k = 10\ \mathrm{m^2\ mol^{-1}}$ and the value of d for the cell, calculate K_p, first in $\mathrm{N\ m^{-2}}$ and then in atm., at each temperature. Hence determine ΔH° for reaction (1). Interpolate the value of K_p at 298·2 K and so calculate ΔG° and ΔS° at this temperature. Estimate the experimental errors in the values of D' and D, and hence in those of K_p, ΔG° and ΔS°. (The value of k will depend somewhat on the blue filter used.) The theory assumes (a) ideal gas behaviour, (b) that k is independent of temperature and (c) that k is independent of pressure, i.e. that the gas is optically thin.

EXPERIMENT
Absorption spectra of bromophenol blue.[2]

If a solution contains several light-absorbing species, then:

$$D = \log_{10}(I_0/I) = d \sum_i c_i k_i \tag{1}$$

Experimentally two minor effects must be taken into account: the reflection of part of the light from the walls of the cell and the absorption of some light by the solvent itself. These effects can be eliminated by subtracting the optical density of the cell filled with solvent only (the 'blank'), so that the D value in Eq. (1) is given by

$$D = D_{\text{solution}} - D_{\text{blank}} \tag{2}$$

In this experiment, use is made of the visible absorption spectrum of the coloured acid-base indicator bromophenol blue to determine its dissociation constant. The formula of the indicator (Fig. 9D.3) shows the presence of two acidic groups. The $-SO_3H$ group, as in other sulphonic acids, is completely dissociated in aqueous solutions. The –OH group, though only weakly acidic in phenols, has a much

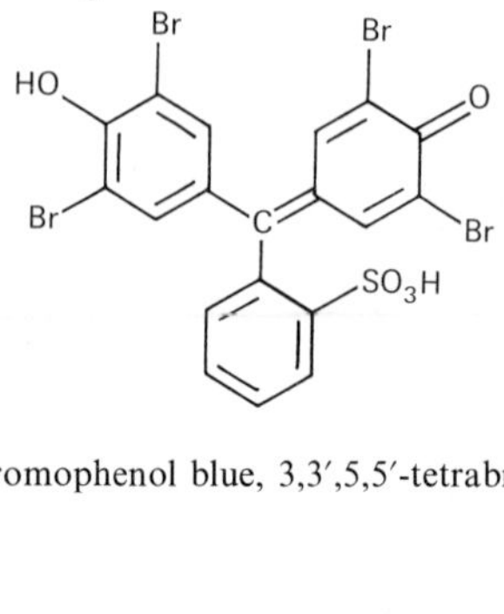

FIG. 9D.3 Structure of bromophenol blue, 3,3′,5,5′-tetrabromophenol sulphonphthalein.

higher dissociation constant in bromophenol blue because of the influence of the bromine substituents and the stabilization of the basic form by resonance. To save space we can write the –OH dissociation

$$HIn^- \rightleftharpoons H^+ + In^{--} \tag{3}$$

The concentrations of HIn^- and of In^{--} can be derived, as shown below, from the absorption spectra (Fig. 9D.4). The spectrum of

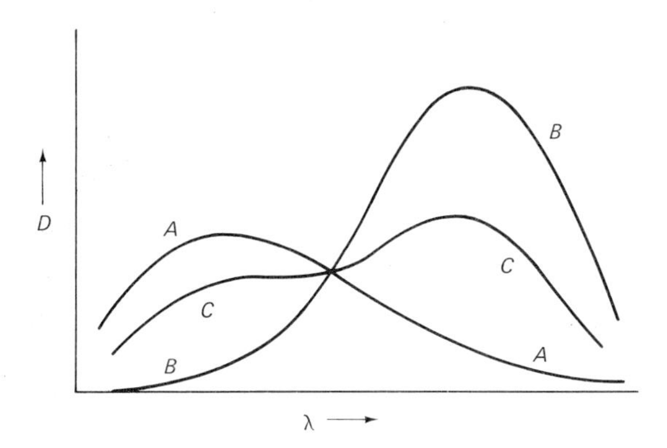

FIG. 9D.4 Absorption spectra of bromophenol blue.

HIn^- is given by curve *A* (indicator dissolved in HCl solution), that of In^{--} by curve *B* (indicator dissolved in NaOH solution), and that of a mixture of HIn^- and In^{--} by curve *C* (indicator dissolved in a buffer mixture of known pH). Before applying Eq. (1), let us assume that the total concentration c of indicator is the same in all three solutions, and that in the buffer solution a fraction α of the indicator is in the basic In^{--} form and a fraction $(1-\alpha)$ in the acidic HIn^{--} form. Then at any specified wavelength:

$$\text{Curve A:} \quad D_A/d = [HIn^-]k_{HIn^-} = ck_{HIn^-} \tag{4}$$

$$\text{Curve B:} \quad D_B/d = [In^{--}]k_{In^{--}} = ck_{In^{--}} \tag{5}$$

$$\begin{aligned} \text{Curve C:} \quad D_C/d &= [HIn^-]k_{HIn^-} + [In^{--}]k_{In^{--}} \\ &= (1-\alpha)ck_{HIn^-} + \alpha ck_{In^{--}} \\ &= ck_{HIn^-} + \alpha(ck_{In^{--}} - ck_{HIn^-}) \end{aligned} \tag{6}$$

Hence

$$\frac{D_C}{d} = \frac{D_A}{d} + \alpha\left(\frac{D_B}{d} - \frac{D_A}{d}\right)$$

from which

$$\alpha = \frac{D_C - D_A}{D_B - D_A} \tag{7}$$

The 'concentration' equilibrium constant K_c of reaction (4) is then given by

$$K_c = \frac{[H^+][In^{--}]}{[HIn^-]} = [H^+]\frac{\alpha}{1-\alpha} \tag{8}$$

with α obtained from Eq. (7). In reality ions are not quite independent entities in solution for cations and anions attract each other. The effective or thermodynamic concentration (the activity) is therefore rather smaller in dilute solution than the analytical concentration and is related to the latter by the activity coefficient f:

$$\text{activity} = \text{concentration} \times f$$

The thermodynamic dissociation constant K of reaction (4), in terms of activities, is then

$$K = \frac{\{H^+\}\{In^{--}\}}{\{HIn^-\}} = \{H^+\}\left(\frac{\alpha}{1-\alpha}\right)\frac{f_{In^{--}}}{f_{HIn^-}} \tag{9}$$

The activity coefficients can be calculated for solutions of known ionic strength from the theory of Debye and Hückel, and numerical values are listed below. The hydrogen ion activity $\{H^+\}$ can be derived from e.m.f. measurements which yield the pH ($= -\log_{10}\{H^+\}$). The pH value for the buffer solution used is also given below.

Procedure. Before commencing, read the instrument makers' instruction booklet for the spectrophotometer used.

A fresh solution of bromophenol blue is made up by dissolving exactly 0·02 g in 100 cm^3 distilled water, stirring well for some time. Measure the absorption spectrum from 700 to 350 nm for each of the following three solutions:

(a) 5 cm^3 exactly (use a pipette) of indicator solution diluted to 100 cm^3 with distilled water and with two drops of conc. HCl added.
(b) As for (a), but with the addition of two drops of conc. NaOH solution instead of HCl.
(c) 5 cm^3 exactly of indicator solution diluted to 100 cm^3 with 50 mol m^{-3} potassium hydrogen phthalate buffer solution.

Because of the inherent instability of the solutions, especially (b), do not prepare these until immediately before use, and keep the indicator solution in the dark if you want to store it.

Test the Beer–Lambert law at any one wavelength for either the acidic or the basic form of the indicator by suitable dilution of either (a) or (b) to give solutions of four different concentrations.

Calculations. Using the spectra of absorbance D versus λ for (a), (b) and (c):

(i) Explain why all the curves pass through one point (the isos-bestic point).

(ii) Determine values of α at two suitable wavelengths. Calculate the thermodynamic dissociation constant K with error limits, taking the pH of the buffer as 4·00 and the activity coefficients $f_{In^{--}}$ and $f_{HIn^{-}}$ as 0·50 and 0·84, respectively.

(iii) Calculate the maximum extinction coefficients of HIn^{-} and of In^{--}. The wavelengths at which the extinction coefficients are a maximum should be given as well.

OTHER EXPERIMENTS

(1) The absorption curve of other coloured substances in solution may be determined and the validity of the laws of Lambert and Beer examined. (Suitable solutions: I_2 in CCl_4, K_2CrO_4, $CuSO_4$, $KMnO_4$ in water.)

(2) The absorption curve for mixtures may be investigated. Provided no chemical interaction occurs, the extinction of the mixture is the sum of the extinctions due to the separate absorbing substances.

(3) The absorption curve of a pH indicator in a series of buffer mixtures of known pH may be studied (cf. section 13F).

The spectrophotometer can be used to investigate the formation and stability of coloured complexes (cf. Section 7E).

(4) The *diffuse reflection spectra* of pigments can be determined.

(5) *Fluorescence spectra* emitted by solutions or solids (phosphors) under excitation by ultra-violet light can be photographed.

BIBLIOGRAPHY 9D: Absorption spectra

Weissberger, Vol. I, Pt 3, chapter XXX.
Calder, *Photometric Methods of Analysis*, 1969 (Hilger, London).

[1] Gray and Yoffe, *Quart. Rev.*, 1955, **9**, p. 362.
[2] Gold, *pH Measurements*, 1956 (Methuen, London).

9E Refractometry

1. REFRACTIVE INDEX: THE CRITICAL ANGLE PRINCIPLE

When a ray of monochromatic light passes from a less dense to a more dense medium, it is bent or refracted towards the normal. Thus, in Fig. 9E.1(a) if *I* is the less dense and *II* the more dense medium, a

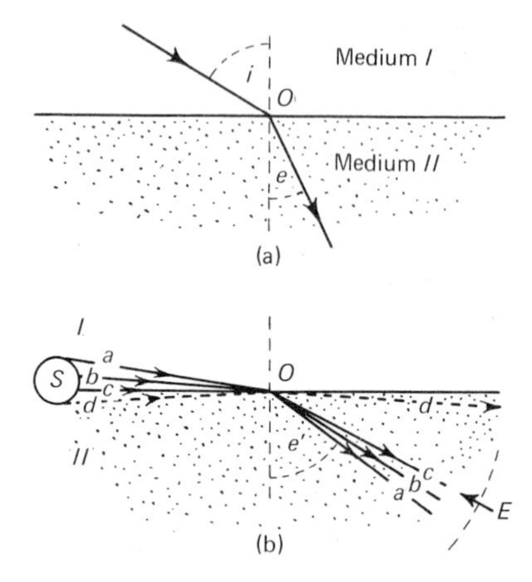

FIG. 9E.1 Refraction and the critical angle principle.

ray of light passing from *I* to *II* will be bent so that the angle of refraction *e* will be less than the angle of incidence *i*; and, according to the law of refraction, the relation between these two angles will be such that

$$\sin i/\sin e = N/n$$

where *n* is the index of refraction of the less dense, and *N* the index of refraction of the more dense medium. As the angle *i* increases, the angle *e* also increases, and reaches its maximum value *e′* when *i* becomes equal to a right angle; that is, when the incident light is horizontal. Since $\sin 90° = 1$, the above equation becomes $1/\sin e' = N/n$, or $\sin e' = n/N$. (If $i > 90°$ the ray is totally internally reflected.)

If, therefore, a narrow source of light (*S*, Fig. 9E.1(b)) is placed on the interface between *I* and *II*, and is observed by means of an eyepiece *E* which can turn about the point *O*, one observes a band of light due to rays such as *a* and *b*, leading to a sharp edge, corresponding to the critical ray *c*, and then a region of darkness. The position of the sharp edge gives the critical angle *e′*. Most instruments for measuring refractive index (*refractometers*) employ this *critical angle principle*.

2. SPECIFIC AND MOLECULAR REFRACTIVITY

Whereas the refractive index n of a substance varies with the temperature, the expression

$$\frac{n^2-1}{n^2+2}\cdot\frac{1}{d}$$

(Lorentz and Lorenz), where d is the density, remains nearly constant at different temperatures. The value of this expression is, therefore, dependent only on the nature of the substance, and is a characteristic of it. It is called the *specific refractive power* or *refractivity* of the substance. If the refractivity is multiplied by the molecular weight of the substance, one obtains the *molecular refractivity*,

$$[R] = \frac{n^2-1}{n^2+2}\cdot\frac{M}{d},$$

where M is the molecular weight. $[R]$ has the dimensions of a volume.

Refractivity and polarizibility. Refraction arises from the fact that the extra-nuclear electrons of atoms tend to follow the oscillations of the electromagnetic field associated with light. It can be shown that, for a particular frequency of the light $[R]=4\pi N\alpha/3$ (Clausius–Mosotti equation), where N is Avogadro's number and α is the *electronic polarizability* of a molecule of the medium, that is to say, the value of the instantaneous dipole moment induced in the molecule when it is placed in an electric field of unit intensity. α is therefore a measure of the 'looseness' of binding of the electrons. It is understandable that the polarizability (and therefore, refractivity) of a molecule is approximately the sum of the polarizabilities (or refractivities) of its constituent atoms, since the electronic condition of a given element is very similar in its different compounds. Tables of *atomic refractivities* $[r]$ have therefore been prepared from which one can calculate molecular refractivities with fair accuracy. These tables include values for double and triple bonds to be added to the atomic refractivities for compounds containing such bonds; these corrections are needed because multiple bonds are more polarizable than single bonds. Refractivities depend on the wavelength of light employed. The accompanying short list gives atomic refractivities $[r]_D$ for the Na–D line.

Element	$[r]_D$	*Element*	$[r]_D$
C (singly bonded)	2·418	Cl	5·967
H	1·100	Br	8·865
O (in OH group)	1·525	I	13·900
O (in ethers)	1·643	Double bond	1·733
O (in CO group)	2·211	Triple bond	2·398

3. THE ABBE REFRACTOMETER

This instrument is the refractometer most commonly used in chemical laboratories. The principle of operation is illustrated diagrammatically in Fig. 9E.2(a), and the appearance of the instrument, as made by Messrs Bellingham and Stanley (London), is shown in Fig. 9E.2(b).

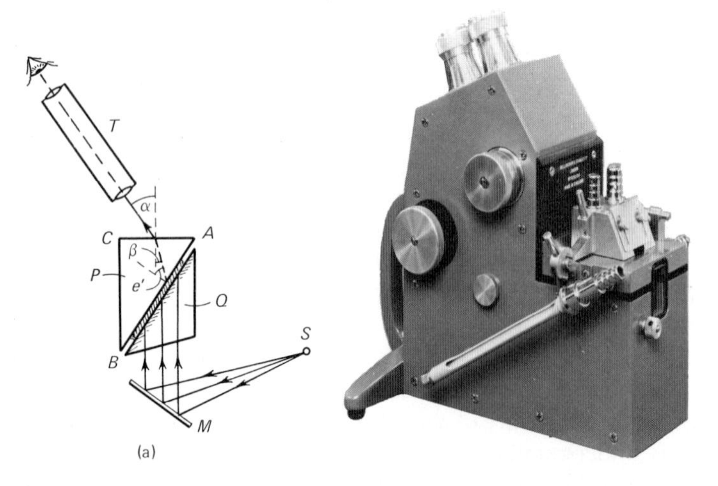

FIG. 9E.2 The Abbe refractometer. (a) Optical principle. (b) General view of the instrument. (Courtesy Bellingham and Stanley Ltd, London.)

The optical system consists of three parts—a mirror *M*, a prism box *PQ* which can be rotated as a whole by means of a milled knob *R*, and a *fixed* telescope, *T*.

Light from a source *S* (a pearl lamp or sunlight) is cast by means of the mirror on to the lower prism *Q*, thus illuminating the upper face of this prism. This face is ground so that it acts as a diffusing screen and provides rays in every direction. The narrow space between the lower (illuminating) prism Q and the upper (refracting) prism P contains a small quantity of the liquid under examination. This liquid must have a refractive index lower than that of the glass of prism P, so that the critical angle phenomenon (explained above) can be observed. Then, on looking through the telescope, one sees a band of light due to rays which pass obliquely through the liquid. This band finishes sharply, the edge of the light band corresponding to rays which pass through the prism face AB at grazing incidence, thus entering the prism at the critical angle e'. Now $\sin e' = n/N$ where n and N are the refractive indices of the liquid and the prism respectively. It is readily shown that the angle α at which the critical ray emerges from the face AC of the prism is related to n by the equation

$$n = \sin B\widehat{A}C \sqrt{N^2 - \sin^2 \alpha} - \cos B\widehat{A}C . \sin \alpha$$

In practice the telescope is fixed but the prism box is rotated until the critical ray is seen to coincide with cross-wires set in the telescope. Each setting of the prism thus corresponds to a definite critical angle and therefore a definite value of n and this can be read off directly on an engraved scale by means of the eyepiece. The instrument requires only a few drops of liquid, and the refractive index can be read very quickly. The usual prism covers the range 1·30 to 1·70 with an accuracy of $\pm 0{\cdot}0002$.

Procedure. In order to carry out a determination of the refractive index of a liquid, open the prism box and place a few drops of the liquid on the ground surface of the lower prism. Close and fasten the prism box again, taking care (by tilting the refractometer forward a little, if necessary) that the liquid does not flow away. A film of liquid will thus be enclosed between the two prisms. Focus the cross-wires of the telescope by rotating the eyepiece and adjust the mirror so as to get good illumination. By means of the lower knob, turn the prism box slowly backwards and forwards until the field of view becomes partly light and partly dark. When white light is used the edge of the light band will show a coloured fringe. By means of the upper knob, rotate the 'compensator' (consisting of two prisms, which rotate in opposite directions, and so form a system of variable dispersion) until the coloured fringe disappears and the light-band shows a sharp edge. Now rotate the prism box until this sharp edge is in coincidence with the intersection of the cross-wires in the telescope, and read off directly the index of refraction on the scale through eyepiece. The third decimal place in the refractive index can be read directly, and the fourth can be estimated with an accuracy of about $\pm 0{\cdot}0002$.

For the purposes of temperature regulation the prisms are enclosed in a metal jacket through which water from a thermostat can be circulated. Before circulating the water, the thermometer must be screwed into place. The temperature of the liquid should be controlled within 0·5°C. The effect of temperature on the calibration of the instrument is small—about 0·0001 per 15°C.

The scale gives values for the refractive index for the D line (n_D). Connected with the compensator, however, there is a divided circle, and by reading the number on this dispersion circle, after the light-band in the telescope has been made sharp and free from colour fringe, the value of the dispersion, $n_F - n_C$, can be calculated from tables supplied with the instrument.

Adjustment of the refractometer. It may be necessary, from time to time, to adjust the setting of the refractometer, and for this purpose a standard glass test-piece is provided, the index of refraction of which is marked on the glass. Open the prism box until the lower prism can be slipped off its hinge. By means of a drop of monobromonaphthalene applied to the polished surface of the test-piece, fix the latter on the surface of the upper prism, the ground edge of the test-piece being directed towards the mirror. Excess of the monobromonaphthalene is to be avoided. Turn the lever so that the reading on the divided arc corresponds with the refractive index of the glass test-piece, and by means of the compensator obtain a sharp band of light, without colour fringe, in the telescope. If this edge coincides with the intersection of the cross-wires, then the instrument is in proper adjustment; but if not, turn the adjustment knob pin until coincidence occurs.

4. THE IMMERSION REFRACTOMETER

When comparatively large quantities of material are available, the dipping or immersion refractometer may be used. In this instrument, the optical principle of which is the same as that of the Abbe refractometer, the prism is rigidly fixed in the telescope tube, in which also there are contained a compensating prism (as in the Abbe apparatus) and a scale. In use, the refractometer is suspended so that the prism dips into the liquid to be investigated, which is continued in a beaker immersed in a thermostat, regulated to a temperature of 17·5°C, the temperature for which the instrument is calibrated. White light is reflected by means of a mirror placed below the beaker, and the sharp edge of the band of light is read off on the scale.

5. THE PULFRICH REFRACTOMETER

This instrument is less used than the Abbe refractometer, but is important for its great accuracy ($\pm 0{\cdot}00002$). The critical angle is observed by means of a right-angled prism (refractive index N) (Fig. 9E.3(a)) which carries a glass ring cell cemented to its upper surface to contain the liquid (refractive index n). The prism is fixed. The observation telescope bearing cross-wires rotates about an axis through the point O, and permits the angle of emergence i of the

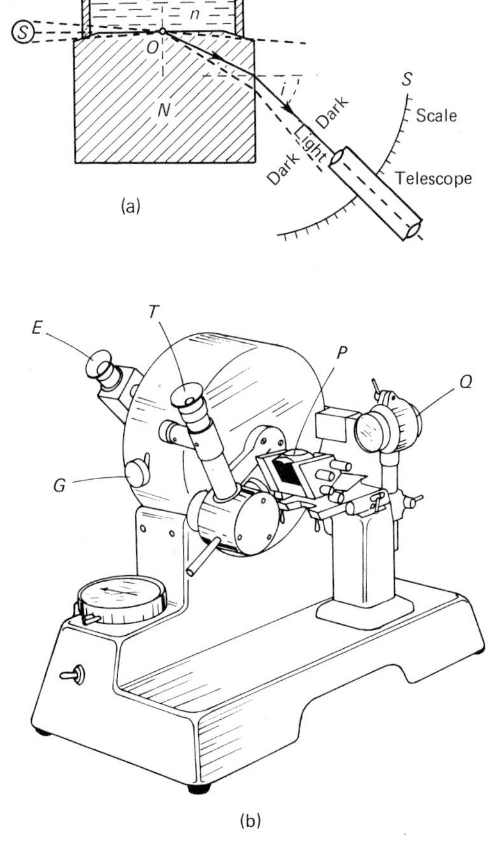

FIG. 9E.3 The Pulfrich refractometer. (a) Optical principle. (b) General view of the instrument. (Courtesy Bellingham and Stanley Ltd, London.)

critical ray from the side of the prism to be determined on a scale S. The theory is as explained above, and it readily follows that $n = \sqrt{(N^2 - \sin^2 i)}$. Tables of n corresponding to the measured angle i are supplied by the makers for each prism. As N is a function of the wavelength concerned, the tables are provided for a number of different wavelengths. A prism with N about 1·6 is suitable for liquids of $n = 1{\cdot}3$ to 1·5, and another prism of $N = 1{\cdot}7$ extends to the range from 1·5 to 1·7.

Details. Figure 9D.3(b) shows the Pulfrich refractometer. The light source used is a discharge tube or lamp (e.g. Na, H_2, Hg or Cd lamp) giving well-separated lines. The source is aligned with the upper surface of the prism P and placed at a short distance from a lens Q which concentrates the light into the required region. Water from a thermostat is circulated round the prism housing. If the full accuracy of the instrument is to be obtained, the temperature of the liquid sample must be controlled to $\pm 0{\cdot}05$°C.

The observation telescope T is mounted on a vertical turntable carrying an accurate angular scale which is read by means of the eyepiece E. The scale is graduated directly to 15 minutes, and, by means of a micrometer screw and divided drum, head readings can be repeated to 5 seconds. (In older models the scale was divided into degrees and half-degrees (30 minutes), and a vernier, read with a lens, permitted readings to be taken to the nearest minute.) An essential detail, of course, is a screw by which the position of the telescope can be adjusted with the corresponding accuracy.

6. OTHER REFRACTOMETERS

The Hilger–Chance refractometer. This recent instrument is of special interest as being a high-precision refractometer which does not employ the customary critical angle principle. Instead, the refraction produced by a right-angle prism of the material is measured by what is in effect, a goniometer arranged in a vertical plane. The means provided for measuring the angular deviation of the ray are extremely refined: a main scale divided in intervals of 10 minutes is supplemented by a micrometer scale reading to 3 seconds. An accuracy of 0·00001 in the refractive index is obtainable over the range 1·30–1·95.

This instrument was originally designed for measurement of the refractive indices of optical glasses, but it is readily adaptable to liquids, 2–3 cm^3 being required. Naturally, the full precision of the instrument is not obtainable without exceptionally good temperature control.

Gases: Rayleigh interference refractometer. The principle of the interferometer was adopted by Lord Rayleigh (1896) for measurement of the refractive indices of gases. A beam of light is divided into two parts which pass through two similar parallel slits and then, after traversing similar cells, are brought together to produce interference bands. Any change in the refractive index of the medium in one of the cells (e.g. if a gas is introduced instead of vacuum) causes a displacement of the bands which can be measured by adjusting a tilted compensating glass plate placed in one of the beams. A movement of one-fortieth of a band can be detected, corresponding to a refractive index difference of 2×10^{-8}. The method is therefore particularly suitable for detecting small changes of gas composition; for instance, 0·01% of hydrogen in air can be detected. The instrument can also be adopted for measurements with liquids; again, it provides an exceptionally sensitive method of measuring small changes of refractive index, and, hence, of composition.

Differential refractometers. The modern method of molecular weight determination based on the scattering of light requires a knowledge of the refractive index of the solute particles. This can be obtained from measurements of the refractive indices of solutions of different concentration.

7. REFRACTOMETRIC DETERMINATION OF THE COMPOSITION OF SOLUTIONS

Refractometric measurements may be used, very advantageously, for the quantitative determination of the composition of binary solutions. For this purpose, the refractive indices of a series of solutions of known composition are first determined (e.g. by means of the Abbe or immersion refractometer), and the values so obtained are plotted on a graph. The composition of an unknown solution can then be ascertained from the graph after the index of refraction of the solution has been determined. The relation between refractive index and composition for a large number of solutions is given in the *International Critical Tables* (Vol. VII).

The composition of a *dilute* solution of two liquids can be calculated approximately from the refractive index of the solution provided one knows the refractive indices and densities of the two components. If n_1 and n_2 are the refractive indices of the two components, and n_3 the refractive index of the solution, and if d_1, d_2 and d_3 are the corresponding densities, the percentage amount p, of component '1' can be calculated approximately by means of the expression,

$$\frac{n_1-1}{d_1}\cdot p = 100\cdot\frac{n_3-1}{d_3}-\frac{n_2-1}{d_2}(100-p)$$

The refractive index of a solution, calculated by means of this expression, should be compared with that determined directly.

BIBLIOGRAPHY 9E: Refractometry

Weissberger, Vol. I, Pt 2, chapter XVIII.

9F Polarimetry

1. When ordinary light is passed through a Nicol prism (made from Iceland spar), the emergent light is plane-polarized, that is, the electromagnetic vibrations take place in one plane. If this polarized light is now examined by means of another Nicol prism, it will be found that, on rotating the latter, the field of view appears alternately light and dark, the minimum of brightness following the maximum as the prism is rotated through an angle of 90°. The prism by which the light is polarized is called the polarizer, and the second prism, by which the light is examined, is called the analyser.

If, when the field of view appears dark (which occurs when the axes of the two prisms are at right angles to each other), a tube containing a solution of cane sugar is placed between the two prisms, the field lights up, and one of the prisms must be turned through a certain angle α before the field becomes dark again. The solution of cane sugar has therefore the power of turning or rotating the plane of polarized light through a certain angle, and is hence said to be *optically active*. When, in order to obtain darkness, the analyser has to be turned to the right, i.e. clockwise, the optically active substance is said to be *dextro-rotatory*, and *laevo-rotatory* when the analyser must be turned to the left.

It will, of course, be possible to obtain a position in which the field of view becomes dark by rotation of the analyser either to the right or the left, because in one complete rotation of the prism through

360°, there are two positions of the analyser, 180° apart, at which the field is dark, and similarly, two positions at which there is a maximum of brightness. In determining the sign of the activity of a substance, one takes the direction in which the rotation required to give extinction is less than 90°.

The angle of rotation depends on (1) the nature of the substance, (2) the length of the layer through which the light passes, (3) the wavelength of the light employed (the shorter the wavelength, the greater the angle of rotation), (4) the temperature. In order, therefore, to obtain a measure of the rotatory power of a substance, these factors must be taken into account, and one then obtains the *specific rotation*. This is defined as the angle of rotation produced by a liquid which in the volume of 1 m^3 contains 1 kg of active substance, when the length of the column through which the light passes is 1 m. The specific rotation is represented by $[\alpha]$, the observed angle of rotation being represented simply by α.

With a pure active liquid, the specific rotation is obtained from

$$[\alpha] = \frac{\alpha}{l.d}$$

where l is the length of the column of liquid and d is the density. If account is taken also of the other factors on which the rotation depends, viz. temperature and wavelength of light, a number is obtained which, for the particular conditions of experiment, is a constant, characteristic of the substance. Thus, $[\alpha]_D^{25^\circ}$ represents the specific rotation for the D line (sodium light) at the temperature of 25°C.

When the active substance is examined in solution, the concentration must be taken into account, in accordance with the expressions:

$$[\alpha] = \frac{\alpha}{lc} = \frac{\alpha}{lpd}$$

Here c is the concentration of active substance in kg m^{-3}, p is the weight fraction in kg (kg soln)$^{-1}$ and d is the density of the solution. In expressing the specific rotation of a substance in solution, the concentration and the solvent (which also has an influence on the rotation) must be stated.

2. THE POLARIMETER

The arrangement of the optical parts of the Lippich polarimeter, the type now generally adopted, is shown diagrammatically in Fig. 9F.1.

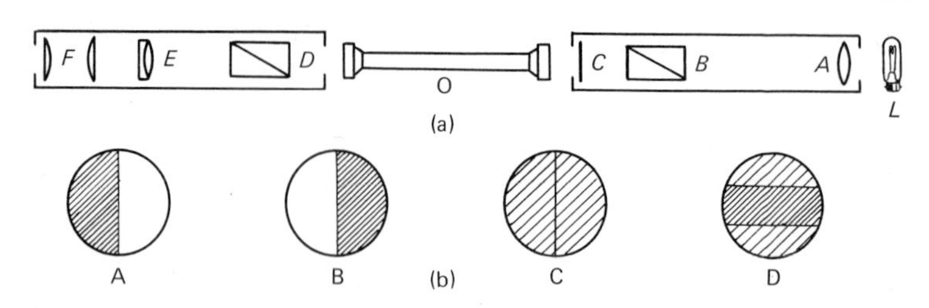

FIG. 9F.1 Principle of the polarimeter. (a) Arrangement of the optical system. (b) Appearance of the field of view.

Monochromatic light from the source *L* passes through the lens *A*, which renders the rays of light parallel, and then through the polarizing prism *B*. It then passes through the observation tube *O*, and thence through the analyser *D*. The field of view is observed through the telescope *EF*. At *C* there is a small Nicol prism which covers half of the opening at the end of the polarizer tube. The light on passing through this prism is altered in phase by half a wave-length, but still remains plane-polarized. In this way, two beams of polarized light are obtained; and if the polarizer is rotated so that the plane of polarization forms an angle (δ) with the optical axis of the Nicol prism, the planes of polarization will also be inclined at an angle, equal to 2δ. This is the half-shadow angle. On rotating the analyser, a position will be found at which the one beam will be completely, the other only partially, extinguished. The one half of the field of view, therefore, will appear dark, while the other half will still remain light (as shown by A in Fig. 9F.1(b). On rotating the analyser still further through the angle 2δ, a second position, B, will be found at which the second beam will be extinguished, while the first is no longer so. In this position of the analyser, the half of the field which was formerly bright will now be dark, and that formerly dark will now be light. When, however, the analyser occupies an intermediate position, C, the field of view will appear of uniform brightness; and this is the position to which the analyser must be set. Many polarimeters give a triple field, D, and this arrangement facilitates matching, especially if the illumination of the field of view is not perfectly uniform.

By diminishing the angle δ (by rotating the polarizer) the sensitiveness of the instrument can be increased, because now the angle 2δ, through which the analyser must be rotated in order to cause the shadow to pass from one half to the other of the field of view, is diminished. By diminishing the angle of half-shadow, however, the uniform illumination of the field of view is also diminished, so that the increased sensitiveness due to diminution of the angle of half-shadow is partly counteracted by the greater difficulty in deciding when the field is uniformly illuminated, unless the light intensity of the source can at the same time be increased. With a source of light of given intensity, therefore, the angle of the half-shadow must be fixed so that the determination of the position of uniform illumination can be made without unduly straining the eyesight.

The visual matching of the two beams can be made more accurately if the eye is dark adapted; either the polarimeter should be used in a darkened room or the observer should use a black cloth to exclude extraneous light. As with all optical instruments, eyestrain is reduced if the observer can be comfortably seated and can look into the eyepiece without stretching.

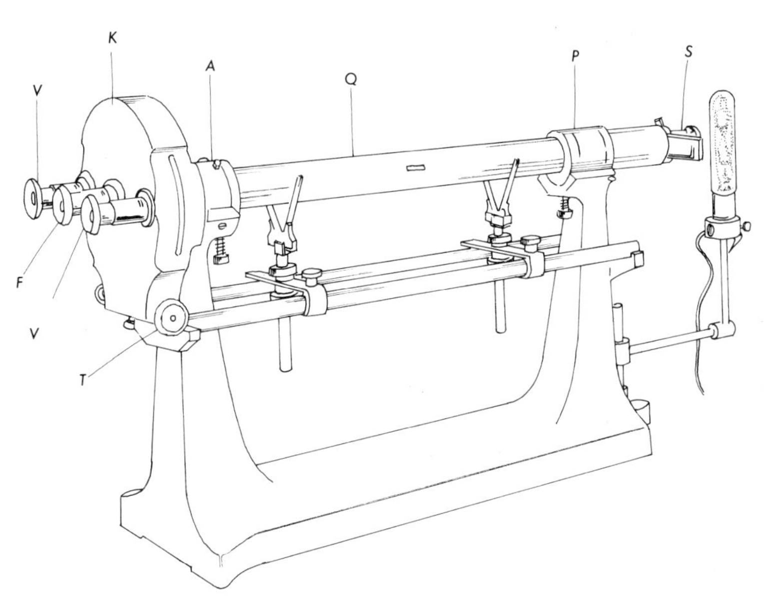

FIG. 9F.2 Standard form of polarimeter. (Courtesy Bellingham and Stanley Ltd, London.)

The complete polarimeter is shown in Fig. 9F.2. At the end *S*, which is directed towards the source of light, are the lens and the light-filter, consisting either of a solution of potassium dichromate or an equivalent gelatine filter. The polarizing prism is at *P*. In most polarimeters there is a lever and scale by means of which the angle of half-shadow can be altered. The observation tube is placed in a trough in the middle part of the instrument, *Q*, and is protected from extraneous light by a cover. The analyser is in the portion of the tube at *A*. *F* is the telescope with eyepiece. *K* contains a graduated disc, which can be caused to rotate, along with the analyser, past fixed verniers situated on opposite sides of *F* by means of a slow-motion screw. The two sides of the scale are read to 0·01° by the verniers with the assistance of a magnifying eyepiece *V*. In the polarimeter made by Adam Hilger Ltd (London), verniers have been replaced by a direct-reading eyepiece scale.

Light sources. Sodium laboratory or street-lighting lamps (see section 9A(2)) are generally used for polarimetry. When measurements are needed at other wavelengths, a mercury lamp may be used in conjunction with special filters, or, better, the image can be examined with a spectroscopic eyepiece or the instrument can be combined with a monochromator (section 9F(4)).

Polarimeter tubes. The observation tube for containing the liquid generally consists of a tube of thick glass with accurately ground ends which are closed by means of circular plates of optical glass. These are held in position by screw-caps with rubber washers (Fig. 9F.3(a)).

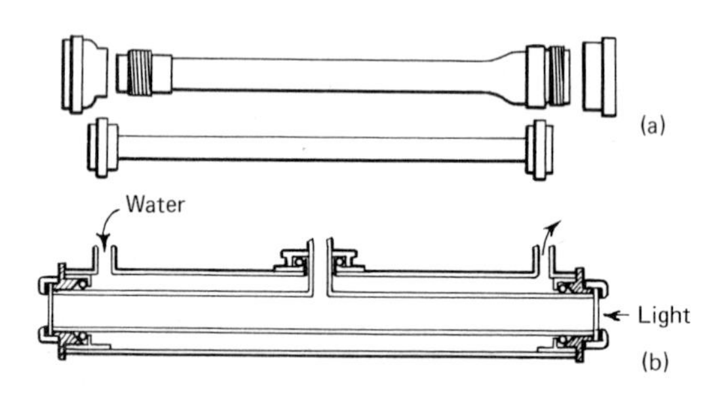

FIG. 9F.3 Polarimeter tubes. (a) Plain tubes. (b) Water-jacketed tube with side opening.

The caps must not be screwed down too tightly as strain in the glass may cause spurious polarization. Polarimeter tubes are generally 10 cm long and about 1 cm diameter, but other sizes are made, including micro-tubes requiring less than 0·5 cm^3 of liquid.

The form of tube shown in Fig. 9F.3(b) has a side opening for filling and a metal jacket through which water from a thermostat can be circulated to control the temperature of the liquid.

3. ADJUSTMENT AND USE OF THE POLARIMETER

Set up the polarimeter so that the polarizer lens is situated at its focal distance from the sodium lamp—i.e. so that parallel light passes along the section which will be occupied by the polarimeter tube. The light source should be adjusted to give as uniform an image as possible in the eyepiece. Clean the polarimeter tube, fill it with distilled water and set it in place between polarizer and analyser. Focus the eyepiece of the telescope on the line dividing the field of view. Now determine the 'zero-point' of the instrument by rotating the analyser until the best match between the parts of the field is obtained. This position should be approached several times from either side, and readings should be taken on the two verniers (in instruments so fitted), and the mean of the readings taken. The object of reading both sides of a circular scale, i.e. at points on the graduated circle 180° apart, is to correct for any eccentricity in the construction of the scale.

Since the position of the analyser required to give equal illumination of the two parts of the field depends on the angle of the half-shadow, the arm which rotates the polarizer must be fixed before the zero point is determined. Some instruments provide means of setting the scale to 0·00° when the analyser has been set to the match-point with distilled water in the cell. With others, the 'zero' reading must be taken and subtracted from the results subsequently obtained with optically active solutions to obtain their true rotations.

EXPERIMENT (see also section 14(B))
Determine the specific rotation of sucrose.

Procedure. Prepare three solutions of sucrose by weighing out about 5, 10 and 20 g of pure sucrose, previously dried in a steam oven, dissolving in distilled water, and making up the volumes to 100 cm^3 at the temperature to be used in the determination (say, 20 or 25°C). Water at this temperature must be circulated from a thermostat round the water-jacket of the polarimeter tube. Determine the 'zero' with distilled water as described above and then the rotation for each solution in turn in the same way. The tube must, of course, be rinsed several times with portions of the new solution each time the solution is changed. Calculate the specific rotation $[\alpha]_D^t$ from the results, and ascertain whether $[\alpha]$ depends on the concentration.

In place of sucrose, the above experiment may be made with any other optically active substance in an appropriate solvent (e.g.

tartaric acid, quinine sulphate, camphor, ethyl malate, etc.). One may verify, also, that the angle of rotation of a given solution is strictly proportional to the path length of the cell employed.

4. OTHER INSTRUMENTS

Simplified polarimeters using Polaroid film instead of prisms are available with an accuracy of 0·05°; these are useful for routine determination. Polarimeters intended for the determination of the concentration of sugar (sucrose) are called *saccharimeters*: they are provided with a scale calibrated directly in terms of sugar concentration.

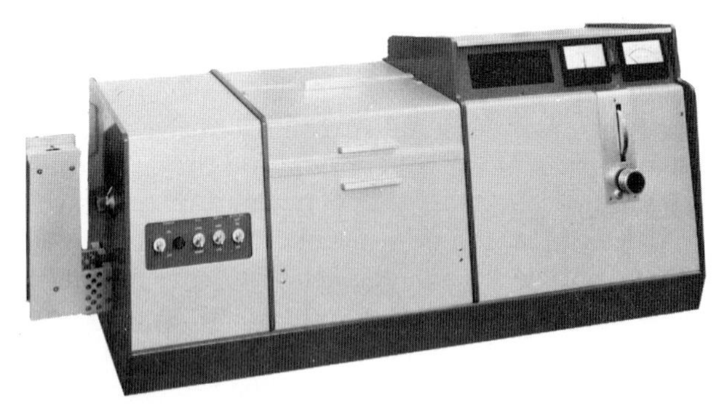

FIG. 9F.4 Self-balancing polarimeter. (Courtesy Bellingham and Stanley Ltd, London.)

A modern version of the polarimeter is shown in Fig. 9F.4. The speed and accuracy of reading can be improved using photoelectric detection of the matching of the beams: the angle of rotation is altered by a servo mechanism until the beams are matched, and the final value displayed digitally. These instruments can be operated continuously on a flowing liquid, and are consequently useful for process control in industry.

The optical rotatory power of a molecule is a function of wavelength; the variation can be determined using a *spectropolarimeter*. Monochromatic light is obtained by passing light from an argon or krypton discharge lamp through a grating monochromator; it is then polarized and passed through the sample in the way described above, and the beams are matched photoelectrically. Curves of optical rotatory power against wavelength are characteristic of the molecule, and the optical rotatory dispersion (*ORD*) curve can also be used for analysis, particularly of organic molecules.[1]

BIBLIOGRAPHY 9F: Polarimetry

Weissberger, Vol. I, Pt 3, chapter XXXIII.

[1] Weissberger, Vol. I, Pt 3, chapter XXXIV; Djerassi, *Optical Rotatory Dispersion*, 1960 (McGraw-Hill, New York).

10
Thermochemistry

Introduction

Most chemical processes are accompanied by a measurable absorption or evolution of heat. It follows from the 1st Law of Thermodynamics that the magnitude of the heat change is proportional to the quantity of substance involved and depends also on the physical state of the reactants and the products, but it is independent of the 'path' by which the reaction is brought about. Consequently, to express unambiguously the thermal process accompanying a given reaction, say, $A+B \rightarrow C+D$, one must state the conditions of A, B, C and D (e.g. gas, liquid, solid or solution; temperature and, if gaseous, pressure), and indicate the amount of substance for which the value is given (e.g. one mole of A).

A heat change measured at constant volume is the change in *internal energy* U for the reaction; at constant pressure the change in *enthalpy* H. The latter is defined by $H=U+PV$, so for a reaction at constant pressure $\Delta U=\Delta H-P\,\Delta V$ where ΔV is the change in volume. ΔV can be ignored for solids and liquids; if the gas laws are obeyed $\Delta V=\Delta n\,RT$ for gas reactions where Δn is the change in the number of moles of gas.

For an exothermic reaction H (products) $<$ H (reactants) and ΔH is negative; conversely ΔH is positive for an endothermic reaction. Similar statements hold for ΔU. We can only measure changes in U and H since their absolute values are undefined.

The heat change accompanying a physical or chemical process is measured by some form of *calorimeter*; the measured rise or fall of temperature multiplied by the total heat capacity of the calorimeter gives the quantity of heat in joules. Many different forms of calorimeter are in use for studying the 'heats' of the various types of physical and chemical processes. For example, it is frequently necessary to know *specific heats** (for gases, liquids and solids), *latent heats** (of fusion, evaporation, sublimation and transition), *heats of solution, dilution and mixing*, and *heats of reaction* (including neutralization, combustion, hydrogenation, etc.). In addition, special forms of calorimeter have been designed for measurements at very high or very low temperature, and for special purposes such as determination of heats of adsorption, wetting, etc.

The technique of calorimetry presents two principal difficulties—firstly, how to determine the heat capacity of the calorimeter and all its contents, and, secondly, how to deal with the inevitable exchange of some heat between the calorimeter and its surroundings. The heat capacity can be computed if all the components of the calorimeter and its contents are weighed separately and their specific heats are known. When this method is not applicable the 'water equivalent' must be found by supplying a known amount of heat to the system and determining the rise of temperature which it produces. The heat is best supplied electrically and measured as the product of volts × amps × seconds (joules), but it should be noted that the accuracy of the product will not be high unless the electrical quantities are measured with high precision—say, by a potentiometric method (section 4C). Other methods of supplying a known number of calories include the 'method of mixtures' using a heated body of known specific heat, and by the use of well-established heats of reaction (e.g. neutralization or combustion).

There are two ways of dealing with the problem of heat exchange between calorimeter and surroundings. The usual method is to apply a correction to the observed change of temperature. An obvious precaution is to arrange that the correction shall be relatively small and the heat exchange regular; this is the principle of the classical 'shielded can' calorimeter. The other solution is the *adiabatic calorimeter* in which the surroundings are automatically maintained at the same temperature as the calorimeter. This eliminates heat exchange entirely, but rather complicated instrumentation is required. For many purposes the heat exchange can be rendered practically negligible by using a Dewar vessel (Thermos flask) as calorimeter. The heat capacity must be measured with a small electrical immersion heater. For endothermic processes one can determine directly how much electrical energy is required to counteract exactly the cooling effect of the process.

The temperature change of the calorimeter must be kept to a few

* A textbook of practical physics should be consulted for details of calorimeters employed for measuring specific and latent heats.

degrees, and consequently sensitive thermometers of the Beckmann type are generally used. Alternatively, platinum resistance thermometers or multiple-junction thermocouples may be employed in work of high precision.

10A Classical calorimetry

1. Figure 10A.1 shows diagrammatically the most common type of calorimeter, suitable for measurements with liquids and solids in the neighbourhood of room temperature. The central vessel *A* is a metal

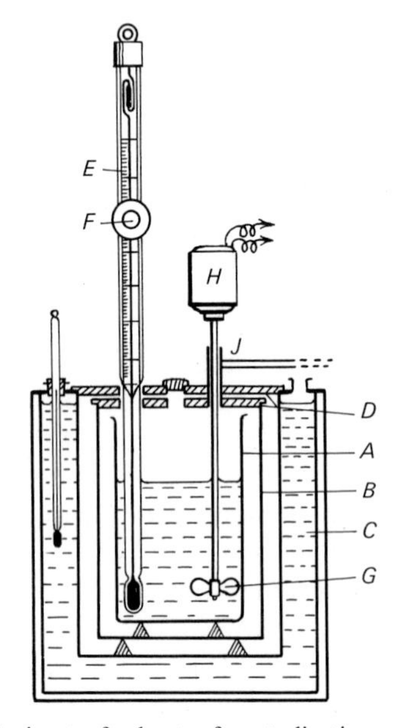

FIG. 10A.1 Calorimeter for heats of neutralization, solution, etc.

can, usually of silver, having a capacity of at least 500 cm^3. It is surrounded by one or more polished metal radiation screens *B*, and finally a water-filled double jacket *C* itself enclosed in felt. The various compartments are closed by ebonite lids which are suitably cut to take a Beckmann thermometer *E* (provided with a lens *F*), and a glass or stainless steel stirrer *G* preferably driven by a small electric motor *H* and guided by a bearing *J*. The speed of the motor is controlled by an external rheostat. The whole arrangement is intended to minimize heat exchange to or from the vessel *A*, and, above all, to make such exchange as does occur as regular as possible so that corrections can be applied (see below). The use of the calorimeter will be illustrated for measurements of heats of neutralization and of solution.

EXPERIMENT

Determine the heat of neutralization of hydrochloric acid with sodium hydroxide.

Procedure. Prepare approximately 250 mol m^{-3} solutions of hydrochloric acid and of sodium hydroxide, free from carbonate, and determine their concentration by titration.

Fit together the calorimeter as shown in Fig. 10A.1, the outer vessel having been filled with water some hours previously in order that it may acquire, as nearly as possible, the temperature of the room. A Beckmann thermometer, previously set (section 7B) so that the mercury stands at the lower end of the scale at the temperature of the room, is passed through the holes in the covers of the calorimeter and supported so that the bulb passes about two-thirds down the calorimeter. 250 cm^3 of the caustic soda solution are placed in the calorimeter, and the stirrer is set in motion at a moderate speed.

Meanwhile, 250 cm^3 of the hydrochloric acid solution are placed in a tap funnel which is set up to deliver into the calorimeter. (The funnel may be rinsed previously with acid and drained for one minute to correct for the liquid which adheres to the glass. A trace of wetting agent may also be added.) The funnel is wrapped with felt to minimize temperature fluctuations. The temperature of the acid is read by another Beckmann or calorimeter thermometer. The readings of this thermometer must be compared with those of the thermometer placed in the alkali, to ascertain whether the temperatures of the acid and alkali are the same at the time of mixing, or, if not, what the difference of temperature is.

The temperatures registered by the thermometers immersed in the acid and alkali should be read, say every minute, for at least five minutes before the mixing of the solutions takes place. During this time the solutions should be stirred quietly. Then, at a particular moment which must be noted, the acid is run as rapidly as possible into the alkali, the two solutions mixed well, and the temperature of the mixture read every half-minute or every minute for five or ten minutes after mixing took place, until it is found that the fall of temperature becomes uniform. At first the temperature rises rapidly, then more slowly, and then begins to fall. As the temperature rises above that of the room, radiation from the calorimeter is taking place; it follows that the highest temperature read will be lower than if no loss of heat by the calorimeter occurred. In order, therefore, to get the true elevation of temperature produced by the heat of neutralization, the temperatures read on the thermometer before and after mixing should be plotted on squared paper, the thermometer readings being represented as ordinates and the time as abscissae. In this way two graphs similar to those shown in Fig. 10A.2 will be obtained.

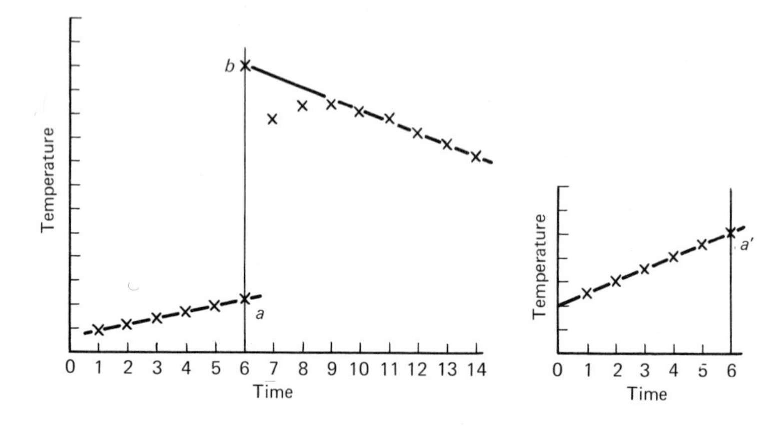

FIG. 10A.2 Simple graphical correction for heat exchange in calorimetry.

In this graph the temperature of acid and alkali is represented as rising slowly previous to mixing, but the reverse may, of course, be found. If the time of mixing was, say, at the sixth minute, the temperature (t_1 or t'_1) which the alkali and acid would have at that moment is obtained by drawing a line through the different temperature readings and producing it to cut the perpendicular at the sixth minute (point *a* or *a′*); and the highest temperature (t_2) which would have been reached in the absence of radiation is obtained by drawing a straight line through the last readings (when the fall of temperature has become uniform), and producing this line back so as to cut the perpendicular at the sixth minute. This gives the point *b*. The distance *ab* then gives the elevation of temperature required, $(t_2 - t_1)$.

Calculation of the heat of neutralization. The heat produced in the reaction must be equal to the heat required to raise the solution, the calorimeter, the thermometer and the stirrer through the range of temperature t_2–t_1 degrees. This, however, is equal to the sum of the masses of the different parts multiplied by their specific heats. For the heat evolved on mixing the acid and alkali, therefore, one has

$$\text{heat evolved} = (m_1s_1 + m_2s_2 + m_3s_3 + m_4s_4)(t_2 - t_1)$$

where m_1, m_2, m_3, m_4 are the masses of the solution, calorimeter, thermometer and stirrer respectively, and s_1, s_2, s_3, s_4, their respective specific heats. The following list gives the specific heats of some materials commonly employed in calorimeters: Ag 234, Ni 456, Cu 381, brass 372, stainless steel 448, glass 670, all in $J\ kg^{-1}\ K^{-1}$.

The specific heat of the solution will vary more or less from that of pure water according to the concentration of the dissolved salt. With solutions of the concentration used above, it will be sufficiently accurate for present purposes to take the water-equivalent of the solution (i.e. its mass multiplied by its specific heat), as being equal to that of the *water* contained in it. Otherwise, tables must be consulted.

In the case of the thermometer, which consists of glass and mercury the weight of which cannot be determined separately, the water-equivalent is obtained by making use of the fact that the specific heat of equal *volumes* of glass and mercury is practically the same and equal to $1{\cdot}797\ J\ K^{-1}\ cm^{-3}$. To obtain the volume, a beaker of water is counterpoised on the balance, and the thermometer then supported on a stand so that the bulb is immersed in the water. The weight which has now to be added in order to obtain equipoise gives the volume of the bulb. In the case of the Beckmann thermometer the stem above the bulb is not solid, so that the external volume does not represent the volume of the glass and mercury. The external volume of the stem, so far as it was immersed in the solution during the experiment, should be determined separately from that of the bulb, and about one-fifth of the volume so obtained taken as the volume of the glass and mercury. Fortunately, this term is small.

Results. Express the heat of neutralization in $k\ J\ mol^{-1}$. It is better practice to write the reaction, followed by the value of ΔH (= −heat of neutralization). For precise work the initial and final states must be specified, e.g. HCl ($200\ mol\ m^{-3}$, 25°C) + NaOH ($200\ mol\ m^{-3}$, 25°C) = NaCl ($99{\cdot}82\ mol\ m^{-3}$, 25°C), $\Delta H = -57{\cdot}4\ k\ J$.

When a strong acid reacts with a strong base, the reaction is $H^+ + OH^- \rightarrow H_2O$. The heat of reaction is therefore approximately a constant, namely, $57{\cdot}3\ k\ J\ mol^{-1}$. In practice different acids and alkalis may give appreciably different values on account of (a) incomplete dissociation, and (b) concentration effects. These may be investigated by repeating the experiment, using (a) a weak acid or base, and (b) a concentrated solution of acid or alkali. One may also investigate heats of dilution directly.

Similar experiments may be made with redox reactions, precipitation reactions, etc.

2. HEATS OF SOLUTION

The heat of solution of a solid or liquid substance can be determined in practically the same manner as that employed for the determination of the heat of neutralization, and the same apparatus can be used. Since the heat which is evolved or absorbed on dissolving a substance depends on the amount of water or other solvent employed, the statement of the heat of solution has a definite meaning only when the concentration of the solution formed is given. If the dilution is so great that further dilution is unaccompanied by any heat effect, then the heat measured per mole of solute is known as the heat of solution at infinite dilution. Usually, however, it will not be possible to determine this heat of solution directly, and one must therefore state the number of moles of water in which one mole of solute is dissolved. Further, a clear distinction must be made between the quantity known as the *integral heat of solution* (which is the heat obtained when 1 mole is dissolved in X moles of solvent), and the *partial molar heat of solution*. The latter is the heat change *per mole of solute* when an infinitesimal amount of substance is dissolved in a large amount of solution of stated concentration. Partial molar quantities are discussed in section 6A(4). This is the 'heat of solution' involved in calculations of the influence of temperature on solubility (section 8B(6)).

EXPERIMENT

Determine the integral heat of solution of potassium nitrate in water.

Procedure. Blow a fragile bulb on the end of a 25 cm length of glass tubing of about 1 cm bore. Dry it, along with a quantity of finely powdered KNO_3, in an oven, and, when cool, weigh about 15 g of the salt into the bulb. Then set up the bulb in the centre of the calorimeter with about 500 g of water (weighed). Set the stirrer going (as fast as permissible without splashing) and take a series of temperature readings as in the previous experiment. When a steady rate of exchange of temperature is attained break the bulb on the bottom of the calorimeter and stir up the contents vigorously. Take temperature–time readings until a steady temperature change is again reached.

Results. Correct the temperature change graphically as before (Fig. 10A.2) and compute the heat capacity from the weights. Estimate the proportion of the glass tube which was actually immersed in the calorimeter. The specific heat of KNO_3 solution containing 1 mole per cent of the salt can be taken as 3 946 J kg^{-1}. The accuracy of the experiment depends largely on whether the salt dissolves rapidly. If more than about 5 min are required, the simple extrapolation for heat exchange becomes inadequate (cf. section 10B(2)).

In determining the heat of solution of salts attention must be paid not only to the amount of water per mole of salt but also to whether the salt is anhydrous or hydrated. The table below shows that the degree of hydration affects the heat of solution considerably. The difference between the heats of solution of the anhydrous and the hydrated salt gives, of course, the heat of hydration of the solid.

Salt	*Number of moles of water per mole of salt*	*Heat of solution evolved (kJ mol^{-1})*
KCl	200	−18·4
KNO_3	200	−35·5
$ZnSO_4$	400	+77·4
$ZnSO_4.7H_2O$	400	−17·74
$CuSO_4$	300	+66·1
$CuSO_4.5H_2O$	300	−11·2

3. HEAT OF MIXING

The heat of mixing, say of methanol and water, can be determined in a similar manner to the heat of solution (section 10B.4). The *integral heat of mixing* n_1 moles of water with n_2 moles of methanol is the change in enthalpy for

$$n_1\ H_2O + n_2\ MeOH \rightarrow \text{solution}$$

For a discussion of the corresponding partial quantities see section 6A(4). Note that

$$\left(\frac{\partial \Delta H}{\partial n_1}\right)_{T_1 P_1 n_2}$$

is always largest when $m = n_1 = 0$ (adding H_2O to pure MeOH or vice versa) and goes to zero as $m(= n_1)$ becomes large (adding H_2O to H_2O or MeOH to MeOH).

EXPERIMENT
The heat of mixing of methanol and water.

The calorimeter is a simple Dewar vacuum flask. Provided the mixing is correctly performed heat losses during the time of the temperature change will be negligible. Consequently the maximum temperature change, which should be attained within 15 s, can be used without correction in calculating the heat of reaction.

The water equivalent of the calorimeter plus stirrer and thermometer, although small (about 16 g), enters into every subsequent calculation and should therefore be determined. Clean the Dewar flask, dry it with a cloth, and put into it an accurately weighed quantity of distilled water (approx. 70 g) that has been warmed to 10–20°C above room temperature. Arrange a mechanical stirrer so that the water is well stirred without splashing and measure the temperature T_w of the water as accurately as possible. Meanwhile place another 70 g or so of distilled water at room temperature into a stoppered vessel, weigh it, and measure the temperature T_c of the water accurately. Add it to the water in the Dewar as quickly as possible and note the temperature of the mixture, T_m. The weight of the cooler water added is found by weighing the empty vessel. Since the calorimeter and the warm water have been cooled from T_w to T_m while the added water has been warmed from T_c to T_m, we can write

$$(C_{\text{cal.}} + C_{\text{warm}})(T_w - T_m) = C_{\text{cool}}(T_m - T_c) \qquad (1)$$

C is the heat capacity at constant pressure, and for the warm or cool water it is equal to the mass of the liquid times its specific heat. The small temperature variation of the specific heat may be neglected and a mean value employed. Equation (1) can then be solved to find the water equivalent of the calorimeter, $C_{\text{cal.}}$.

The heat of mixing experiments are carried out in a similar manner. Both the water and the methanol are initially at around room temperature. For the first run put into the Dewar a known weight (approx. 100 g) of distilled water, measure its temperature T_1 accurately, and add to it sufficient methanol at a known temperature T_2 to give an approximately 10 mol % solution. Note the temperature rise ($T_3 - T_1$) and find the amount of methanol added by re-weighing the empty stoppered vessel. Since the calorimeter and water have been heated from T_1 to T_3 while the methanol has been raised from T_2 to T_3, the experimental heat of mixing (ΔH_1) is given by

$$-\Delta H_1 = (C_{\text{cal.}} + C_{\text{water}})(T_3 - T_1) + C_{\text{MeOH}}(T_3 - T_2) \qquad (2)$$

ΔH_1 refers strictly speaking to T_3; it could be corrected to 20°C by applying Kirchhoff's relation

$$\frac{\partial \Delta H}{\partial T} = \Delta C = C_{\text{mixture}} - C_{\text{components}} \qquad (3)$$

but the correction is negligible in the present case.

After this first addition, further quantities of methanol should be added to the solution to give successively 20, 30 and 40 mol % mixtures and the corresponding heats of mixing (ΔH_2, ΔH_3, ΔH_4) measured. For these, Eq. (2) must be modified to

$$-\Delta H_n = (C_{\text{cal.}} + C_{\text{solution}})(T_3 - T_1) + C_{\text{MeOH}}(T_3 - T_2) \qquad (4)$$

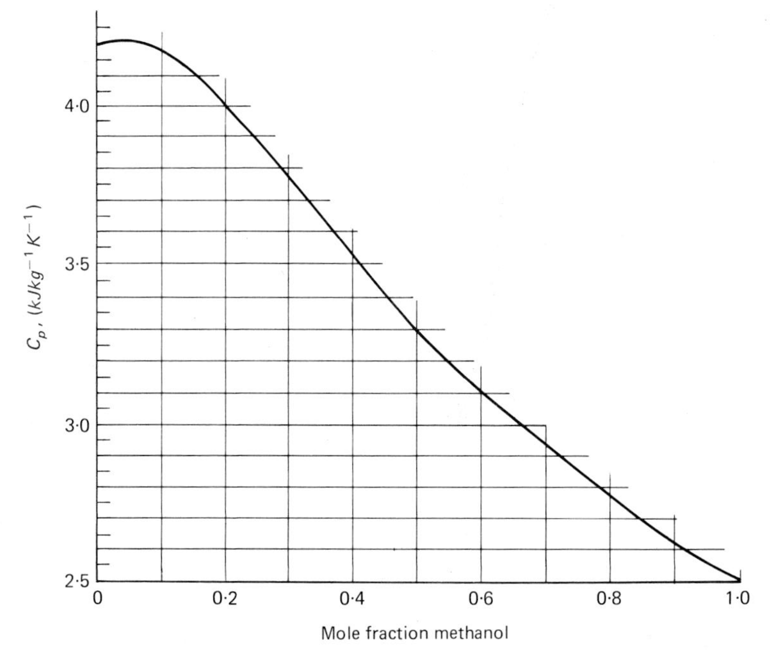

FIG. 10A.3 Heat capacity of water and methanol mixtures at 20°C.

where T_1 is now the temperature of the solution in the Dewar flask before the methanol has been added. The specific heat of methanol–water mixtures is given in Fig. 10A.3. It should be noted that ΔH, the *integral* heat of mixing for a solution of given composition, is the change in enthalpy when pure water is added to pure methanol to form that solution; if the solution has been formed in stages ΔH is therefore obtained by adding together the contributing experimental heats of mixing ΔH_n in accordance with Hess's law.

The calorimeter should now be emptied, dried, a weighed quantity of methanol added, and then the appropriate amounts of water added to give solutions approximately 90, 80, 70 and 60 mol % in methanol. The experimental and the integral heats of mixing are obtained in a manner analogous to that used above.

4. CALCULATION

Scale the values of the integral heat of mixing ΔH so they correspond to solutions containing 1 kg water (for the addition of methanol to water) or 1 kg methanol (water to methanol): e.g. in the former case multiply the values of ΔH by $1/W_2$ where W_2 kg is the original weight of water in the dewar. Calculate the corresponding values of the molality: for the first case if a total of W_1 kg of methanol had been added, the molality would be W_1/M_1W_2 where M_1 is the molecular weight of methanol (in kg). Plot ΔH_n against m in the two cases, and draw a smooth curve. Remember (0, 0) is a point on both graphs! The slope of the curve at any point is the value of the partial molar heat of mixing

$$\overline{\Delta H_1} \equiv \left(\frac{\partial \Delta H_1}{\partial n_1}\right)_{T,P,n_2}$$

at that point (addition of 1 to 2). On each graph draw targets at (0, 0) and the experimental points, and hence obtain values of $\overline{\Delta H_1}$ (first plot) and $\overline{\Delta H_2}$ (second plot). For the values of $\overline{\Delta H_1}$ find the corresponding values of $\overline{\Delta H_2}$ from

$$\begin{aligned}\Delta H &= n_1\,\overline{\Delta H_1} + n_2\,\overline{\Delta H_2} \\ &= m_1\,\overline{\Delta H_1} + \frac{1}{M_2}\,\overline{\Delta H_2}\end{aligned}$$

Conversely find values of $\overline{\Delta H_1}$ from the value of $\overline{\Delta H_2}$ from the second plot. Calculate the mole fraction of methanol in each of the experimental mixtures:

$$x_1 = \frac{W_1/M_1}{(W_1/M_1)+(W_2/M_2)}$$

Values of $\overline{\Delta H}_1$ and $\overline{\Delta H}_2$ can now be plotted against mole fraction over the whole composition range. Show the integral heat of mixing *per mole of solution* ($x_1 \overline{\Delta H}_1 + x_2 \overline{\Delta H}_2$) on the same plot. Estimate the heat evolved on mixing 1 mole of methanol with 1 mole of water.

BIBLIOGRAPHY 10A: Classical calorimetry

Weissberger, Vol. I, Pt 1, chapter X.
Rossini (ed.), *Experimental Thermochemistry*, 1956 (Interscience, New York).
Eitel, *Thermochemical Methods in Silicate Investigation*, 1952 (Rutgers Univ. Press, New Jersey).
Swietoslawski, *Microcalorimetry*, 1946 (Reinhold, New York).
Roberts (revised Miller), *Heat and Thermodynamics*, 4th edn, 1951 (Blackie and Son, London).
Rossini *et al.*, *Selected Values of the Classical Thermodynamic Properties*, 1952 (U.S. Govt Printing Office); National Bureau of Standards Circular No. 500.

10B The bomb calorimeter

1. A large number of thermochemical data have been obtained from measurements of *heats of combustion*. These data are not restricted in usefulness to the directly-measured combustion reactions, for by combining several such reactions according to Hess's law one can deduce the heats of other reactions. For example, *the heat of formation* ΔH_f of methane CH_4 cannot be determined directly, but it can be calculated from measurements of the separate heats of combustion of carbon, hydrogen and methane; thus:

$$
\begin{array}{lll}
(1)\ C+O_2 \rightarrow CO_2 & : \Delta H_1 & = -393{\cdot}42\ kJ \\
(2)\ 2H_2+O_2 \rightarrow 2H_2O & : \Delta H_2 & = -571{\cdot}69\ kJ \\
(3)\ CH_4+2O_2 \rightarrow CO_2+2H_2O & : \Delta H_3 & = -890{\cdot}31\ kJ \\
(4)\ = (1)+(2)-(3) & : \Delta H_4 & = \Delta H_1+\Delta H_2-\Delta H_3 \\
\qquad C+2H_2 \rightarrow CH_4 & & = -74{\cdot}80\ kJ
\end{array}
$$

The heat of combustion is best determined by the method due to Berthelot, which consists in burning the substance in an atmosphere of compressed oxygen. The original design of the autoclave in which the combustion takes place (the Berthelot bomb) has been modified in various ways. The form to be described here is the modification due to Mahler, as made by Messrs C. W. Cook & Sons Ltd, Birmingham.

It should be noted that, unlike other calorimeters, a bomb calorimeter measures the heat of a reaction *at constant volume* (instead of at constant pressure). The heat evolved is therefore identified with the decrease of internal energy, $-\Delta U$.

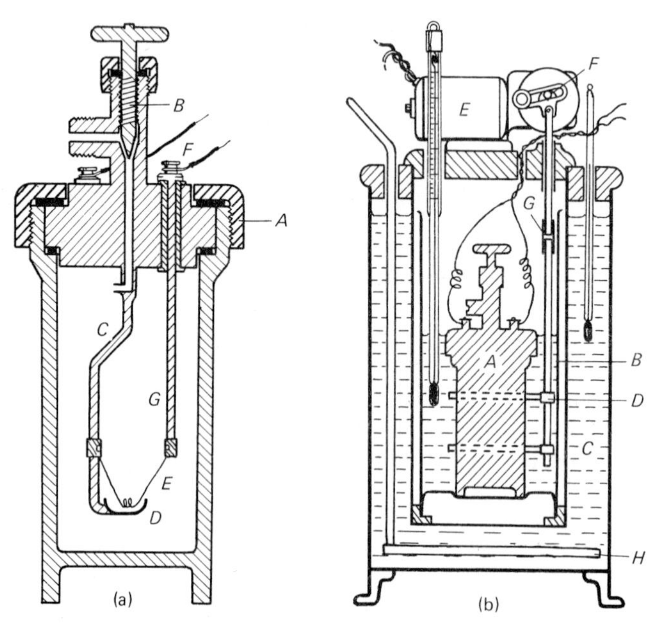

FIG. 10B.1 The bomb calorimeter. (a) Construction of the bomb. (b) Arrangement of apparatus in the calorimeter.

2. THE BOMB

The bomb (Fig. 10B.1(a)) consists of a stainless steel pressure vessel about 7 cm in diameter. The lid is held in place by a strong locking-nut *A* and rendered gas-tight by washers. Oxygen is charged into the bomb through the needle-valve *B*. A rod *C* supports a crucible of silica or platinum to contain the substance which is to be burnt. Combustion is initiated by means of a fine platinum or iron wire *E* which is heated momentarily to redness by passage of an electric current, the current being introduced through an insulated terminal *F* and the rod *C* (which is in electrical connection with the lid).

3. THE CALORIMETER

Figure 10B.1(b) shows the rest of the apparatus. The bomb A is placed in water in a copper calorimeter vessel B which is surrounded by a double-walled copper water-jacket C. The water in the inner vessel is stirred automatically by a reciprocating stirrer D driven by a small electric motor E and reduction gear and eccentric F. In order to minimize heat conduction along the stirrer, an ebonite rod G is interposed between the upper and lower parts. The top of the apparatus is closed by two wooden lids which are provided with holes to take a Beckmann thermometer for the inner vessel, and an ordinary thermometer graduated in tenths of a degree for the outer water-jacket. The latter should be provided with a simple stirrer H.

The water-equivalent of the calorimeter must be determined by burning a known weight of a substance of well-established heat of combustion. Pure benzoic acid is usually used, its heat of combustion being $2{\cdot}643 \times 10^7$ J kg^{-1}. As the water-equivalent is about 3 kg, combustion of 1 g of benzoic acid gives a temperature rise of about 2°C. The same amount of water must, of course, be used in the calorimeter for the experiment with benzoic acid as with the test substance, or allowance must be made.

Procedure. The substance to be burned is first compressed into a pellet, and, after being weighed, is placed in the crucible D. A piece of fine iron wire, about one-tenth of a millimetre in diameter and 6–7 cm in length, is weighed, and its ends connected with the rods C and G. The middle portion of the wire should be formed into a narrow spiral by twisting it round a pin. When the wire is attached, raise the crucible until the middle portion of the iron ignition wire touches the pellet of compressed substance. After having coated the washer and the screw on the bomb with vaseline, the cover is tightly screwed down with the help of a large spanner, the bomb being meanwhile fixed in its holder. Be very careful that no grit gets either on the washer or in the screw of the bomb. The latter is now ready to be filled with oxygen.

A cylinder of compressed oxygen fitted with a needle valve and pressure gauge is connected to the valve B by the metal pipe provided with the calorimeter. Open the valve B slightly, then the main cylinder valve, and then *cautiously* open the slow-release valve attached to the cylinder and allow oxygen to stream slowly into the bomb until the pressure registered on the manometer is 20–25 atm. Now close the valves and disconnect the tube from the bomb. If there is any leakage of gas it will generally be detectable by a slight hissing sound.

The calorimeter may now be prepared. The water-mantle having been filled with water (preferably some hours previously), a thermometer is hung in the air-space inside. After it has taken the temperature of the enclosure, the temperature is read. A Beckmann thermometer is now set (section 7B) so that the lower end of the scale represents a temperature of about 1·5–2 degrees *below* that found in the enclosure of the water-mantle.

The calorimeter vessel is then tared, 2·5 dm^3 of water placed in it, and the weight of the water determined on a balance accurate to about 1 g. The readings of the Beckmann thermometer and the outer-jacket thermometer should be noted when they are at the same temperature in a vessel of water; this is for the 'heat-loss correction' (see below).

In order to reduce the error due to radiation, the temperature of the water in the calorimeter should be such that, before the combustion in the bomb occurs, it is about as much below the surrounding temperature (temperature of the air-space) as it will be above it after the combustion has taken place. As the rise of temperature should be about 2·5–3°C, the temperature of the water should be made about 1·5°C lower than that of its surroundings.

The charged bomb is lowered cautiously into the water of the calorimeter. Flexible wires are then connected to the terminals, and the lid of the calorimeter placed in position. Insert the thermometer through the cover of the calorimeter and set the stirrer in motion at such a rate that it rises and falls about once per second. After the bomb has been in the water about 5 min, commence reading the temperature on the thermometer, readings being made every minute for about 10 min. At the tenth minute fire the ignition wire by means of a switch key, in series with the mains and a rheostat—a suitable potential must be found by preliminary experiments. The iron wire will thereby be caused to burn and will ignite the substance in the crucible. The temperature of the water in the calorimeter will now begin to rise rapidly. Again take readings of the temperature, minute by minute (it will probably be impossible to make a reading at the first minute after ignition), until the highest temperature is reached, a point which must be carefully noted. The temperature will now begin to fall slowly, and readings at intervals of a minute must again be made for about 10 min. During the experiment the outer vessel should be stirred occasionally and its temperature taken.

This completes the series of observations. The bomb is removed from the calorimeter and carefully dried. The needle valve *B* is slowly opened so as to allow the gases to escape from the bomb. When the pressure has fallen again to that of the atmosphere the cover is removed and the interior of the bomb cleaned and dried. Any of the iron wire which has not been oxidized should be detached and weighed, and the weight subtracted from that originally taken.

Calculations. For approximate work the 'heat-loss correction' can be made by the simple graphical method previously described, but this is not accurate when the passage of heat is slow; a correction method should be used, such as the '3-period method' described below.

'*Heat-loss*' *correction.* For the purpose of this correction one can assume that the heat exchange between calorimeter and surroundings is due to two factors—(a) conduction, convection and radiation to the outer vessel, and (b), Joule heating by the working of the stirrer. The rate of rise of temperature due to (a) is approximately proportional to (temperature of the outer jacket − temperature of the calorimeter), according to Newton's Law of Cooling, while the rate of rise due to the stirrer is constant. Thus, the rate of gain of heat should be represented by the equation, $dQ/dt = k_1(t_s - t_c) + k_2$, where dQ is the heat in joules gained in time dt when the temperature of the calorimeter is t_c and that of

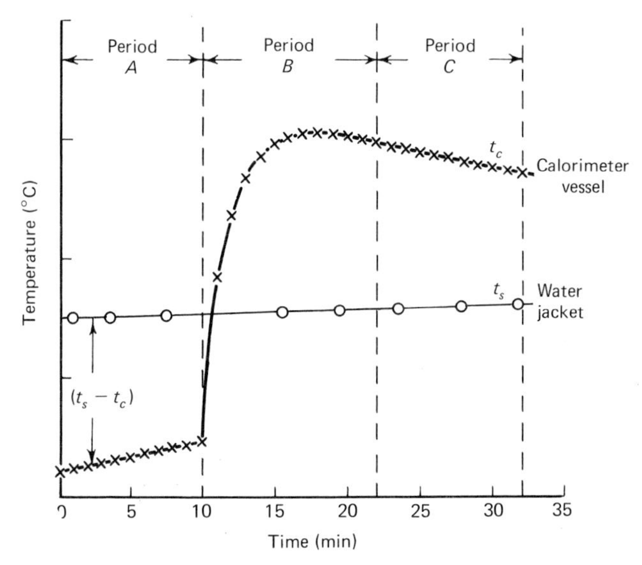

FIG. 10B.2 Temperature record for heat-exchange correction by the three-period method.

the surrounding jacket is t_s. To apply this theory one must, in effect, determine the constants k_1 and k_2, and this can be done from the temperature–time data.

First, complete temperature–time graphs for the calorimeter (t_c) and the water-jacket (t_s) must be plotted; clearly, t_c and t_s must be in the same units, and therefore the Beckmann readings should be converted to °C by making use of the preliminary comparison readings which have been made for this purpose. Figure 10B.2 shows the kind of graphs which should be obtained. The chart can clearly be divided into three distinct periods, namely, a preliminary rating period (*A*) during which t_c is rising slowly and regularly, the combustion period (*B*) during which heat is being transmitted from the bomb to the calorimeter, and a final rating period (*C*) during which t_c falls regularly. The temperature of the outer jacket, t_s, may perhaps rise slightly during the course of the experiment, as indicated in Fig. 10B.2.

The above heat exchange equation can now be tested for periods *A* and *C*. The rate of change of temperature (per minute) is obtained by subtracting consecutive readings of t_c, and the difference between the temperature of the calorimeter and its surroundings can be measured on the graph for the time half-way between the two temperature readings. A graph of δt_c against $(t_s - t_c)$ can then

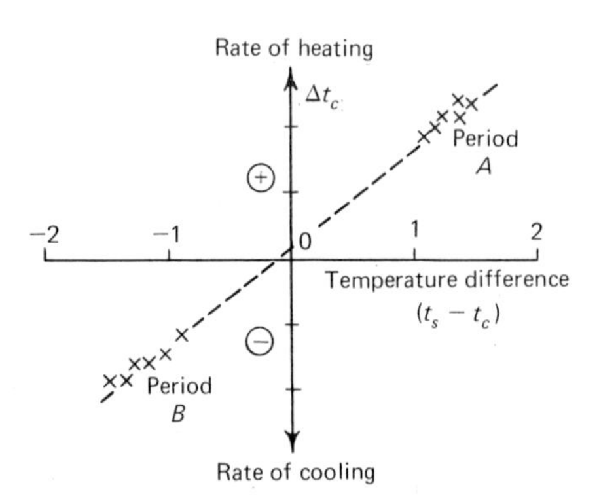

FIG. 10B.3 Calculation of heat-exchange correction.

be plotted. The data shown in Fig. 10B.2 would lead to a 'heat correction graph' like that shown in Fig. 10B.3, a number of points being obtained from period *A* with positive values of $(t_s - t_c)$ and of δt_c, and other points from period *C*, during which $(t_s - t_c)$ is negative. Between the two sets of points no data are available, but, if the above equation is correct, the δt_c against

$(t_s - t_c)$ graph should be a straight line; consequently, linear interpolation is permissible. (If the stirrer produces appreciable heating effect, a positive value of δt_c will be found for $(t_s - t_c) = 0$.)

The 'heat-loss' graph can now be used to calculate the heat loss or gain which the calorimeter must have experienced during the period *B*. For each minute one calculates the mean calorimeter temperature t_c, and then the difference $(t_s - t_c)$—paying due regard to sign. The heat *gain* δt_c is then read from the correction graph; values will be positive while $t_s > t_c$, and then become negative when $t_s < t_c$. Positive values of δt_c mean that the calorimeter has been gaining heat from the surroundings, equal to $\delta t_c \times W$ (where W is the total water-equivalent) during that minute; negative values show that heat loss must have been occurring. Since W is a constant, $W \sum \delta t_c$, taken over period *B*, gives the total number of joules which the calorimeter has *gained* from the outside during period *B*. If the overall change of temperature from the beginning to the end of period *B* is T°C, the corrected rise due to the heat of combustion alone is $(T - \sum \delta t_c)$. Of course, if on the whole there has been a loss of heat (as, for example, is true for the hypothetical experiment illustrated in Fig. 10B.2), one would find $\sum \delta t_c$ negative, and therefore the corrected rise of temperature would come out greater than the observed—as it must.

As a check on the calculation, one should calculate the corrected temperature rise for several points subsequent to the maximum in the curve; a constant corrected value should be found for the heat due to the combustion.

4. WATER-EQUIVALENT OF THE CALORIMETER

As explained above, it is necessary to make preliminary determinations with benzoic acid in order to find the water-equivalent of the calorimeter. Since the amount of water used in different experiments may, perhaps, be varied, it is usual to calculate the water-equivalent of the bomb and calorimeter vessel separately. Allowance should be made for the small amount of heat produced by combustion of the iron wire (if used), the heat of combustion of iron being 6.7×10^6 J kg^{-1}. It follows that the heat balance for the calibration experiment gives

$$\begin{pmatrix}\text{wt of}\\\text{benzoic}\\\text{acid}\end{pmatrix} \times 2{\cdot}643 \times 10^7 + \begin{pmatrix}\text{wt of}\\\text{iron}\\\text{burnt}\end{pmatrix} \times 6{\cdot}7 \times 10^6 = \begin{pmatrix}\text{corrected}\\\text{rise of}\\\text{temperature}\end{pmatrix}\begin{pmatrix}\text{wt of}\\\text{water}\end{pmatrix} + \begin{pmatrix}\text{water-equivalent}\\\text{of calorimeter}\end{pmatrix}$$

Two or more calibration experiments should be carried out until concordant values are obtained for the water-equivalent of the calorimeter.

5. HEATS OF COMBUSTION

When the apparatus has been calibrated, as described above, exactly similar determinations are made with the substance for which the heat of combustion is required (e.g. sucrose, anthracene, coals, etc.). Volatile liquids can be burned by enclosing them in a gelatine capsule of known weight and calorific value.

The method of calculating the heat of combustion from the corrected temperature rise will be apparent.

Bomb calorimetry is applicable to a wide variety of substances, but complications arise for substances containing sulphur, nitrogen, phosphorus, etc., since one must know the products formed in the reaction.

BIBLIOGRAPHY 10B: The bomb calorimeter

Weissberger, Vol. I, Pt 1, chapter X.

[1] Jessop, *J. Res. Nat. Bur. Standards*, 1942, **29**, 247.

11
Conductivity of electrolytes

11A Theory and technique

1. TERMINOLOGY

An electric current may be carried either by electrons—as in metals and most 'semi-conductors'—or by movement of ions; conduction in *liquid* solutions is entirely *electrolytic*. The conduction obeys Ohm's law, i.e. the current (in amperes) = difference of potential (volts)/resistance (ohms). The resistance between opposite faces of a unit cube of a conductor is called the *specific resistance* or *resistivity*. The reciprocal of resistance is *conductance*, and the reciprocal of specific resistance is *specific conductance* or *conductivity* (symbol κ; units $\Omega^{-1}\ m^{-1}$).

With electrolytic solutions conductance is a more natural quantity to employ than resistance, since it is related directly to the number of ions present and to their rate of movement. Generally the pure solvent is practically non-conducting, and consequently the observed conductance is due entirely to dissolved electrolyte. In comparing the conducting properties of different substances one should compare chemically comparable quantities of the substances, i.e. equimolar quantities. If the concentration of the solution is c mol m^{-3}, the specific conductance (κ) is clearly due to c moles: the *molar conductivity* $\Lambda = K/c$ ($\Omega^{-1}\ m^2\ mol^{-1}$) is the theoretical conductivity per mole of electrolyte.

2. THEORY

The molar conductivity Λ of a solution is a measure of the number and rate of migration of anions and cations from 1 mole of the solute. Experiment shows that Λ increases as the solution is diluted (whereas κ, of course, decreases). The increase of Λ with dilution could arise from either an increase in the number of ions present (i.e. increased dissociation) or from a greater speed of movement of the individual ions. Arrhenius (1883) put forward the former explanation and assumed that the true ionic mobilities were independent of concentration so that Λ is proportional to the number of ions present. According to the law of mass action all dissociation processes should tend to completion as the concentration tends to zero, and hence the equivalent conductivity at zero concentration, Λ_0, represents the current-carrying capacity of the solute in a completely dissociated state.

For a 1:1 electrolyte $\Lambda_0 = \Lambda_+ + \Lambda_-$ where Λ_+, Λ_- are the ionic molar conductances, associated with the cation and anion respectively. For the electrolyte $M_{\nu^+}X_{\nu^-}$ (one mole contains ν^+ mole of cation M and ν^- mole of cation X),

$$\Lambda_0 = \nu^+\Lambda_+ + \nu^-\Lambda_-$$

For example $CaCl_2$: $\nu^+ = 1$, $\nu^- = 2$ and $\Lambda_0 = \Lambda_+ + 2\Lambda_-$.

The conductivity at finite concentrations, Λ_c, should therefore give the degree of dissociation $\alpha = \Lambda_c/\Lambda_0$. The Arrhenius theory is approximately correct for *weak electrolytes*, i.e. solutes which are only slightly ionized, and for such substances conductivity measurements can be used to obtain the degree of dissociation and hence the dissociation constant (section 11B).

Salts of strong acids and bases also show Λ increasing with dilution, but this cannot be ascribed to increasing ionization as salts are known to be ionized even in the solid state. The effect is due to interaction between the anions and cations, tending to retard their movements. In extreme cases, particularly with polyvalent ions, temporary 'ion-pairs' may form owing to the strong coulombic forces, but otherwise the interaction is a statistical 'ionic atmosphere' effect. The theory was developed by Debye and Hückel (1923) and Onsager (1926, 1927), who obtained an expression for the conductivity of very dilute solutions which is of the form

$$\Lambda_c = \Lambda_0 - (a\Lambda_0 + b)\sqrt{c}$$

where c is the concentration in mol m^{-3} and a and b are constants which can be evaluated from the theory. (For a uni-univalent electrolyte at 25°C, $a = 7{\cdot}25 \times 10^{-3}$ $m^{\frac{3}{2}}$ $mol^{-\frac{1}{2}}$ and $b = 1{\cdot}917 \times 10^{-4}$ $m^{\frac{7}{2}}$ Ω^{-1} $mol^{-\frac{3}{2}}$.) In practice, graphs of Λ_c against $\sqrt{c}$ are approximately linear, and do approach the correct slope of the Onsager theory at great dilutions ($c \to 0$). Appreciable departures from the theory occur, however, at quite moderate concentrations, and Λ_0 cannot be obtained accurately by graphical extrapolation of data. Instead, it is better to assume the constants of the Onsager equation and calculate Λ_0 from the equation in the form

$$\Lambda_0 = \frac{\Lambda_c + b\sqrt{c}}{1 - a\sqrt{c}}$$

At the present time there is no quantitative theory for the conductance of concentrated solutions of strong electrolytes.

3. OUTLINE OF EXPERIMENTAL METHODS

Resistance is almost always measured by means of a Wheatstone bridge circuit (section 4D.1). With an electrolytic conductor there are several complications that do not arise with electronic conductors. The principal one is that the passage of a current causes electrolysis, and a back e.m.f. of 'polarization' is set up. This difficulty is overcome by using an alternating current so that whatever electrolysis occurs during the first half cycle is reversed during the second and the net electrolysis is negligible. The Wheatstone bridge (d.c. and a.c.) is described in section 4D. All connections to the bridge should be made with stiff straight wire to maintain the capacitance constant; the resistance boxes should be wound non-inductively. For the most accurate work the resistance of the cell is measured as a function of bridge frequency and then extrapolated to infinite frequency.

4. CAPACITY BALANCE

It has been mentioned that a perfect balance on an a.c. bridge cannot be expected unless reactance is balanced as well as resistance, and this condition is usually achieved by adding a variable condenser to one arm of the bridge. A convenient arrangement is to use a *differential* variable air condenser Z of capacity about 100 $\mu\mu$F, the fixed sides being connected to points 1 and 3 (Fig. 11A.1) and the movable vanes being connected to point 4. Capacity can then be applied in parallel with either R_3 or C as required by simply turning the insulated knob of the condenser. Further fixed condensers may be required if the cell capacitance is high. The procedure is to balance the bridge first by adjusting R_3 as well as possible, then alternately adjust the condenser and readjust R_3 until the best null is achieved.

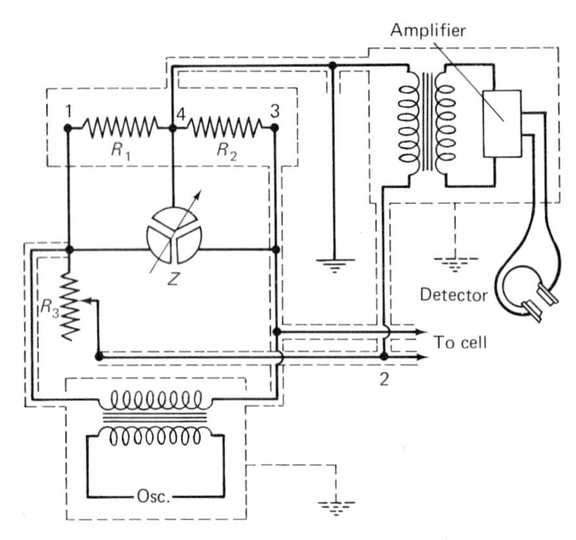

FIG. 11A.1 Earthing of components in the conductivity bridge.

5. EARTHING

The following points in the circuit should be connected to earth: the chassis of the oscillator and amplifier, all screens surrounding leads or resistance boxes, the thermostat, and the point 4 on the bridge. (Fig. 11A.1.) This earthing arrangement should be found quite satisfactory if measurements are to be made of resistances ranging from about 100 to 10 000 Ω with a sensitivity of 0·1%. For work of higher precision it is essential that the Wagner earthing circuit be employed.[1]

6. TEMPERATURE CONTROL

As the conductivity of solutions varies by about 2 per cent per degree, the temperature must be controlled to within 0·05–0·01°C if an accuracy of 0·1% is to be obtained on κ. Consequently, the conductivity cell must be immersed in an efficient thermostat bath. Oil thermostats are preferred for research work on conductivity because electrical leakage is thereby minimized, but an ordinary water thermostat is more convenient. A simple method of supporting the cell in the thermostat is to place it in a small 'box' made of copper sheet which can be hung on the side of the tank. The 'box' is partially open at the ends so that water from the thermostat can pass through it.

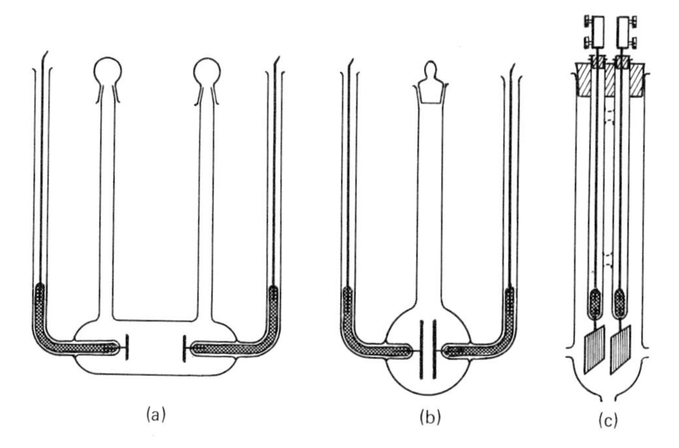

FIG. 11A.2 Conductivity cells. (a) For liquids of high conductance. (b) For liquids of low conductance. (c) Dipping cell.

7. CONDUCTIVITY CELLS

Figure 11A.2 shows a number of conductivity cells for various purposes: (a) and (b) are suitable for accurate measurements, (a) for solutions of high conductance, and (b) for those of low conductance; (c) is a simple dipping cell, which is convenient for approximate work and conductimetric titrations. The cells should be made of Pyrex or other resistance glass, and the electrodes of platinum sufficiently thick to be quite rigid. Connection to the electrodes is made by a drop of mercury, but it is convenient to seal the ends of the tubes with wax and attach brass connectors (as in Fig. 11A.2(c)) so that the cells can be inverted during washing, etc.

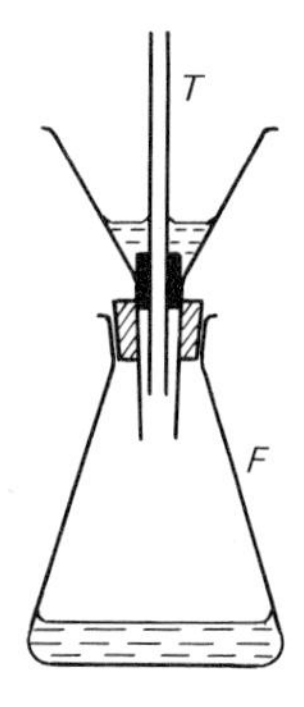

FIG. 11A.3 Flask for steaming glass vessels.

All glass vessels in conductivity work should be very thoroughly washed free from electrolyte. After the usual cleansing with chromic acid and repeated rinsing with distilled water, they should be steamed out by being inverted over the tube *T* of the flask shown in Fig. 11A.3. Water is boiled in the flask, steam passes up through the tube and the condensate collects in the funnel. Finally, the vessel may be dried by drawing a current of air through it while it is still warm.

8. PLATINIZING THE ELECTRODES[2]

Even with a.c. current some polarization occurs if *smooth* platinum electrodes are used, but it can be eliminated by coating the electrodes with a thin layer of platinum black. This has the effect of increasing the effective area for current discharge, thus reducing the local current density, and it also catalyses the electrode processes.

The electrodes are platinized by electro-deposition from a solution consisting of 3 g chloroplatinic acid and 5 mg lead acetate in 100 cm^3 of water. The current from two accumulators is regulated by a rheostat so that only a moderate stream of gas is formed at the electrodes; a current density of 30 mA per cm^2 of platinum surface is recommended. The current is reversed every half-minute, and after about 10 min the electrodes should be sufficiently coated. For work with dilute solutions only a very thin coating should be applied.

After the electrolysis, great care must be taken to remove all traces of the platinizing solution, which is liable to be adsorbed by the electrodes. It is advisable to replace the platinizing solution by a dilute solution of sulphuric acid and pass a current for some time in each direction. This treatment should be followed by repeated rinsing with warm distilled water and then with conductivity water (see below). Cells should be kept filled with pure water when not in use.

9. CONDUCTIVITY WATER

Ordinary laboratory distilled water generally has a conductivity of about $3\text{–}5 \times 10^{-4}\ \Omega^{-1}\ m^{-1}$. Most of this conductivity is due to impurities, and very little to ionization of the water itself, as the value has been reduced to $4 \times 10^{-6}\ \Omega^{-1}\ m^{-1}$ by extreme purification. Dissolved carbon dioxide in equilibrium with that in the air produces a conductivity of about $8 \times 10^{-5}\ \Omega^{-1}\ m^{-1}$ and the remainder is due to traces of salts, ammonia, etc. Elaborate precautions are required in order to produce water of conductivity less than $10^{-4}\ \Omega^{-1}\ m^{-1}$ by distillation; the principle usually employed is to redistil good distilled water with the addition of alkaline permanganate and to condense and reject part of the steam, the remainder being condensed in a tube of pure tin or silica in a stream of purified air.

However, the simplest method of obtaining water of low conductivity is by passing it slowly through a column packed with a mixture of anion- and cation-exchange resins, the former being in the –OH form and the latter in the –H form. Such a mixture is supplied by the Permutit Co. Ltd, London, under the name 'Bio-Deminrolit'. Anions present in the water are replaced by OH^- ions, and cations by H^+ ions, and since these are in equivalent proportions, the water is 'deionized' completely. After two or three 'bed-volumes' of water have been passed through the column, the specific conductivity of the effluent falls to about $10^{-5}\ \Omega^{-1}\ m^{-1}$, or less. If the column is fed with ordinary distilled water, it will treat a very large volume of water before it requires regenerating. When necessary, this is effected by first separating the two types of resin (by an upward flow of water) and then treating the anion-exchanger with excess of NaOH in a column and the cation-exchanger with excess of HCl. The two resins are then washed and mixed again for use.

10. PREPARATION OF SOLUTIONS

Since much of the theory of conductivity is concerned with very dilute solutions and since their conductivity can be measured with considerable accuracy, it is necessary to make sure that the concentration of the solutions studied is known with similar accuracy. The usual procedure is to prepare first the most concentrated solution required and to dilute it successively. Dilution by volume is hardly accurate enough, and therefore the solutions should be made up and diluted *by weight*.

A number of wide-necked stoppered Pyrex flasks of about 150 cm^3 capacity are cleaned, steamed out, and dried as described above. A suitable quantity of the substance is weighed accurately from a weighing bottle into one of them. The flask is then weighed to the nearest centigram on a large balance, conductivity water is added, and the flask weighed again. More dilute solutions are subsequently prepared from this solution by pouring some of it into another (weighed) flask, adding conductivity water, and weighing again. Buoyancy corrections should be applied where significant (section 2A).

11. DETERMINATION OF THE CELL CONSTANT

The cells used for conductivity measurements are not of accurately known dimensions, and consequently they need to be calibrated before it is possible to deduce values of specific conductance from measurements of cell resistance. This is done by means of solutions of accurately known conductivity. Suppose a given cell, when filled with a solution of known specific conductivity κ is found to have a resistance R, then the *cell constant* K is defined by $K=\kappa R$. The cell constant may therefore be thought of as the resistance that the cell would exhibit when filled with liquid of unit conductivity. If another solution shows a resistance R' when measured in the same cell, its conductivity κ' is obtained from the same equation, i.e. $\kappa'=K/R'$.

Conductivity cells are generally calibrated with solutions of potassium chloride of carefully adjusted concentration and at a temperature of 25·0°C. In the following table[3] are given the values of the conductivities of solutions containing 0·1 and 0·01 moles of potassium chloride (1 mole=74·553 g) in 1 kg *of solution*, i.e. 0·1 and 0·01 *demal* (D).

These solutions can be prepared by dissolving 7·41913 and 0·745263 g KCl in 1 kg water respectively; these weights require correction for the buoyancy of air (section 2A(3)). The unit of resistance is the international ohm (=1·0005 absolute ohm).

Temperature (°C)	*Specific conductivity* (Ω^{-1} m^{-1})	
	0·1D	0·01D
0	0·7138	0·07736
18	1·1167	0·12205
25	1·2856	0·14088

A quantity of pure, dry KCl is weighed out, and the necessary quantity of water to produce exactly 0·1 or 0·01 D is calculated and then obtained by adjustment (using a robust balance). Alternatively, if an accuracy of 1 part in 1 000 is sufficient, the solution may be made by dissolving any weighed quantity of KCl in a graduated flask of water and, after bringing the temperature to 25°C, making up by volume to the mark. The concentration of the solution is then known, and the conductivity can be calculated from an empirical equation given by Shedlovsky[4] for KCl solutions at 25°C, namely:

$$\Lambda = 1{\cdot}4982 \times 10^{-2} - 2{\cdot}9678 \times 10^{-4}\sqrt{c} + 9{\cdot}49 \times 10^{-6} c(1 - 7{\cdot}191 \times 10^{-3}\sqrt{c})\ \Omega^{-1}\ \mathrm{m}^2\ \mathrm{mol}^{-1}$$

where c is the concentration in mol m^{-3}.

Another procedure is to make up a solution by weight (without, however, adjusting the quantities) and then use Shedlovsky's formula together with the density, which for KCl solutions up to 10 mol m^{-3} is given by $\rho_c = 997{\cdot}1 + 0{\cdot}47c$ kg m^{-3}.

BIBLIOGRAPHY 11A: Conductivity; theory and technique

Weissberger, Vol. I, Pt 4, chapter XLV.
Robinson and Stokes, *Electrolyte Solutions*, 2nd edn, 1959 (Butterworths, London).
Potter, *Electrochemistry: Principles and Applications*, 1956 (Cleaver-Hume, London).
Davies, *Electrochemistry*, 1967 (Newnes, London).

[1] Jones and Josephs, *J. Amer. Chem. Soc.*, 1928, **50**, 1049; Shedlovsky, *J. Amer. Chem. Soc.*, 1930, **52**, 1793; Luder, *ibid.*, 1940, **62**, 89; Foy and Martell, *Rev. Sci. Instrum.*, 1948, **19**, 628.
[2] Feltham and Spiro, *Chem. Rev.*, 1971, **71**, 177.
[3] Jones and Bradshaw, *J. Amer. Chem. Soc.*, 1933, **55**, 1780.
[4] Shedlovsky, *J. Amer. Chem. Soc.*, 1932, **54**, 1411.

11B Weak electrolytes

EXPERIMENT

Determine the equivalent conductivity of a weak acid at several concentrations and hence calculate the dissociation constant of the acid.

Suitable acids for this experiment are succinic, benzoic and mandelic acids. Weak dibasic acids such as succinic behave in conductance measurements as monobasic acids since only the first hydrogen is ionized at the concentrations employed.

Procedure. If an accuracy of better than 1% is required, the solutions must be made up by weight rather than volumetrically. The acid should be weighed out into a steamed Pyrex flask fitted with a stopper, and a quantity of conductivity water sufficient to make a solution of about 50 mol m^{-3} should be added and weighed. More dilute solutions are then prepared from the stock solution by repeated dilution by weight in other Pyrex flasks, the concentration being approximately halved each time. The flasks are suspended in the thermostat at 25·0°C in readiness for the measurements of the conductivity of the solutions.

For weak electrolytes a cell having a small cell constant should be used. To make a determination, the cell is rinsed with a little of the solution and then filled and placed in the thermostat. The leads are connected and the balance point of the bridge is determined. The cell is left in the thermostat until the resistance becomes constant, showing that the solution has come to 25·0°C. The cell is then emptied, and the measurements are repeated with a further sample of the same solution as a check. This check is particularly necessary with very dilute solutions where appreciable amounts of substance derived from a previous solution may be slowly desorbed from the electrodes. When concordant results have been obtained with the first (most concentrated) solution, the cell is washed with water and then rinsed twice with the next solution, the conductivity of which is then determined in the same way. A series of solutions of increasing dilution down to about 1 mol m^{-3} is studied, and the conductance of the water alone is also obtained. The constant of the cell used must, of course, be determined with a standard KCl solution as described, but a rather dilute solution is necessary if the resistance of the cell is not to be too low for accurate measurement. (Alternatively, the conductance of the most concentrated acid solution employed may be measured both in a cell of small constant and also in one of larger constant, and the absolute cell constant of the latter may be determined with one of the standard KCl solutions referred to above (section 11A(11)). Hence the cell constant of the first cell can be calculated.)

Treatment of results (*elementary theory*). The molar concentration of each solution is calculated, and then its specific conductance and hence molar conductivity. If the conductivity of the water is less than 10^{-4} Ω^{-1} m^{-1} it may be ignored, but if larger, it should be subtracted from that of the solutions before calculating Λ. In order to calculate the degree of dissociation one needs Λ_0, but, in the case of weak electrolytes, this cannot be obtained reliably by extrapolation, and it is therefore calculated from the sum of the ionic conductivities. The following values may be taken for Λ_0 at 25°C: acetic acid 0·3908, benzoic acid 0·383, mandelic acid 0·378, succinic acid 0·382 Ω^{-1} m^2 mol^{-1}. The degree of dissociation is calculated for each solution as $\alpha = \Lambda_c/\Lambda_0$. The values of α are then substituted in the Ostwald equation to give the dissociation constant, $K_c = c\alpha^2/(1-\alpha)$, where c is the concentration. The resulting values of K_c should be approximately constant over a considerable range of concentration. The numerical value of K_c will, of course, depend on the units of concentration used, and these should always be stated.

Improved treatment for weak electrolytes. There are two distinct reasons why the elementary theory does not give absolutely constant values for the dissociation constant.

(a) The degree of dissociation is not accurately given by $\alpha = \Lambda_c/\Lambda_0$ because the ionic mobilities change with ionic concentration. A correction can be applied as follows, using the method of successive approximations. First the 'classical' value of α is obtained as above. This is then substituted in the Onsager equation to calculate the value of Λ (say Λ') which the ions would exhibit at the actual ionic concentration present in the solution (i.e. at αc). Thus,

$$\Lambda' = \Lambda_0 - \sqrt{\alpha c}(a\Lambda_0 + b)$$
$$= \Lambda_0 - \sqrt{(\Lambda_c/\Lambda_0)c}\,(a\Lambda_0 + b)$$

The new value Λ' is now used in place of Λ_0 to calculate an improved value of α, say, $\alpha' = \Lambda_c/\Lambda'$. A second value of Λ (say Λ'') may be calculated in the same way, giving a further refinement of α (α'') and so on. Usually two recalculations give a substantially constant value of α, which may then be used to calculate $K(=\alpha''^2/(1-\alpha'')V)$. Values of a and b are given in the next section.

(b) In calculating K one ought to use activities rather than concentrations. For very weak acids little error will be introduced since the ionic strength is very low, and the ions and the undissociated acid will have activity coefficients very close to unity. For moderately weak acids, however, a correction becomes necessary for the ionic species—say if the ionic strength exceeds 0·1 mol m^{-3}. Thus, if c_i is the concentration of hydrogen ions (and of anions) and c_u is the concentration of undissociated acid, the thermodynamic dissociation constant is given by $K_{Th} = (c_i \cdot f_\pm)^2/c_u$, where $f_\pm$ is the mean ion activity coefficient. Such activity coefficients are known to follow the law $\log f_\pm = -A\sqrt{c_i}$, where A is a constant (about 0·0161, according to the Debye-Hückel theory). Hence,

$$\log K_{Th} = \log(c_i^2/c_u) - 2A\sqrt{c_i} = \log K_c - 2A\sqrt{c_i}$$

where K_c is the concentration equilibrium constant calculated as explained under (a). Thus, a graph of $\log K_c$ against $\sqrt{c_i}$ should give a straight line from which $\log K_{Th}$ can be obtained as the intercept at $c=0$, and the slope gives $2A$.

BIBLIOGRAPHY 11B: Weak electrolytes

Robinson and Stokes, *Electrolyte Solutions*, 2nd edn, 1959 (Butterworths, London).

11C Strong electrolytes

EXPERIMENT

Determine the equivalent conductivity of solutions of a strong electrolyte, and hence examine the validity of the Onsager theory as a limiting law at great dilutions.

Suitable electrolytes are HCl, KNO_3, $AgNO_3$, NaCl. (Uni-divalent salts may also be studied, but di-divalent salts such as $CuSO_4$ and $MgSO_4$ are only partially dissociated on account of ion-pair formation.[1])

Procedure. The solutions are prepared and their conductivity measured in the manner described above for weak acids. Concentrations from about 100 mol m^{-3} to about 0·1 mol m^{-3} should be studied. This necessitates the use of two cells of high and low cell constants to cover the whole range accurately. The conductivities should be corrected for that of the water.

Treatment of results. Since the Onsager equation gives $\Lambda_c=\Lambda_0-(a\Lambda_0+b)\sqrt{c}$, a graph should be plotted of Λ_c against $\sqrt{c}$. This should tend to linearity as $c\to 0$. The limiting straight line should be drawn and its slope determined and compared with that given by the Onsager theory, namely, $-(a\Lambda_0+b)$ where $a=7{\cdot}25\times10^{-3}$ $m^{\frac{3}{2}}$ $mol^{-\frac{1}{2}}$ and $b=1{\cdot}917\times10^{-4}$ $m^{\frac{7}{2}}$ Ω^{-1} $mol^{-\frac{3}{2}}$ for a uni-univalent electrolyte at 25°C. Λ_0 is obtained by extrapolation of the limiting line to $c=0$. If the limiting slope is in reasonable agreement with the Onsager equation, the theory may be assumed correct and a more reliable value of Λ_0 may be calculated from the experimental values of Λ_c for the most dilute solutions by means of the expression $\Lambda_0=(\Lambda_c+b\sqrt{c})/(1-a\sqrt{c})$ (i.e. a theoretical extrapolation rather than a graphical one).

Shedlovsky[2] showed that conductivity data for strong electrolytes can be represented within an accuracy of 0·1% up to concentrations of 100 mol m^{-3} by an *empirical* extension formula from the Onsager equation, namely, $\Lambda_0 = [(\Lambda_c + b\sqrt{c})/(1 - a\sqrt{c})] - Ac$, where A is a constant to be determined empirically. Such formulae are useful for interpolation and for representing data in a concise form. (See the case of KCl solutions, mentioned in section 11A(11))

BIBLIOGRAPHY 11C: Strong electrolytes

1 Robinson and Stokes, *Electrolyte Solutions*, 2nd edn, 1959 (Butterworths, London).

2 Shedlovsky, *J. Amer. Chem. Soc.*, 1932, **54**, 1405.

11D Applications of conductance measurements

1. Since the measurement of electrical conductance of solutions can be made accurately and conveniently, it provides a valuable means of studying many problems in physical chemistry, as may be seen from the following examples. The use of conductance to determine the concentration of a solution of a pure substance by comparison with previously prepared calibration curves is sufficiently obvious to need no further exemplification. The method can frequently be used to advantage when the substances to be determined are not readily estimated by simple analytical methods.

2. SOLUBILITY OF SPARINGLY SOLUBLE SALTS

The conductance method is applicable for finding the concentration of the saturated solution, provided that the salt is not hydrolysed, and that the solubility is not too high. It may then be assumed that the ions possess their limiting conductivities, and Λ_0 may be taken as the sum of the ionic conductivities (see section 11A(2)): these must, of course, be known for the same temperature. Hence, measurement of the specific conductivity of the saturated solution leads to a value for the concentration. A refinement of the calculation is to correct Λ_0 by use of the Onsager equation to allow for the finite concentration.

EXPERIMENT

Determine the solubility of lead sulphate, or of silver chromate, in water at 20°C.

Procedure. The conductivity of the water employed should first be determined at 20°C. A quantity of finely powdered lead sulphate or silver chromate is then shaken repeatedly with the conductivity water in order to remove any impurities of a comparatively soluble nature. The well-washed salt is then placed with conductivity water in a hard-glass vessel, which is placed in a thermostat at 20°C, and shaken from time to time. After intervals of about 15 min, a quantity of the solution is transferred to a conductivity cell, and the conductivity determined. A conductivity cell with large electrodes placed a short distance apart should be used. This is repeated with fresh samples of the solution, until constant values are obtained.

The conductivity so determined is corrected by subtracting the conductivity of the water employed, and the solubility then calculated by means of the equation $c = \kappa/\Lambda_0$. (For the solubilities of some sparingly soluble salts, see Appendix, Table A9.)

3. HYDROLYSIS OF SALTS

When the salt of a weak acid or base is dissolved in water, hydrolysis occurs, so that the conductance of the solution is now due partly to the ions of the salt and partly to the ions of the acid or base formed by hydrolysis (more especially the hydrogen ion and hydroxide ion). If, however, a quantity of the weak acid or base which, in the presence of its salt, can be regarded as completely un-ionized, is added to the solution, the hydrolysis of the salt will be diminished, but the ionization will be unaffected. From measurements of the conductivity of pure solutions of the salt (in which, therefore, hydrolysis occurs), and of solutions containing excess of the weak base or acid, the degree of hydrolysis can readily be calculated.

Considering here only the simplest case of a binary salt, say of a strong monobasic acid with a weak monoacid base, the hydrolytic equilibrium is given by the expression

$$\frac{K_b}{K_w} = \frac{(1-\alpha)}{\alpha^2 c}$$

where c is the concentration in mol m^{-3}, α the degree of hydrolysis, K_b the affinity constant of the weak base and K_w the ionic product of water. The amount of the unhydrolysed salt is represented by $(1-\alpha)$, and of the free acid by α. The molar conductivity Λ_c, of the solution of hydrolysed salt is therefore given by

$$\Lambda_c = (1-\alpha)\Lambda'_c + \alpha\Lambda''_c$$

where Λ'_c and Λ''_c are the molar conductivities at concentration c of the unhydrolysed salt and of the strong acid respectively. The former, as has been pointed out, can be determined by measuring the molar conductivity of the salt in presence of excess of the weak base. The degree of hydrolysis of the salt at the given dilution is then given by the expression

$$\alpha = \frac{\Lambda_c - \Lambda'_c}{\Lambda''_c - \Lambda^c}$$

EXPERIMENT

Determine the degree of hydrolysis of aniline hydrochloride in aqueous solution at 25°C.

Procedure. Make a solution (say 30 mol m^{-3}) of aniline hydrochloride in water. Place 20 cm^3 of this solution in a conductivity cell and determine the conductivity at 25°C. Dilute the solution with water to 15 mol m^{-3} and 7·5 mol m^{-3} as described in section 11A(10) and determine the conductivity at each dilution. Now make a solution of aniline hydrochloride, not in pure water, but in a 30 mol m^{-3} solution of aniline, and determine the conductivity at the same dilutions as before, the dilution of the original solution being carried out with the 30 mol m^{-3} solution of aniline. From the conductivities so determined, calculate the degree of hydrolysis of aniline hydrochloride at each of the dilutions, the molar conductivities of hydrochloric acid at $c=30$, 15 and 7·5 mol m^{-3} being taken as 0·393, 0·399 and 0·401 Ω^{-1} m^2 mol^{-1} respectively.

Calculation. From the value of the degree of hydrolysis found calculate the affinity constant of aniline, assuming the ionic product of water at 25°C to be $1{\cdot}02 \times 10^{-8}$ mol^2 m^{-6}.

4. RATES OF REACTION, SOLUTION, DIFFUSION

If a reaction in solution involves the formation or removal of electrolytes, or even a change in the number or type of ions present in the solution, the rate of the reaction can often be followed very conveniently by consecutive measurements of the conductance of the solution. Examples are the rate of hydrolysis of esters (see section 14B for an experiment) or esterification of alcohols, or the rate of ion exchange between a salt solution and an ion-exchange resin where the exchanging ions have different mobilities. The rate of diffusion of salts in water can also be determined from the rate of change of resistance exhibited between electrodes placed in the diffusion column.

5. CONDUCTIMETRIC TITRATIONS

Measurements of the electrical conductance can also be employed in order to determine the end-point of reactions between electrolytes, e.g. neutralization of acids and alkalis, precipitation reactions, etc. The method is of especial value for coloured or turbid solutions, in which the change of colour of an indicator would be more or less masked, or with very dilute solutions, or for reactions for which no convenient indicator is available.

Acid–alkali titrations. When a strong acid is added to an alkali, the conductance of the solution will decrease owing to the disappearance of the hydroxide ion, and its replacement by the less mobile anion of the acid; but when all the hydroxide ion has been removed, by combination with hydrogen ion from the acid added, then any further addition of acid will cause the conductance to increase, owing to the addition to the solution of free hydrogen ion. Since the hydrogen ion has a much greater mobility than any other ion, the presence of a slight excess of free acid will cause a marked increase in the conductance. When, therefore, the conductance of the solution is plotted against the volume of acid added, a curve of the form shown in Fig. 11D.1 is obtained. The point of intersection of the two curves gives the volume of acid required to exactly neutralize the solution of the alkali.

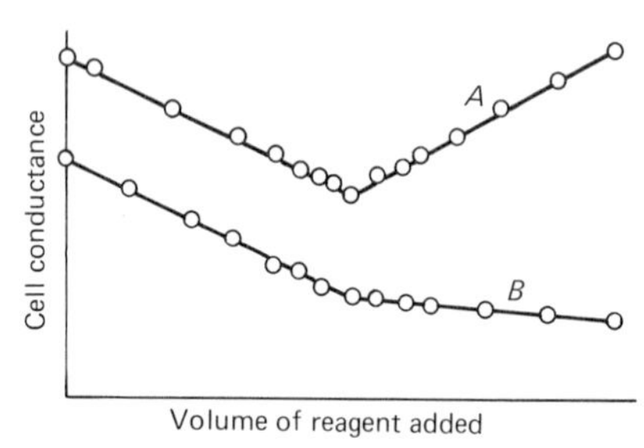

FIG. 11D.1 Determination of the equivalence point in conductimetric titrations. *A*, titration of strong base with strong acid. *B*, titration of strong base with weak acid.

When both acid and alkali are strong, one may carry out the titration from either side; the acid may be added to the alkali, or the alkali to the acid. In the latter case, there is first a diminution of the conductance owing to the replacement of hydrogen ion by a less mobile cation, and then, after the neutral point has been reached, an increase in the conductance, which is sharply defined owing to the fact that OH^- is also a highly mobile ion.

When, however, the acid is weak, it is necessary to add the acid to the alkali (*and in this case a strong alkali must be chosen*), and not the alkali to the acid. If the alkali is added to a weak acid, the minimum is not sharp owing to the fact that the change in conductance is due not so much to the disappearance of the fast-moving hydrogen ion (which is present in comparatively small concentration), as to the replacement of the un-ionized acid molecules by the ions of the salt formed. When, however, the acid is added to the alkali, the hydroxide ion is replaced by the much slower anion of the acid, and a consequent diminution of the conductance of the solution is produced. As the acid is, however, only slightly ionized, and the ionization is further reduced by the presence of the neutral salt, the addition of excess of acid does not, in general, lead to an increase of the conductance, but only to a sharp change in the direction of the conductance curves, as shown in the lower graph in Fig. 11D.1.

Strong acids, also, may be titrated in presence of weak acids, and the basicity of an acid may be determined.

Other electrolytic reactions. The conductimetric method of titration may also be used to determine the end-point of many other types of electrolytic reaction.[1] Thus, in the reactions

$$\mathrm{NaCl + AgNO_3 = AgCl + NaNO_3}$$

$$\mathrm{CH_3.COONa + HCl = CH_3.COOH + NaCl}$$

the conductance of the solution remains almost unchanged on addition of the silver nitrate or of the hydrochloric acid, because the conductance of sodium nitrate is similar to that of sodium chloride, and the conductance of sodium chloride is similar to that of sodium acetate. When, however, excess of silver nitrate or of hydrochloric acid is added, the conductance shows a sudden increase. In the reaction $\mathrm{MgSO_4 + Ba(OH)_2 = Mg(OH)_2 + BaSO_4}$ the resultants are both sparingly soluble and separate out from the solution. There is therefore a marked decrease of the conductance. On adding excess of barium hydroxide, the conductance rapidly increases.

Practical procedure. A known volume of the solution to be titrated is pipetted into a suitable conductivity cell. The dipping type shown in Fig. 11A.2(c) will be suitable for many dilute solutions and can be immersed in a small beaker. For solutions of about 10–50 mol $\mathrm{m^{-3}}$ it is better to use a similar cell with smaller electrodes; a pair of platinum wires (platinized), about half an inch in length and sealed into glass tubes containing a little mercury, are suitable. If the solutions are at room temperature it will not be necessary to place the cell in a thermostat for the titration unless the end-point happens to be rather indistinct. The resistance of the cell may be determined in the usual manner. The titrating solution is then run in from a burette in small quantities at a time, and after each addition the solution is well mixed and the resistance redetermined. To avoid the disturbing effects due to dilution, the concentration of the titrating solution should be five or ten times greater than the concentration of the solution to be titrated, the titrating solution being delivered from a microburette graduated to 0·01 cm^3. A graph is drawn of resistance against volume of titrating solution added, and the point of intersection of the curves gives the equivalence point.

Simplified apparatus. Conductimetric titrations can be greatly simplified and expedited by replacing the Wheatstone bridge by a direct reading conductance meter.

EXPERIMENTS

Titrate the following solutions conductimetrically:

(a) 50 cm^3 of 10 mol m^{-3} NaOH with 1 000 mol m^{-3} HCl from a microburette, (b) 50 cm^3 of 10 mol m^{-3} Na_2CO_3 with 1 000 mol m^{-3} HCl, (c) 50 cm^3 of 20 mol m^{-3} Na_2SO_4 with 1 000 mol m^{-3} $BaCl_2$.

BIBLIOGRAPHY 11D: Applications of conductance measurements

Delahay, *New Instrumental Methods in Electrochemistry*, 1954 (Wiley, New York).

[1] Lingane, *Electroanalytical Chemistry*, 2nd edn, 1958 (Interscience, New York); Wilson, *Ann. Repts. Chem. Soc.*, 1952, **49**, 325; Sand, *Electrochemistry and Electrochemical Analysis*, 1941, Vol. 3 (Blackie and Sons, London); Kolthoff and Laitenin, *pH and Electrotitrations*, 1941 (John Wiley, New York).

12
Transport numbers and electrode processes

12A Transport numbers

1. INTRODUCTION

When a current of electricity is passed through an electrolyte, the change of concentration at the electrodes is, in general, not the same. Although equivalent amounts of positive and negative ions are necessarily discharged at the electrodes (Faraday's law), the velocity with which the ions move in the potential gradient is different. Consequently, the amount of electricity carried by cations in one direction is different from that carried by anions in the other direction. These two amounts are, indeed, in the ratio of the velocities of migration (mobilities) of the cation and anion, u_+ and u_-, respectively. It is of interest to mention that in the extreme case of certain ionic *crystals* (e.g. silver chloride) only one of the ions can move through the lattice and therefore carries *all* the current. In solution, however, both ions can move, but their velocities differ because of their different (hydrated) radii.

The total current passing through a solution under a given potential is proportional to the sum of the ionic mobilities $(u_+ + u_-)$. The cation carries a fraction equal to $u_+/(u_+ + u_-)$ and the anion $u_-/(u_+ + u_-)$. These fractions are called the *transport* (*or transference*) *numbers* of cation (t_+) and anion (t_-), and clearly $t_+ + t_- = 1$.

The relationship between transport numbers and ionic mobilities is readily derived, as follows. If a univalent ion at concentration c mol m^{-3} has a transport number of t, and the specific conductivity of the solution is $\kappa\ \Omega^{-1}\ m^{-1}$, then in a unit cube of solution subjected to a potential gradient of 1 V m^{-1}, a current of κ A would flow. The quantity of electricity transported by the given ions in 1 s would therefore be $t \times \kappa$ C, which corresponds to $t\kappa/96487$ mol. The solution contains c mol m^{-3}, and therefore the rate of migration of the ions, u, is given by $u = (t\kappa/96\,487c)$ m s^{-1}.

Transport numbers can be determined by (a) measurement of the quantity of electrolyte transported to the cathode or anode during electrolysis (Hittorf's method, 1853), or (b) by measurement of the rate of movement of a boundary between two electrolytes during electrolysis (Lodge 1886, Whetham 1893), or (c) in certain cases, from measurement of e.m.f. of concentration cells with and without transport (as described in chapter 13).

2. HITTORF'S METHOD

In this method one measures the total quantity of electricity passed through a cell and the amount of one of the ions which has passed away from (or into) the space around one of the electrodes. It is assumed that the change of concentration takes place only in the neighbourhood of the electrodes, and that the intermediate portion of the solution remains unaltered. It is convenient, therefore, to have an apparatus in which the anode region, cathode region and intermediate region are well separated so that the solutions can be withdrawn from the sections separately after electrolysis and analysed. It is evident that too great a quantity of electricity must not be passed through the cell. Further, the current must not be, on the one hand, too intense, as that might cause heating and mixing by convection, or, on the other hand, so feeble that diffusion could partially annul the concentration changes produced by electrolysis.

Hittorf's method, although simple in principle and widely applicable, suffers from the experimental disadvantage that it depends on measuring small *changes* of concentration, and therefore very accurate analytical methods must be employed if reliable values of the transport number are to be obtained.

3. MOVING BOUNDARY METHOD

A sharp boundary is formed in a tube between solutions of two salts having a common ion (say, A^+X^- and B^+X^-, Fig. 12A.1). It must be possible to observe the position of the boundary by having A and B of different colour, or by having an indicator substance in the solutions or by an optical method depending on a difference of refractive index or ultra-violet absorption spectrum between the two solutions. If now a potential gradient is set up along the tube by applying a voltage to electrodes situated at the ends of the tube,

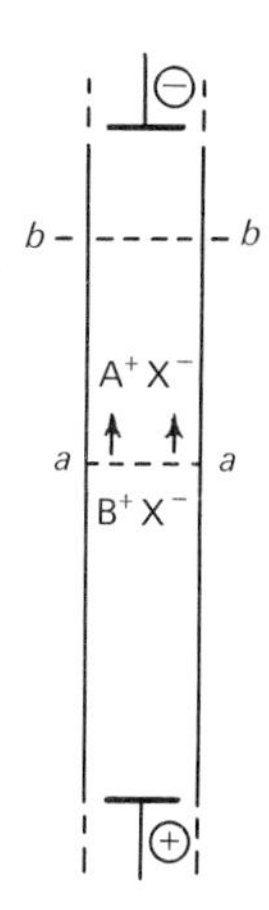

FIG. 12A.1 Basis of the moving boundary method for determining transport numbers.

current will flow by migration of the cations towards the cathode and anions towards the anode. At the boundary the A^+ and B^+ ions will move in the same direction and hence the boundary will be observed to move. In general A^+ and B^+ ions will not move at the same rate. Suppose the cathode is situated in the AX solution. If B^+ ions move faster than A^+ ions, they will overtake the latter and pass them, causing the boundary to become more and more diffuse. If, on the other hand, the A^+ ions are the faster, they will tend to move away from the B^+ ions at the boundary. Of course, a separation of ions cannot occur to any appreciable extent, but the effect prevents the A^+ and B^+ ions from mixing by diffusion. The result is the formation of a self-sharpening boundary *aa* which migrates along the tube with a velocity equal to that of the *faster* ion (A^+), and therefore the experiment leads to a value for the transport number of A^+ in the solution of A^+X^- (concentration $= c$ mol m^{-3}). The tube is graduated so that the volume, V m^3, swept through in time t s during the passage of Q C of electricity can be determined. The amount of current carried by A^+ ions across any section *bb* in the AX solution is clearly equal to the number of moles of A^+ ions which would have passed that section (namely, $c \times V$) multiplied by the charge (in coulombs) carried by each mole (namely, 1 faraday $= 96\,487$ C). Hence, the fraction of the current carried by A^+ ions ($= t_+$) is given by $t_+ = (cV \times 96\,487)/Q$. c is known from the original concentration of AX solution, and V and Q are measured during the experiment. The moving boundary method is therefore a direct one, and is capable of high accuracy.

In carrying out a moving-boundary determination, the following conditions must be observed. (1) A suitable 'following ion' B^+ must be employed, having a mobility less than that of A^+. (2) The conductance of the 'indicator' solution should be slightly less than that of the leading solution. (3) The denser solution must be the lower one in the tube, to obviate mixing. (4) The current of electricity passed through the tube must not be so large as to cause appreciable heating, expansion and convection of the liquid. An obvious precaution is to keep the tube fairly narrow and immerse the apparatus in a bath of water. (5) As there must be no *flow* of liquid along the tube, one end of the apparatus must be closed, and a non-gassing electrode must be employed at this end. These conditions are met in the experiment described below. Another technique sometimes applicable is the 'autogenic boundary' device of Franklin and Cady;[1] the 'indicator' solution is produced *in situ*—for example, by the anodic solution of cadmium, giving $CdCl_2$.

Concentration of following ion. While the boundary is moving, the theory given above for the A^+ ion must also hold for the B^+ ion, and hence $t_{A^+}/c_{A^+} = t_{B^+}/c_{B^+}$. It appears therefore that the 'indicator' solution must be of exactly the right concentration if the experiment is to succeed, and that one must know t_{A^+}/t_{B^+} *a priori* to arrange c_{A^+}/c_{B^+}. In practice, however, the concentration of the 'indicator' solution is not highly critical as it automatically adjusts itself in the neighbourhood of the boundary, so that t_{A^+}/t_{B^+} need only be known approximately (within 10%). If BX is the lower solution, it should be made slightly *more* concentrated than the value of c_{A^+} calculated from the above equation so that the density changes near the boundary during electrolysis do not cause convection. If the 'indicator' solution is *above* the leading solution, then it should be *less* concentrated than that calculated.

4. MEASUREMENT OF QUANTITY OF ELECTRICITY

The simplest method of measuring the quantity of electricity passed during a transport number determination is by means of a reliable milliammeter in series with the apparatus. If the current changes slowly, one can note the reading at frequent intervals and calculate the total electricity passed by graphical summation of current × time. Alternatively, the area under the current–time graph can be computed by Simpson's Rule. For greater accuracy, the milliammeter can be replaced by a standard resistance, and the potential across it determined by an accurate potentiometer or digital voltmeter (section 4C(2–3)). The value of the resistance should be chosen to suit the full range of the potentiometer; for example, if the current is 10 mA, the potential drop across a standard 100 Ω resistance would be 1 V, and this could be determined to better than 1 mV.

In determinations by the moving-boundary method it is convenient to maintain a *constant current* through the apparatus so that the boundary moves at constant velocity. Errors in the current–time integration are thereby reduced. The resistance of the cell increases during the measurement and hence the current tends to fall. For work of moderate accuracy the current can be regulated by manual adjustment of a rheostat (preferably a large one with rotary adjustment) used as potential-divider across the supply voltage. Numerous electronic circuits have been devised for current control, e.g. Fig. 12A.2.

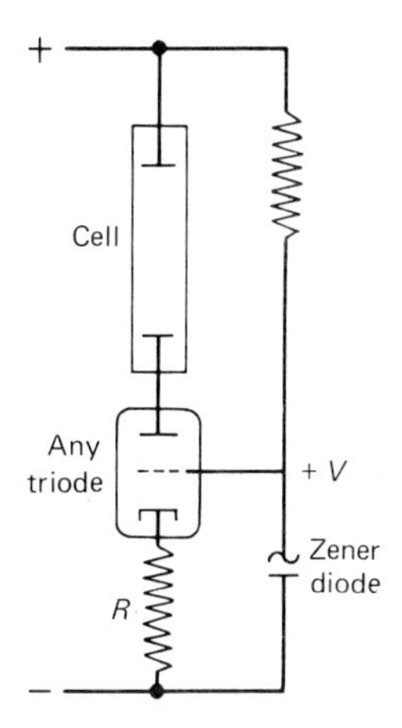

FIG. 12A.2 Constant current regulator. Current through cell ≈ V/R where V is the Zener diode voltage.

In determinations of transport numbers by Hittorf's method it is traditional to use a *coulometer* for measuring the quantity of electricity passed. This is particularly suitable since only one value of Q is needed and a coulometer gives the integrated number of coulombs directly irrespective of changes of current. The silver or iodine coulometers are capable of giving very high accuracy, but the copper coulometer, formerly much used, is less satisfactory, and has nothing but cheapness to recommend it.

5. THE SILVER COULOMETER

A convenient form of apparatus is shown in Fig. 12A.3. A glass or porcelain crucible *a* having a porous base of No. 2 or 3 porosity is supported inside a small beaker by a glass tripod *b* so that its base is clear of the bottom of the beaker. The cathode *c* is a sheet of platinum

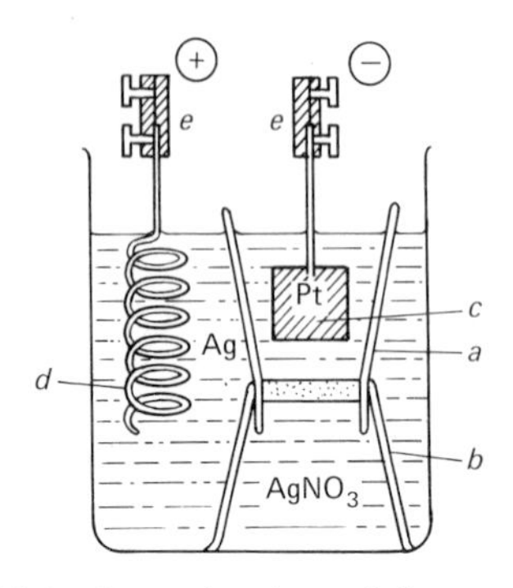

FIG. 12A.3 Convenient form of silver coulometer.

foil, about 1 cm^2 in area, with a platinum wire welded to it. The crucible + cathode are cleaned, washed, and dried to constant weight in an oven at 150°C before the experiment. The usual precautions of cooling for sufficient time in a desiccator must, of course, be observed. The anode *d* is a coil of stout silver wire (or a piece of silver foil with wire attached). Both electrodes are connected by small brass connections *e*. The electrolyte is approximately 15% $AgNO_3$ solution; the $AgNO_3$ should be very pure, and in particular, free from organic matter. The current passed through the coulometer should not exceed 10 mA cm^{-2} of cathode surface. After the electrolysis, the cathode is carefully disconnected and left in the crucible, together with any loose crystals of silver. The crucible is lifted out of the beaker with platinum-tipped tongs and carefully washed externally with a jet of water and then internally with suction in the usual manner until the washings give no reaction for silver. The crucible and contents are then dried to constant weight at 150°C, cooled in a desiccator and weighed again. The accuracy obtainable with this coulometer will generally be limited only by the accuracy of weighing. 1 C deposits $1{\cdot}118 \times 10^{-6}$ kg of silver.

6. THE IODINE COULOMETER

The apparatus is shown in Fig. 12A.4. The electrodes are platinum

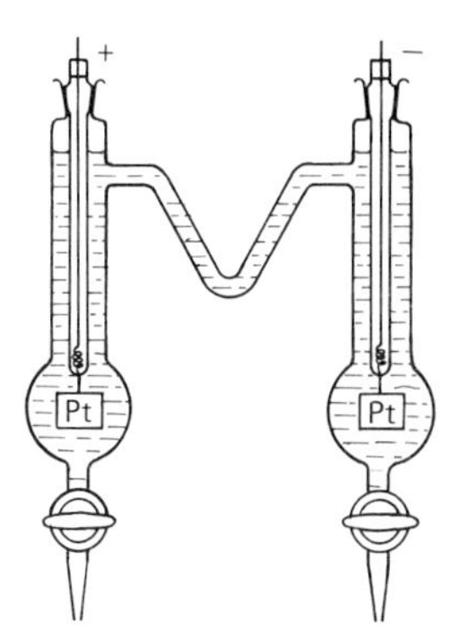

FIG. 12A.4 Iodine coulometer.

foils. The electrolyte *in the bulb* of the anode compartment is concentrated KI solution; that around the cathode is a concentrated solution of I_2 in KI. The upper parts of the apparatus contain 10% KI solution. During electrolysis iodine is formed at the anode and consumed at the cathode. After electrolysis the anode solution is run off and titrated for iodine in the usual way.

EXPERIMENT

Determine the transport number of Ag^+ *in a solution of silver nitrate by Hittorf's method.*

The type of apparatus traditionally used for determinations by Hittorf's method is shown in Fig. 12A.5. *A*, *B* and *C* are the cathode, anode and intermediate compartments. The electrodes consist of stout silver wires, cemented into glass tubes by means of wax, and soldered to copper leads inside the tubes. The whole apparatus is filled with a solution of silver nitrate, about 50 mol m^{-3}, made up accurately *by weight*. As the solution in the cathode compartment becomes more dilute during electrolysis and is therefore liable to undergo mixing by convection, one can insert a thin *loose-fitting* disc of porous glass to minimize this effect. The apparatus should be protected from direct sunlight or heat which might cause convectional mixing.

A current of about 10–20 mA is suitable. A potential of about 30–40 V is required (depending on the dimensions of the apparatus), and can be obtained from batteries, a rectifier or d.c. mains with variable resistance. It is convenient to have a 0–50 milliammeter in series with the apparatus, but the quantity of electricity passed is usually measured by a coulometer.

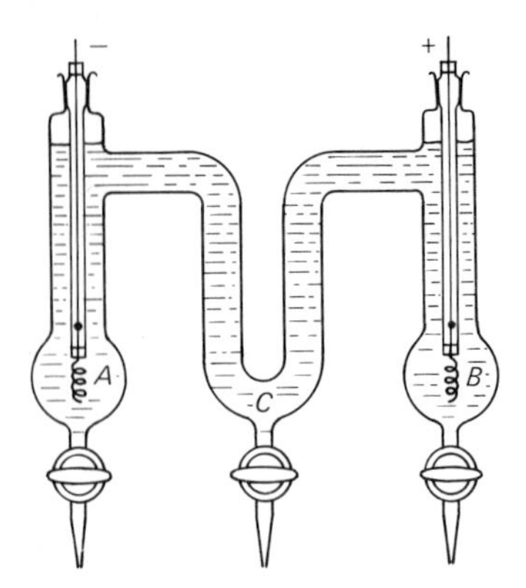

FIG. 12A.5 Apparatus for determination of transport number of silver nitrate by Hittorf's method.

About 100 C are passed through the cell (2–3 hours at 10 mA), and *A*, *B*, and finally *C* are drained into separate, dry, weighed flasks. Each compartment is rinsed with a little of the original solution, the rinsings being added to the appropriate flask. The flasks are then weighed, and their contents titrated with standard 50 mol m^{-3} potassium thiocyanate solution, using ferric alum as indicator. (Alternatively, the silver may be determined gravimetrically or by electrometric titration with KCl solution.) The analysis must be made as accurately as possible. The composition of the intermediate compartment *C* should be found identical with that of the original solution; if it has changed, the current has been passed too long, and the experiment must therefore be repeated for a shorter time.

Calculation. Let Q be the number of coulombs passed through the apparatus. Consider the processes occurring in the anode compartment. Suppose the initial concentration of *silver* in the solution was w kg per kg of *water*. Let the amount of Ag found by analysis after electrolysis be w' kg in W kg of *solution*. The anolyte after electrolysis therefore consisted of w' kg of Ag$+(W-w')$ kg of water. If no concentration changes had taken place, this quantity of water would have contained $w \times (W-w')$ kg of Ag. There has therefore been a gain of silver equal to $w'-w(W-w')=X$ kg. This increase is the net result of a gain by solution of the anode and loss by migration of Ag^+ out of the anode compartment. By Faraday's law, the amount of silver dissolved at the anode must be $Q \times 0{\cdot}10788/96\,487 = Y$ kg. Hence, the amount that has migrated away is $(Y-X)$ kg. This amount has therefore carried a fraction of the current given by $(Y-X)/Y$, which is therefore the transport number of the silver ion, t_{Ag^+}, in the $AgNO_3$ solution.

The calculation for the cathode compartment is entirely analogous, and should, of course, lead to the same value of t_{Ag^+}.

EXPERIMENT

Determine the transport number of H^+ *in* HCl *solution by the moving-boundary method.*

A simple form of moving-boundary apparatus is shown in Fig. 12A.6. A sharp boundary is produced by means of a stopcock A as described later. The boundary is caused to move down the tube BC, which consists of half of a 1 cm^3 pipette, graduated in 0·01 cm^3. The cathode F is a silver–silver chloride electrode of considerable capacity so that no gases are evolved, the electrode reaction being $AgCl + e \rightarrow Cl^- + Ag$. The electrode is constructed from platinum gauze attached to a platinum wire sealed in the glass tube D; the gauze is first electro-

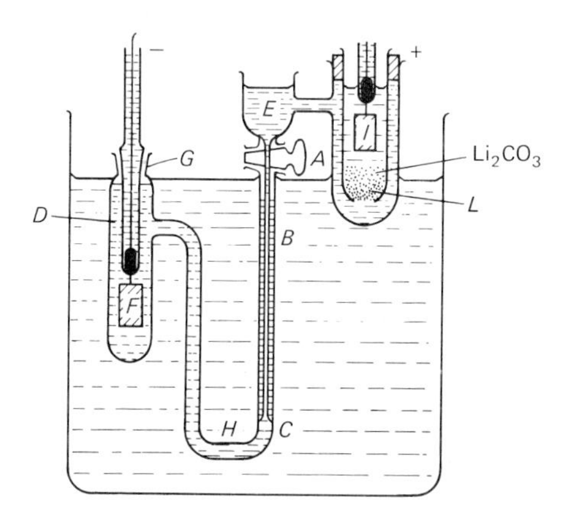

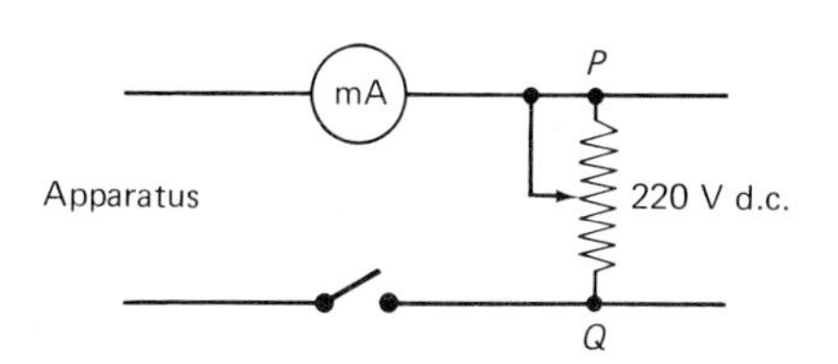

FIG. 12A.6 Apparatus for determination of transport number of HCl by the moving-boundary method. (a) Glass cell. (b) Circuit diagram.

plated with silver which is then chloridized by being made the anode in conc. HCl. The cathode compartment is sealed by a greased ground glass joint *G*. The 'indicator solution' in compartment *E* consists of LiCl, and the anode is a platinum foil electrode *I* which is placed inside a tube containing some solid Li_2CO_3 (retained in position by a plug of glass wool, *L*). The object of this arrangement is to ensure that no hydrogen ions escape from the neighbourhood of the anode.

A current of about 2–5 mA produces a convenient rate of movement. A potential of 100–400 V is needed, and may be obtained from batteries or a d.c. supply. The current is best measured by an accurate 0–100 milliammeter. It is convenient to maintain a constant current during the experiment by manual adjustment of a large sliding rheostat *PQ* of about 5 000 Ω, used as a potential divider (Fig. 12A.7) or better by using a triode to maintain a constant current as shown in Fig. 12A.2.

Procedure. Prepare a litre of approximately 100 mol m^{-3} HCl, and add to it sufficient solid methyl orange to give it enough colour to be seen in the moving boundary tube. Standardize this solution accurately by titration with standard alkali or borax. Take a portion of the solution and add to it solid Li_2CO_3 a little at a time, allowing each lot to dissolve before adding the next, until the solution is just definitely yellow. A slight excess is permissible, but too much would make the LiCl solution so prepared too dense. The concentration of this LiCl indicator solution must be adjusted approximately to conform to the relation $(t_+)_{HCl}/c_{HCl} = (t_+)_{LiCl}/c_{LiCl}$. Since $(t_+)_{HCl}$ is roughly 0·8, and $(t_+)_{LiCl}$ is about 0·3, the theoretical concentration of LiCl must be about $0{\cdot}1 \times 0{\cdot}3/0{\cdot}8 = 0{\cdot}04$. For reasons explained above, a slightly less concentrated solution is preferable. The 100 mol m^{-3} LiCl should therefore be diluted with about three times its own volume of distilled water.

The apparatus is first cleaned and rinsed twice with the HCl solution. It is then entirely filled with the HCl solution, and the stopper *G* is inserted and the stopcock *A* is closed. The bore of the stopcock will therefore contain HCl. The HCl is then poured off from vessel *E*, and this part of the apparatus is rinsed with water and then filled with the LiCl solution. Particular attention must be given to see that the section of tube immediately above stopcock *A* is well rinsed with LiCl solution; this part can be washed out with the aid of a glass tube drawn out to a fine jet. The anode assembly is then inserted, as shown. The whole apparatus is set up in a glass-sided tank of water to prevent disturbances of the boundary by temperature change. (The vessel D would act like a thermometer.) The electrical circuit is wired up, and checked for polarity.

To form the boundary, the stopcock A is turned very slowly until fully open. The current should then be switched on at once. A current of ≈ 10 mA is convenient to form the boundary and bring it into the calibrated volume rapidly. It may then be reduced to a constant value between 2 and 5 mA. During the experiment the resistance of the cell increases because HCl in the tube BC is gradually replaced by LiCl which has a lower conductivity; the potentiometer is adjusted to maintain a constant current if electronic regulation is not available.

The boundary—yellow above, red below—should be very sharp. It can be seen best by placing a thin sheet of white paper behind the apparatus and an electric light bulb behind the paper. As soon as the boundary reaches the first graduation on the tube BC, a stop-clock should be started, and subsequently the time should be noted (without stopping the clock) when the boundary reaches each graduation down to the bottom of the tube.

Calculation. Plot a graph of volume reading on the tube BC against time. If the current has been kept constant, this graph should be a straight line, as the theory of the method gives the equation $t_+ = 96\,487\, cV/Q$, where c = conc. of HCl in mol m^{-3}, V = volume in m^3 swept through by the boundary during the passage of Q C (current I in A $\times$ time in seconds). Draw the best straight line through the points and work out the mean rate of movement of the boundary, $\overline{V}$, in $m^3\ s^{-1}$; then $t_+ = 96\,487\, c\overline{V}/I$.

The small correction for volume change at the Ag–AgCl electrode, and the correction for movement of solvent necessary to make transport numbers determined by the moving boundary method strictly comparable with those obtained by the Hittorf method, can be neglected in the present instance.[2]

BIBLIOGRAPHY 12A: Transport numbers

Weissberger, Vol. I, Pt 4, chapter XLV.
MacInnes, *The Principles of Electrochemistry*, 1939 (Reinhold, New York).
Robinson and Stokes, *Electrolyte Solutions*, 2nd edn, 1959 (Butterworths, London).
Davies, *Electrochemistry*, 1967 (Newnes, London).

[1] Franklin and Cady, *J. Amer. Chem. Soc.*, 1904, **26**, 499.
[2] Cf. Robinson and Stokes, *op. cit.*

12B Electrode processes

1. INTRODUCTION

When dilute sulphuric acid is electrolysed with platinum electrodes, the net reaction is the decomposition of water with liberation of hydrogen at the cathode and oxygen at the anode. Since the reaction

$$H_2O(l) \rightarrow H_2(g, 1\text{ atm.}) + \tfrac{1}{2}O_2(g, 1\text{ atm.})$$

is accompanied by an increase of free energy of 237 k J, this is the *minimum* amount of electrical energy theoretically capable of causing the reaction to proceed in a *reversible* manner. Rather more energy is needed to produce electrolysis at a finite rate. Since $\Delta G = -zEF$ (section 13A.1) and $z=2$ for this reaction, it follows that *at least* 1·23 V must be applied. In practice, about 1·67 V is needed to effect electrolysis of 1 000 mol m^{-3} H_2SO_4 with smooth platinum electrodes because the electrodes are not 'reversible'. At lower voltages no gas is evolved, and very little current passes through the cell, as shown diagrammatically in Fig. 12B.1. The point X is called the *decomposition voltage* (or potential). The small current that flows at

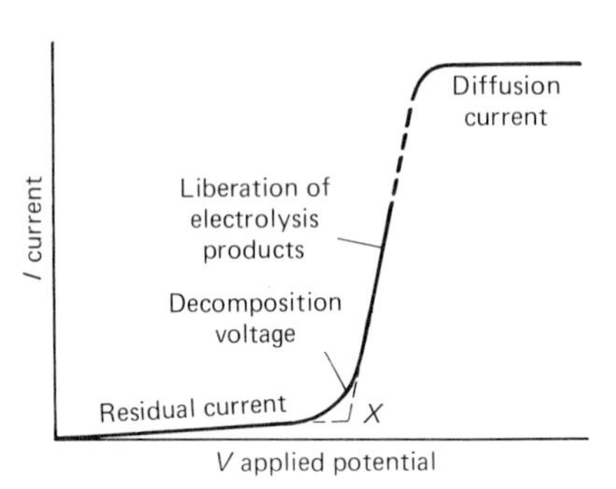

FIG. 12B.1 Current–voltage curve for electrolysis.

applied voltages below X is the *residual current*. It is associated with formation *in solution* of small quantities of hydrogen and oxygen which diffuse away from the electrodes or, in the case of hydrogen, react with dissolved oxygen. When a potential (less than X) is first applied to the electrodes, a relatively large current passes momentarily, but it quickly falls off to the small value indicated in Fig. 12B.1. The cell is then said to be *polarized*. Polarization is caused by accumulation of the products of electrolysis at the surface of the electrodes with the result that the system constitutes a voltaic cell with a *back e.m.f. of polarization* which opposes the applied voltage. If no mechanism existed for removal of polarization products, zero current would then pass. On the other hand, the residual current can be increased by stirring the solution to assist removal of these substances. Similarly, if easily oxidizable or reducible substances (e.g. Fe^{++}, Fe^{+++}) are added to the cell, a progressive electrode reaction can take place with consequent increase of current; this phenomenon is *depolarization* of the electrodes.

When polarization is due solely to accumulation of substance in the neighbourhood of the electrodes it is called *concentration polarization*. Another type of polarization is observed when a gas such as H_2 or O_2 would be the product of electrolysis; a potential markedly greater than the reversible potential must be applied before any gas is liberated. This is the phenomenon of *overvoltage*. Its origin is still controversial, but its practical importance is extensive. For example, the overvoltage of hydrogen on *smooth* platinum is about 0·09 V and that of oxygen about 0·4 V; this is why the decomposition voltage of water is 1·67 V instead of the reversible value of 1·23 V. With a lead or mercury cathode it would be considerably higher because the hydrogen overvoltages of these metals are high. In contrast, the decomposition voltage of HCl with platinized platinum electrodes is close to the reversible value because both H_2 and Cl_2 show negligible overvoltage on platinized platinum. Overvoltages are not easily reproducible as they depend greatly on electrode structure, on current density, and on traces of impurities.

2. DISCHARGE POTENTIALS

As polarization of a cell is the sum of the independent polarization of the two electrodes, it is more informative to study the current-potential curves for each electrode separately rather than that for the cell as a whole. The arrangement shown in Fig. 12B.2 is used. While a current I is being passed through the cell, the potential of the electrode A *with respect to the adjacent solution* is measured by means of a calomel electrode C (section 13B.1) and potentiometer (section 4C(2.3)). The calomel vessel has a special form with a tip L ('Luggin capillary') which can be brought close to the electrode surface so that the measured potential is not complicated by the potential drop ($I \times R$) which necessarily exists between the two electrodes A and B while a current is passing. The tube connecting L to the calomel electrode is filled with KCl soln (1 000 mol m^{-3}) made into a gel (salt bridge), to prevent diffusion into or out of the calomel

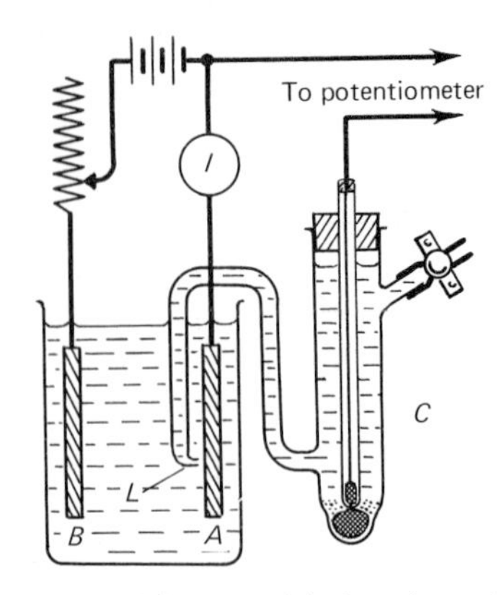

FIG. 12B.2 Determination of the potential of an electrode during electrolysis.

electrode (section 13B). This arrangement permits the determination of *discharge* or *deposition potentials* for single electrode processes.

The two electrodes and solution form a cell whose e.m.f. E normally opposes the applied voltage V so that $V = E + IR$. If an overpotential η is required for electrolysis, E is replaced by the effective decomposition potential for the whole cell, i.e.

$$V = V_{\text{decomp}} + IR = E + \eta_{\text{anode}} + \eta_{\text{cathode}} + IR$$

The e.m.f. of the cell is made up of the e.m.f. of each electrode relative to the solution, i.e. $E = E_{\text{anode}} - E_{\text{cathode}}$. Correspondingly we can write

$$\begin{aligned} V &= (E_{\text{anode}} + \eta_{\text{anode}}) - (E_{\text{cathode}} - \eta_{\text{cathode}}) + IR \\ &= V_{\text{dis, anode}} - V_{\text{dis, cathode}} + IR \end{aligned}$$

where V_{dis} is the discharge potential of each electrode. The voltage measured by the potentiometer in Fig. 12B.2 is $V_{\text{dis}} + V_{\text{calomel}}$; subtraction of the latter (see section 13B) gives V_{dis} relative to a standard hydrogen electrode, whose e.m.f. is conventionally taken as zero (see section 13A).

For a given electrode reaction the overpotential increases with the current density i (A per cm^2 of electrode surface) according to the Tafel equation

$$\eta = b \log_{10} (i/i_0)$$

provided $-0{\cdot}1\ \text{V} \leqslant \eta \leqslant 0{\cdot}1\ \text{V}$ where b and i_0 are constants. The overpotential thus increases exponentially with current density: a plot of

η against $\log_{10} i$ gives i_0 as intercept. Couples with i_0 above about 10^{-6} A cm^{-2} are 'fast' and hence electrochemically reversible; below this value the couples are 'slow', irreversible and show large over-potentials.

By means of an electronic circuit known as the *potentiostat* it is possible to conduct electrolysis at constant electrode potential, measuring the changing current. Equally the current can be maintained constant and the change in voltage followed as electrolysis proceeds. These techniques have facilitated the study of electrode processes, corrosion reactions and electroanalytical techniques.

3. LIMITING CURRENT

When a potential greater than the decomposition voltage is applied to the cell, electrolysis takes place at a rate which increases rapidly with increase of potential (Fig. 12B.2). A limit is set only by the rate at which the proper ions can get to (or the products get away from) the electrode. At relatively high potentials (easily reached with small electrodes and very dilute solutions) the *limiting current* (*diffusion current*, Fig. 12B.2) becomes constant, independent of potential, because the overall rate is controlled by diffusion. The magnitude of the diffusion current is proportional to the concentration of substance which takes part in the electrode reaction. This fact provides the basis of *polarography*, a valuable method of analysing solutions by determination of current–potential curves with a minute platinum electrode or (more often) a dropping-mercury electrode.

4. PASSIVITY

Many practical electrode processes are, of course, complicated by secondary reactions (e.g. sulphate → persulphate, $2Hg + 2Cl^- \rightarrow Hg_2Cl_2$). However, the only other phenomenon that need be mentioned here is that of *passivity*. Fe, Co, Ni, Cr, Mo, W, etc., are metals which will dissolve anodically in dilute HCl provided they are in the *active* condition, but if made the anode in H_2SO_4, HNO_3, etc., they are liable to become *passive* and then subsequently behave as unattackable electrodes. Passivity is due to the formation of a thin, coherent oxide film which is impermeable to the metal ions. The metal can be restored to the active condition by scraping off this film or removing it cathodically or with hot HCl.

EXPERIMENT

Determine current–voltage curves for the following electrodes and electrolytes

	Solution	Cathode	Anode
(*1*)	$CuSO_4$ *100 mol m*$^{-3}$	*copper*	*copper*
(*2*)	H_2SO_4 *100 mol m*$^{-3}$	*platinum*	*platinum*
(*3*)	H_2SO_4 *100 mol m*$^{-3}$	*lead*	platinum
(*4*)	H_2SO_4 *100 mol m*$^{-3}$	platinum	*nickel*

This experiment should be performed with the apparatus shown in Fig. 12B.2. See section 13B for the preparation of the calomel electrode and salt bridge. The electrodes should consist of 1 cm squares of stiff foil, bearing a stout wire by which they can be mounted about 2 cm apart in a small beaker of the solution. The platinum should be freshly platinized, the lead scraped with a knife, and the nickel made active by cleaning with hot conc. HCl and treatment as cathode in dilute H_2SO_4.

Measure the discharge potential V_{dis} as a function of current for each of the electrodes written in italics. It will *usually* be found that the cathode has to be connected to the negative terminal of the potentiometer and the anode to the positive terminal. The mode of connection should be noted.

The current is increased in steps of 5 mA to 20 mA, and then in steps of 20 mA to full-scale deflection of 100 mA. The cell is left for 1–2 min at each stage for the current to become steady, and the potential between the electrode under observation and the reference electrode is then measured. After adjusting the rheostat for the next current, V_{dis} can be calculated and plotted against current whilst the cell is equilibrating. The curve should be retraced reducing the current from its maximum value. Any electrode reactions observed should be noted on the current–voltage curve: record the points at which visible reactions start or cease. Because this experiment is not carried out under conditions of ultra-purity and also because the electrode reaction gradually changes the surface condition of several of the electrodes employed, the potentials will not be reproducible to better than a few millivolts.

Results. Compare the discharge potentials with those expected if no overpotential occurred, i.e. with the values of E obtained from the appropriate equation (section 13A(1)), and hence find the overpotential concerned.

Standard electrode potentials at 25°C:

$$Ni^{++} + 2e \rightleftharpoons Ni \qquad E_0 = -0{\cdot}25\ V$$
$$Cu^{++} + 2e \rightleftharpoons Cu \qquad E_0 = +0{\cdot}34\ V$$
$$O_2 + 4H^+ + 4e \rightleftharpoons 2H_2O \qquad E_0 = +1{\cdot}23\ V$$

Activity coefficients:

$$\gamma_{Cu^{++}}\ (100\ mol\ m^{-3}\ CuSO_4) = 0{\cdot}15$$
$$\gamma_{H^+}\ (100\ mol\ m^{-3}\ H_2SO_4) = 0{\cdot}80$$

Test the Tafel equation for a current–voltage curve showing a large overpotential but little hysteresis.

Since soft metals are generally deposited with negligible overvoltage, the cathode potentials in experiment (1) should correspond to the reversible potential.

Experiment (4) shows initial solution of the Ni anode (confirm with dimethylglyoxime), and after a while a decrease of current as the nickel assumes the passive state; this is subsequently retained at all potentials. Solution of Ni ceases (check) and the only electrode process is evolution of oxygen (with overvoltage) as for a platinum electrode (experiment (2)).

EXPERIMENT

Investigate the electro-oxidation of potassium ferrocyanide; (1) show that the limiting current is controlled by diffusion, and (2) determine the diffusion coefficient of the ferrocyanide ion.

Procedure.[1] Prepare a large silver–silver chloride electrode (section 12A(7)), to be used as unpolarizable cathode of well-defined, constant potential, and two platinum micro-electrodes of the form shown in Fig. 12B.3(a) and (b). The platinum electrodes should be cleaned using cotton wool soaked in carbon tetrachloride, and must not be touched with the fingers. The electrical circuit is shown in Fig. 12B.3(c). The voltmeter V should be capable of reading up to 1 V with a sensitivity of 0·01 V. The maximum sensitivity of the galvanometer should be at least 0·1 μA per mm. A 0–10 microammeter can be used, if available. Another method of measuring these small currents is to place a large standard resistance, e.g. 10 000 Ω, in series with the cell and measure the potential drop across it with a potentiometer or digital voltmeter. The voltmeter reading must be corrected for the potential drop across the resistance.

Prepare 500 cm^3 of approximately 100 mol m^{-3} KCl solution, and

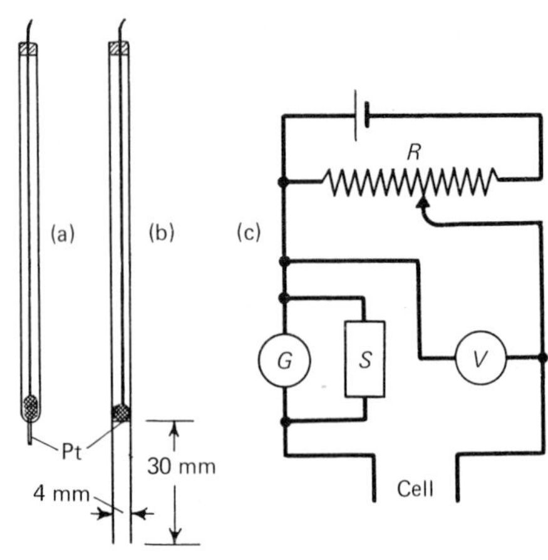

FIG. 12B.3 Apparatus for study of diffusion currents at polarized electrodes. (a), (b) Platinum micro-electrodes. (c) Circuit diagram.

use it to make up 100 cm^3 of solutions 10, 5 and 2·5 mol m^{-3} with respect to potassium ferrocyanide.

(1) Determine current–voltage curves for the KCl base solution, and the three ferrocyanide solutions, using electrode (*a*). Allow 2 min for each reading to become steady, and take readings in steps of 0·1 V up to 1 V. Draw the (corrected) curves, and note the magnitude of the limiting current (at, say, 0·7 V) for each solution, and show that it is proportional to ferrocyanide concentration. Verify that the limiting current is increased by stirring the solution near the electrode. Also, verify by micro-chemical tests that ferrocyanide is oxidized to ferricyanide.

(2) Electrode (*b*) is designed for measuring the *linear* diffusion of ferrocyanide from the bulk of the solution at the lower end of the tube (where the concentration is supposed constant) to the electrode surface where, during electrolysis, the concentration is zero. A diffusion gradient is gradually established along the tube. The ferricyanide solution produced near the electrode has a slightly lower density than the ferrocyanide and therefore tends to keep the column of solution from mixing by convection.

Using the 5 mol m^{-3} ferrocyanide solution, determine a current–time curve with a *constant applied potential* of 0·7 V, i.e. follow the change of diffusion current with time. Start a clock when the current is switched on and take readings every 10 s at first, but then at longer intervals as the current steadies off, continuing the readings for about 30 min. Measure the area of cross-section of the tube.

Make similar determinations with the base solution containing KCl only, and subtract these readings from the limiting current values to obtain the current carried by the ferrocyanide only.

Calculations. Diffusion is controlled by Fick's law, according to which the flow or 'flux' (f) of substance across a given plane is proportional to the concentration gradient. Considering only *linear* diffusion, the position of the plane being denoted by x, the number of moles δN of solute crossing the plane in time δt is given by

$$f = \delta N/\delta t = DA\,\mathrm{d}c/\mathrm{d}x$$

where D is the *diffusion coefficient* (units $m^2\ s^{-1}$), A is the area of cross-section of the diffusion column, and $\mathrm{d}c/\mathrm{d}x$ is the concentration gradient along the direction of x.

In the present system, the concentration gradient is changing with time and is not linear along the tube. It can be shown that if at $t=0$ the concentration at plane $x=0$ is suddenly maintained at $c=0$, then the concentration distribution along a tube, which is long compared with the extent of depletion of the solution, is given by

$$c_{x,t} = c\frac{2}{\sqrt{\pi}}\int_0^z e^{-y^2}\,\mathrm{d}y$$

where c is the original concentration of the solution, y is an integration variable and $z=x/2\sqrt{Dt}$. Now, in the present experiment, the diffusion current is controlled by the rate at which ferrocyanide arrives at the platinum electrode surface, which is the plane $x=0$. Every mole arriving accounts for 1 faraday of current. Therefore, by Fick's law, the current at time t is given by

$$I_t = FAD\left(\frac{\partial c}{\partial x}\right)_{x=0,\,t}$$

The gradient near the electrode is obtained by differentiating the above expression. Thus,

$$\left(\frac{\partial c}{\partial x}\right)_{x=0} = \frac{c}{\sqrt{\pi Dt}}$$

and hence,

$$I_t = FcA\sqrt{D/\pi t}$$

This last expression should represent the change of diffusion current with time. The data can be tested for conformity to the diffusion theory by plotting I_t against $\sqrt{1/t}$ which should give a straight line of slope equal to $\boldsymbol{F}cA\sqrt{D/\pi}$. Hence D can be calculated. (Note that c should be expressed in mol m^{-3}, A in m^2, and I in A to obtain D in $m^2\ s^{-1}$.)

For comparison, an approximate value of the diffusion coefficient of an oxidizable or reducible ion in presence of excess of indifferent salt can be deduced from its equivalent conductance at infinite dilution, Λ_0, since

$$D \approx \frac{\boldsymbol{R}T\Lambda_0}{z\boldsymbol{F}^2} \quad \text{(Nernst, 1888)}$$

In the case of the ferrocyanide ion, $z=4$, and $\Lambda_0 = 0{\cdot}1105\ \Omega^{-1}\ m^2\ mol^{-1}$ at 25°C.

5. THE ROTATING DISC ELECTRODE[2]

If the electrode is rotated in an electrolyte under conditions where the current is diffusion controlled, diffusion occurs only across a thin *diffusion layer* of thickness δ on the surface of the electrode, of the order of 10^{-3} cm. Diffusion across this layer is governed by $dc/dx = (c_0 - c_s)/\delta$ where c_0 and c_s are the bulk solution concentration and the concentration at the surface of the electrode. From Faraday's laws

$$I = DA\boldsymbol{F}\frac{di}{dx} = DA\boldsymbol{F}(c_0 - c_s)/\delta$$

The maximum diffusion current occurs when $c_s = 0$ and I (diff.) = $DA\boldsymbol{F}c_0/\delta$. The limiting current at high potentials is thus proportional to the concentration in the bulk solution; this is used in polarography, amperometry and other analytical techniques.

Levich has shown that δ is uniform over the surface of a rotating disc electrode, and is given by $1{\cdot}613D^{\frac{1}{3}}v^{\frac{1}{6}}\omega^{-\frac{1}{2}}$ where v is the kinematic viscosity, i.e. the viscosity divided by the density, and ω is the angular velocity of the disc in rad s^{-1}. Then

$$I\,(\text{diff.}) = 0{\cdot}62D^{\frac{2}{3}}A\boldsymbol{F}c_0v^{-\frac{1}{6}}\omega^{\frac{1}{2}} \qquad (1)$$

EXPERIMENT

A suitable rotating disc electrode is a 1 mm platinum wire set in a teflon former and rotated at 200 r.p.m. This acts as anode. The circuit, *cleaning of the electrode, and solutions are all described in the preceding experiment (section 12B(4)). Determine the current–voltage curve for $V=0$ to 1·2 V in 0·1 V steps for (i) 100 mol m*$^{-3}$ KCl *soln and (ii) 10, 5 and 2·5 mol m*$^{-3}$ *potassium ferrocyanide in 100 mol m*$^{-3}$ KCl *solution, using a single speed of rotation, and single galvanometer scale for all the readings. Repeat with one ferrocyanide solution at a different speed.*

Results. The limiting (plateau) currents increase slightly with applied potential due to impurities: correct the observed values by subtracting the corresponding current observed with the 100 mol m^{-3} KCl solution alone. Check that the limiting current (a) varies with the square root of the speed of rotation and (b) is proportional to the concentration of potassium ferrocyanide at constant speed. Calculate the diffusion coefficient from the last equation above (1), taking the viscosity and density as those of water at the same temperature. This result may be compared with those obtained in the preceding experiment (section 12B(4)) both experimentally and from Nernst's equation. Also calculate the the diffusion thickness δ.

6. THE DROPPING-MERCURY ELECTRODE (POLAROGRAPH)

In place of (or in addition to) the above experiments, one may study all the principal electrode phenomena with a dropping-mercury electrode. The apparatus and the theory are more complicated than with a stationary electrode, but the dropping-mercury electrode is used in polarography because it gives a constantly renewed electrode surface, and a reproducible average diffusion current. Also, as the hydrogen overvoltage on mercury is high, electroreductions can be carried out at very negative cathode potentials. Further, even alkali metals can be deposited because they amalgamate with the mercury.

It is not necessary to have an elaborate recording polarograph, although, of course, this instrument is very advantageous for routine analytical work. Current–voltage curves ('polarograms') can be obtained satisfactorily with the simple electrical apparatus used in the preceding experiment together with an electrode assembly such as that shown in Fig. 12B.4.

A is a reservoir containing specially purified mercury. *B* is a platinum electrode for contact. *C* is a length of polyvinyl chloride tubing (which is preferable to rubber). The jet *D* consists of about

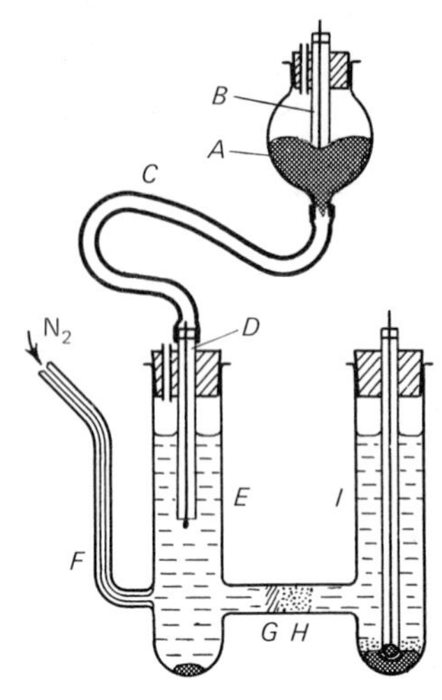

FIG. 12B.4 Cell for determination of polarograms with a dropping-mercury electrode.

7–8 cm of fine-bore thermometer tubing. The apparatus must give a drop-time in dilute KCl of 3–5 s. The cell must provide a tube *F* for bubbling nitrogen to remove dissolved oxygen, and a connection to a satd. KCl calomel electrode *I* of large capacity. The cell illustrated in Fig. 12B.4 has a sintered glass disc *G* and agar plug *H* to separate the test solution in the limb *E* from the KCl of the calomel electrode. It is usually necessary to add a trace of pure gelatine to the solution in *E* in order to obtain smooth diffusion current curves. The oscillations of the current as the drops form can be greatly reduced by connecting an electrolytic condenser of large capacity, e.g. 2 000 μF, in parallel with the galvanometer.

A large number of phenomena can be studied with the dropping-mercury electrode, but for details of special techniques of polarography reference should be made to the monograph by Kolthoff and Lingane.[3] The following polarograms may be determined as examples, using 100 mol m^{-3} KCl + 0·1% gelatine as base solution: (a) deoxygenated base solution, (b) base solution satd with air, (c) satd with oxygen, (d) +1 mol m^{-3} $Pb(NO_3)_2$, (e) +0·5, 1 and 2 mol m^{-3} $CdSO_4$, (f) +1 mol m^{-3} $Pb(NO_3)_2$ +1 mol m^{-3} $CdSO_4$, (g) +1 mol m^{-3} $FeCl_3$ (gives double 'wave'), (h) +1 mol m^{-3} saccharin, (i) amperometric titration of 50 cm^3 of 10 mol m^{-3} lead nitrate with 50 mol m^{-3} potassium dichromate at a fixed potential of −1·0 V with respect to satd calomel electrode.

BIBLIOGRAPHY 12B: Electrode processes

Weissberger, Vol. I, Pt 3, chapter XXXIII.
Davies, *Electrochemistry*, 1967, chapters 15–17 (Newnes, London).
Stock, *Amperometric Titrations*, 1965 (Wiley-interscience, New York).
Delahay, *New Instrumental Methods in Electrochemistry*, 1954 (Wiley, New York).
Potter, *Electrochemistry: Principles and Applications*, 1956 (Cleaver-Hume, London).

[1] Cf. Laitinen and Kolthoff, *J. Amer. Chem. Soc.*, 1939, **61**, 3344.
[2] Riddiford, *Adv. Electrochem. Electrochem. Engng.*, 1966, **4**, 47.
[3] Kolthoff and Lingane, *Polarography*, 2nd edn. 2 vols., 1952 (Interscience, New York).

13
Electromotive force of cells

13A Theory of cells

1. INTRODUCTION

Voltaic cells play an important part in physical chemistry because they provide a convenient means of measuring directly the *free energy change* accompanying certain reactions. When a voltaic cell is discharged, the free energy of the *cell reaction* becomes available for performing useful work. The *maximum useful work* is obtainable in theory only by discharging the cell infinitely slowly so that its maximum electromotive force (e.m.f.) is exerted. In practice, this means that the open-circuit e.m.f. of the cell must be measured; this can be done by determining the e.m.f. that must be applied in opposition to that of the cell in order that no current shall flow when the circuit is closed.

If the e.m.f. of a cell (on open circuit) is E volts (V) and Q coulombs (C) of charge pass, the maximum amount of electrical work done is EQ volt-coulombs, i.e. EQ joules (J). If the cell reaction involves the passage of n moles of electrons, $Q=n\boldsymbol{F}$, where $\boldsymbol{F}$ the Faraday, is the charge of 1 mole of electrons. The decrease in free energy of the chemical reaction, $-\Delta G$, is equal to the maximum electrical work obtained, $n\boldsymbol{F}E$.

A study of the e.m.f. of the cell at different temperatures provides a value for the entropy change in the reaction (ΔS), since $\partial \Delta G/\partial T = -\Delta S$; the heat of the reaction can also be calculated, using the equation $\Delta G=\Delta H-T\,\Delta S$. This method of obtaining thermodynamic quantities is far more convenient than the laborious method of thermal data or the method of studying chemical equilibria, but is restricted, of course, to reactions involving an electron transfer. Also, in practice, relatively few satisfactory ('reversible') electrodes are available for employment in cells.

In many applications of voltaic cells the e.m.f. measurements are used to obtain information about the constitution of solutions—for example, the concentration of hydrogen ions. The relationship between constitution and e.m.f. is of great importance, and may be derived by thermodynamics in various ways; one derivation is given in the section that follows. The theory of electrode potentials was first given in approximate form by Nernst in 1889, who obtained the formula

$$E_M = \text{constant} + \frac{RT}{zF} \log_e C_M$$

for the e.m.f. of a metal electrode (z=valency, C_M=concentration of metal ions), and

$$E_\mathrm{H} = \text{constant} + \frac{RT}{F} \log_e \frac{[\mathrm{H}^+]}{p_{\mathrm{H}_2}^{\frac{1}{2}}}$$

for the e.m.f. of a hydrogen electrode. ($[\mathrm{H}^+]$=concn of hydrogen ions, p_{H_2}=pressure of hydrogen gas.) The Nernst formulae were used for many years, but, as will be seen below, the exact theory involves thermodynamic *activities* rather than concentrations.

2. GENERAL THERMODYNAMIC THEORY OF ELECTRODE POTENTIALS

Every voltaic cell consists of two *electrodes* or 'half-cells' which are connected internally by the electrolyte solution and externally by the leads and potential-measuring instruments. A potential difference exists between each electrode and the solution. Although these 'single electrode potentials' E_1 and E_2 are not susceptible to measurement separately, their algebraic difference is the measured e.m.f. of the cell, i.e. $E = E_1 - E_2$. E_1 and E_2 may be either positive or negative quantities with respect to the solution, depending on the nature of the electrode. Their e.m.f.'s may, therefore, either oppose one another (if they are of the same sign) or reinforce (if they are of opposite sign), and the algebraic difference will determine which electrode becomes the positive pole of the cell.

At each electrode an electrochemical equilibrium is set up of the type ox.$+ne \leftrightharpoons$ red., e.g. $\mathrm{Fe}^{+++} + e \leftrightharpoons \mathrm{Fe}^{++}$. If the half cell is imagined to discharge infinitely slowly so the system remains at equilibrium, the electrical energy produced equals the chemical free energy released by the reaction. The electrical work is $nE_1\boldsymbol{F}$(J). The free energy produced by converting 1 mole of ox. into red. is $(\mu_{\text{ox.}}+n\mu_e-\mu_{\text{red.}})$ where μ is the *chemical potential* (partial molar free energy) of the species. The chemical potentials are related to the thermodynamic activities (a) by the equation $\mu=\mu^\circ+\boldsymbol{R}T\log_e a$, where μ° is the chemical potential of the substance in its *standard state*, defined as that for which the activity is unity. The chemical potential of the electrons on the electrode, μ_e, depends only on temperature for a given electrode, so

$$(\mu_{\text{ox.}}+n\mu_e-\mu_{\text{red.}}) = nE_1\boldsymbol{F},$$

and

$$E_1 = \left(\frac{(\mu_{\text{ox.}}+n\mu_e-\mu_{\text{red.}})}{n\boldsymbol{F}}\right)+\frac{\boldsymbol{R}T}{n\boldsymbol{F}}\log_e\left(\frac{a_{\text{ox.}}}{a_{\text{red.}}}\right)$$

The term $(\mu_{\text{ox.}}+n\mu_e-\mu_{\text{red.}})/n\boldsymbol{F}$ is a constant, and clearly equal to the value that E_1 would assume for an electrode in which $a_{\text{ox.}}=1$ and $a_{\text{red.}}=1$, i.e. with the reactant and product in their standard states. However, μ_e is unknown so the value of E_1 cannot be calculated; nor can its value be measured in isolation from another half cell. This difficulty is overcome by arbitrarily taking the e.m.f. of the standard hydrogen electrode (see below) as zero. The standard electrode potential of a half cell, E_1°, is then the e.m.f. of the electrode with all the species at unit activity relative to that of a standard hydrogen electrode; E_1° is the e.m.f. of a cell with the half cell concerned as one electrode and a standard hydrogen electrode (sections 13A(6), 13D) as the other.

With other concentrations, these electrodes have an e.m.f. given by

$$E_1 = E_1^\circ+\frac{\boldsymbol{R}T}{n\boldsymbol{F}}\log_e\left(\frac{a_{\text{ox.}}}{a_{\text{red.}}}\right)$$

Specific examples of the application of this formula to various practical electrodes are given later.

It will be noticed that e.m.f. measurements primarily give information about thermodynamic *activities*; however, these can often be related to the *concentrations* of the substances concerned (see below).

3. SIGN CONVENTIONS

No difficulty arises for the potential of an actual electrode with respect to a standard hydrogen electrode placed in the solution. If for example, the electrode is a zinc rod in a solution of zinc ions of unit activity, the zinc has a potential of $-0{\cdot}76$ V with respect to a standard hydrogen electrode, i.e. it must be connected to the negative terminal of a potentiometer and the hydrogen electrode to the positive electrode.

However standard electrode potentials can be defined in two ways, according to the way the half-reaction is written. The accepted way is to write ox. $+ne \rightarrow$ red., or for the zinc electrode Zn^{++} $(a=1)+H_2$ $(g,$ 1 atm$)=$Zn (metal)$+2H^+$ $(a=1)$. If ΔG° is the free energy change for this reaction, i.e. G° (products)$-G^\circ$ (reactants), then the standard electrode potential is defined as $E_1^\circ = -\Delta G^\circ/n\boldsymbol{F}$. This is a reduction potential. The corresponding Nernst equation is

$$E_1 = E_1^\circ + \frac{\boldsymbol{R}T}{n\boldsymbol{F}} \log_e \left(\frac{\text{activity of oxidant}}{\text{activity of reductant}}\right)$$

(as given above). In this case

$$E_1 = E_1^\circ + \frac{\boldsymbol{R}T}{2\boldsymbol{F}} \log_e \frac{a(\mathrm{Zn}^{++})}{a(\mathrm{Zn})} = E_0 + \frac{\boldsymbol{R}T}{2\boldsymbol{F}} \log_e a(\mathrm{Zn}^{++})$$

since the activity of a pure solid is taken as unity. The reduction potential has the same sign as that of the actual electrode relative to a standard hydrogen electrode.

The alternative definition of electrode potential is the oxidation potential red. $\rightarrow$ ox. $+ne$, or for our example Zn $\rightarrow$ $Zn^{++}+2e$, i.e. Zn (metal)$+2H^+$ $(a=1)=Zn^{++}$ $(a=1)+H_2$ $(g,$ 1 atm). Both ΔG° for this reaction and $E_1^\circ = \Delta G^\circ/n\boldsymbol{F}$ are equal in magnitude but of opposite sign to ΔG° and E° defined by reduction potentials. This convention was used in the U.S.A. and still appears in some older textbooks.

For the overall e.m.f. of a cell, see section 13A(7).

4. ACTIVITY COEFFICIENTS OF SALTS

It has been mentioned above that electrode potentials are dependent upon the thermodynamic activity of the ions in solution. In general the activity of any species i differs somewhat from its concentration by a factor f_i called the activity coefficient; thus, $a_i = m_i f_i$ and $f_i \rightarrow 1$ as $m_i \rightarrow 0$. For e.m.f. measurements the concentration is expressed as the molality $m=$ mol (kg solvent)$^{-1}$ rather than $c=$ mol m^{-3}. For dilute solutions $c \approx md$ where d is the density of the solution (kg m^{-3}).

Activity coefficients can be measured in a number of independent ways e.g. from f.p. depressions and from the e.m.f. of cells. In the case of a non-electrolyte, e.g. alcohol in water, the factor f is unambiguous, but with electrolyte solutions care is needed in defining and applying activity coefficients. The usual treatment is therefore given here briefly.

The chemical potential of a salt is regarded as the sum of the chemical potentials of its constituent ions. Thus, for a uni-univalent (1:1) electrolyte $\mu_{salt} = \mu_+ + \mu_-$. The activity of a salt is therefore the *product* of the activities of its ions, $a_{salt} = a_+ \times a_-$. Since a_{salt} is a quantity that can be measured (for example, from f.p. depression determinations), whereas there is no sound method of splitting it into a_+ and a_-, it is usual to work in terms of the *mean ion activity* $a_\pm$ defined by $a_\pm = \sqrt{a_{salt}}$. Similarly, although individual ions are regarded as having individual activity coefficients, f_+ ($= a_+/m_+$) and f_- ($= a_-/m_-$), these are not separately determinable, but a mean value $f_\pm$, defined by $\sqrt{f_+ f_-}$, is obtainable, since $a_{salt} = a_+ a_- = (f_+ m_+)(f_- m_-) = f_\pm^2 m_+ m_- = f_\pm^2 m_\pm^2$, where $m_\pm$ is the mean molality ($= \sqrt{m_+ m_-}$). The formulae for salts of other valence types are analogous although somewhat more complicated.[1] *In the case of electrolytes the activity coefficients generally tabulated are the mean ion activity coefficients* $f_\pm$. Many experimentally determined activity coefficients are now available.[2] Some values for a number of common electrolytes are given in Table A10 in the Appendix. The values are quoted for 25°C, but they do not change much with temperature.

5. ESTIMATION OF ACTIVITY COEFFICIENTS

A theory of the activity coefficients of salts has been given by Debye and Hückel (1923). The theory leads to the following approximate expression for the activity coefficient f of an ion in a *very dilute* electrolyte solution:

$$-\log_{10} f \approx 0{\cdot}51 z^2 \sqrt{I_m}$$

where z is the valency of the ion and I_m is the *ionic strength* of the solution, defined by $I_m = \frac{1}{2} \sum m_i z_i^2$ mol (kg solvent)$^{-1}$; the summation is for every type of ion (molality m_i and valency z_i) present in the solution. In the case of a pure 1:1 salt, ionic strength is equal to molality, but not in other cases. Note that if I is defined by $I_c = \frac{1}{2} \sum c_i z_i^2$ where c_i is in mol m^{-3}, the activity coefficient is given by $-\log_{10} f \approx (0{\cdot}51/\sqrt{1\,000}) z^2/\sqrt{I_c} = 0{\cdot}0161 z^2/\sqrt{I_c}$.

The simple Debye–Hückel formula holds up to about 10 mol m^{-3} for 1:1 salts. For higher concentrations more complicated expressions have been derived, but they are of very limited usefulness. The theoretical formulae are not satisfactory for trivalent and polyvalent ions.

It is often necessary to make an estimate of the value of the activity coefficient of an ion, in the absence of experimental values for it. If the ionic strength is less than 10 mol m^{-3} the simple Debye–Hückel formula may be used. A more reliable value is obtainable from a formula of Güntelberg[3] namely, for a salt of valence type $z_+:z_-$, $\log f_\pm = -0{\cdot}51z_+|z_-|\sqrt{I_m}/(1+\sqrt{I_m})$. This gives a useful representation of the *average* behaviour of salts up to $I=0{\cdot}1$, and is as good a *guess* as any other.

A still better value can be obtained if it is possible to calculate the empirical constant β in the extended Güntelberg equation:[4] namely

$$\log f = -0{\cdot}51z_+|z_-|\sqrt{I_m}/(1+\sqrt{I_m})+2\beta\bar{v}$$

where $\bar{v}$ is given by $2/\bar{v}=1/v_+ + 1/v_-$. (Here v_+ and v_- are the molalities of cation and anion respectively.) The constant β can be calculated provided the activity coefficient of at least one solution of the salt has been determined. A list of values recommended by Guggenheim[5] is given in Table 13A.1.

The Güntelberg–Guggenheim formula can be used for estimating activity coefficients in salt mixtures.

TABLE 13A.1 Empirical β constants for calculation of activity coefficients from the Güntelberg–Guggenheim equation*

Electrolyte		*Electrolyte*	
HCl	0·24	KNO_3	−0·21
NaCl	0·12	Na_2SO_4	−0·39
KCl	0·07	K_2SO_4	0·00
LiCl	0·20	$MgSO_4$	0·00
TlCl	0·35	$CuSO_4$	−1·5
$KClO_4$	−0·44	$La(NO_3)_3$	2·25
$NaNO_3$	0·00	$CaCl_2$	0·75

* Guggenheim, *Phil. Mag.*, 1935, **19**, 588; 1936, **22**, 322.

6. TYPES OF ELECTRODE AND CELL

Although in principle any reaction involving transference of electrons is capable of providing an electrode, comparatively few reliable electrode systems are available. To be satisfactory an electrode must be *electrochemically reversible*; that is to say, if an external potential *slightly* greater than that of the electrode is applied in one direction the electrode reaction must proceed quantitatively from left to right and if the potential is reversed the reaction must be quantitatively reversed. Some possible electrodes are imperfectly reversible because side reactions occur. The following list includes most of the more satisfactory practical electrodes.

6.1. *Metal electrodes.* A metal dipping into a solution of one of its salts. The electrode reaction is $M \rightleftharpoons M^{z+} + ze$ and the electrode potential

$$E_M = E^\circ_M + \frac{RT}{zF}\log_e \frac{a_{M^{z+}}}{a_M}.$$

If the electrode consists of pure metal M, $a_M = 1$. (Examples: silver, mercury, lead, zinc, cadmium, copper.)

In *alloy* electrodes, the metal M is in solution (solid or liquid) at an activity a_M less than unity. (Examples: cadmium amalgam—used in Weston cells, alkali metal amalgams.)

A modified type of metal electrode is one in which the concentration of metal ion in solution is controlled by keeping the solution saturated with a sparingly soluble salt of the metal, M^+A^-. (Examples: the calomel electrode, the silver–silver chloride electrode.) By the common ion effect, the concentration (or, rather, activity) of the metal ion is inversely proportional to the concentration (activity) of the anion of the salt. Such electrodes are therefore effectively reversible with respect to the anion, their potential being given by an expression of the form

$$E = E^\circ + \frac{RT}{zF}\log_e \frac{1}{a_{A^-}}$$

where A^- is the relevant anion.

6.2. *Gas electrodes.* The reaction $H_2 \rightleftharpoons 2H^+ + 2e$ is catalysed by the surface of platinum, palladium or gold, the metal acting as acceptor of the electrons and therefore assuming a potential given by

$$E_H = E^\circ_H + \frac{RT}{2F}\log_e \frac{a^2_{H^+}}{p_{H_2}} = E^\circ_H + \frac{RT}{F}\log_e \frac{a_{H^+}}{p^{\frac{1}{2}}_{H_2}}$$

As mentioned above, it is the recognized convention to take the standard hydrogen electrode (having $p_{H_2} = 1$ atm., $a_{H^+} = 1$) as the arbitrary zero of electrode potential. Thus, on this convention,

$$E_{\mathrm{H}}^{\circ} = 0 \quad \text{and} \quad E_{\mathrm{H}} = \frac{\boldsymbol{RT}}{\boldsymbol{F}} \log_{\mathrm{e}} \frac{a_{\mathrm{H}^+}}{p_{\mathrm{H}_2}^{\frac{1}{2}}}.$$

In practice, other gas electrodes (chlorine, oxygen) are less satisfactory than the hydrogen electrode.

6.3. *Oxidation-reduction* (*'redox'*) *electrodes*. Chemically unreactive metals such as platinum can also serve as electron acceptors (and therefore potential-indicators) for reactions involving solutes in two states of oxidation. For example, if the electrode reaction is $X_n{}^+ \leftrightarrows X^{(n+z)+} + z\,\mathrm{e}$ the electrode potential is given by

$$E_r = E_r^{\circ} + \frac{\boldsymbol{RT}}{z\boldsymbol{F}} \log_{\mathrm{e}} \frac{a_{X^{(n+z)+}}}{a_{X^{n+}}}.$$

Examples:

$Fe^{++} \rightleftarrows Fe^{3+} + e$; $Sn^{++} \rightleftarrows Sn^{4+} + 2\,e$;
$Ce^{3+} \rightleftarrows Ce^{4+} + e$; quinone–hydroquinone (see below).

6.4. *Membrane electrodes*. If a membrane is selectively permeable to one kind of ion only, it can be used to compare the activities of the ion in two solutions placed on opposite sides of the membrane by means of measurements of the 'membrane potential'. The most important example is the *glass electrode*, which is selectively permeable to hydrogen ions and is therefore used for comparing the pH values of solutions. The theoretical membrane potential is given by

$$E = \frac{\boldsymbol{RT}}{\boldsymbol{F}} \log_{\mathrm{e}} \frac{(a_{\mathrm{H}^+})_1}{(a_{\mathrm{H}^+})_2}$$

7. CELLS

A cell may be constructed by joining the electrolytes of any two half-cells. Clearly two types of cell may be distinguished: (a) those in which the two half-cells employ the same electrolyte solution, but, of course, have different electrodes, e.g. $H_2(Pt)|H^+$ (in 100 mol m^{-3} HCl), with $Hg|Hg_2Cl_2$ in 100 mol m^{-3} HCl, and (b) those in which the two half-cells contain *different* electrolyte solutions, which therefore have to be connected together at a *liquid junction*, e.g. the cell $H_2(Pt)|H^+$ (in 100 mol m^{-3} H_2SO_4) $\|$ 100 mol m^{-3} KCl, $Hg_2Cl_2|Hg$. The double vertical line denotes the liquid junction.

The sign convention for half cells has been described above. The sign of a complete cell is given as $E = E_{\text{r.h.s.}} - E_{\text{l.h.s.}}$ where $E_{\text{r.h.s.}}$ is the e.m.f. of the second (right-hand) electrode of the cell as written down (section 13A.3). If E is positive, the r.h. electrode is then positive with respect to the l.h. electrode and vice versa.

8. LIQUID JUNCTION POTENTIALS

Whereas the e.m.f. of cells 'without junction' (sometimes termed 'without transport') is unambiguously given by the algebraic difference between the separate electrode potentials, the presence of a liquid junction may introduce an additional source of potential known as the *liquid junction potential* E_j (or 'diffusion potential', since it arises from the unequal rates of diffusion of the ions in the electrolytes at the junction). The net e.m.f. of the cell is then $E = E_1 - E_2 + E_j$.

The magnitude of liquid junction potentials can be calculated in certain simple cases. For example, at a junction between two solutions consisting of different concentrations of the same uni-univalent salt, it is given by

$$E_j = \frac{RT}{F} \cdot \frac{(u-v)}{(u+v)} \log_e \frac{a_1}{a_2}$$

where u and v are the *mobilities* of the cation and anion respectively, and a_1 and a_2 are the activities of the salt in the two solutions. In many cases, however—notably when the two solutions contain different electrolytes—the liquid junction potential cannot be calculated. If it were neglected it might lead to unknown errors in the results. The usual practice is to reduce, 'swamp' or 'eliminate' the liquid junction potential by joining the solution via a 'salt bridge' consisting of a concentrated solution of a salt having $u \approx v$, usually KCl or KNO_3. Since the theory indicates that E_j arises through unequal diffusion of anion and cation, the presence of a high concentration of ions of equal mobility is thought to reduce the junction potential to a negligible value. Normally E_j is only a few millivolts, but in the important case of junctions involving *acids* it may be much more, because H^+ has a much higher mobility than other ions. The degree of success attending the use of salt bridges with rather concentrated acid solutions is not known, and this uncertainty limits the accuracy and significance of pH measurements.

BIBLIOGRAPHY 13A: Theory of cells

Potter, *Electrochemistry: Principles and Applications*, 1956 (Cleaver-Hume, London).
Davies, *Electrochemistry*, 1967 (Newnes, London).
Kortüm, *Treatise of Electrochemistry* (translated from the German), 2nd ed, 1965 (Elsevier, Amsterdam).
Harned and Owen, *Physical Chemistry of Electrolytic Solutions*, 3rd edn, 1958 (Reinhold, New York).

[1] See Harned and Owen, *op. cit.*
[2] Landolt-Bornstein Tables; Stokes and Robinson, *Trans. Faraday Soc.*, 1949, **45**, 612.
[3] Güntelberg, *Z. physik. Chem.*, 1926, **123**, 243.
[4] Guggenheim, *Phil. Mag.*, 1935, **19**, 588.
[5] Guggenheim, *Thermodynamics*, 1957 (North Holland, Amsterdam).

13B Reference electrodes and the salt bridge

1. MEASUREMENT OF ELECTRODE POTENTIALS

In order to measure the conventional potential between an electrode and a solution, it is necessary to have another electrode and solution, the potential difference between which is known. As a convenient sub-standard electrode the normal calomel electrode is generally used, because the standard hydrogen electrode is not easy to set up (section 13D). The calomel electrode consists of mercury in contact with a solution of potassium chloride saturated with mercurous chloride. As a vessel to contain the mercury and the solution, one may use a small, wide-mouthed bottle but one of the most convenient forms of vessel is shown in Fig. 13B.1.

Preparation of the calomel electrode. First prepare the 1 000 mol m^{-3} solution of potassium chloride, using pure recrystallized potassium

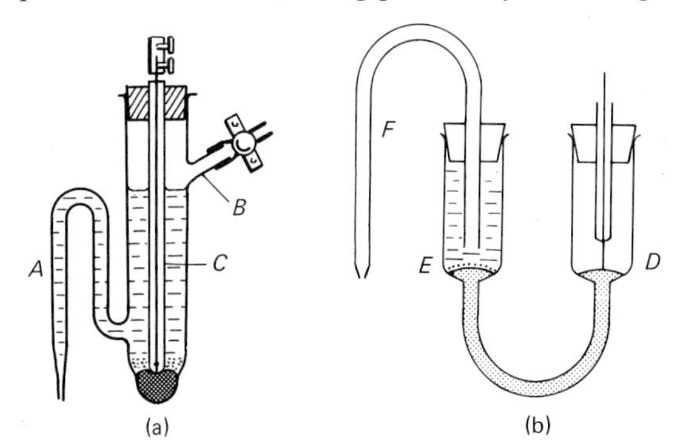

FIG. 13B.1 Convenient forms of the calomel electrode, (a) without and (b) with a salt bridge.

chloride (dried) for the purpose. In the bottom of the electrode vessel, previously thoroughly dried or washed out with the solution of potassium chloride, place a small quantity (1–2 cm^3) of pure mercury, and over this a *thin* layer of calomel paste prepared by rubbing together in a mortar calomel and mercury with some of the solution of potassium chloride. It is then washed two or three times with a quantity of the potassium chloride solution, the mixture being allowed each time to stand until the calomel has settled, and the solution then decanted off.

Having placed the mercury and the calomel paste in the tube insert a rubber stopper or paraffined cork carrying the glass tube *C* (Fig. 13B.1) with platinum wire, which must be immersed in the mercury at the bottom of the tube. The vessel is then filled with the solution of potassium chloride by sucking in the solution through the bent side tube *A*, and then closing the rubber tube on *B* with a clip.

Figure 13B.1(b) shows a form of calomel electrode with salt bridge (see below). The U-tube is filled with mercury. One limb has a dry platinum contact *D*; the other contains KCl solution poured over calomel powder (*E*). A salt bridge (*F*) containing KCl–agar (see below) is placed in the KCl solution.

Since the usual convention is to take the standard hydrogen electrode rather than the calomel as the zero of electrode potential, it is necessary to know the electrode potential of the calomel half-cell on the hydrogen scale. This has been determined from cells consisting of a calomel electrode coupled with a hydrogen electrode. The following values[1] may be adopted for calomel electrodes containing three different concentrations of KCl. (Since earlier workers gave somewhat different values, it is desirable to state the value which is adopted for the e.m.f. of the calomel electrode when reporting measurements in which calomel half-cells have been used.) t is in °C.

$$100 \text{ mol m}^{-3} \text{ KCl calomel electrode: } +0{\cdot}336-0{\cdot}00006\,(t-25)$$

$$1\,000 \text{ mol m}^{-3} \text{ KCl calomel electrode: } +0{\cdot}283-0{\cdot}0002\,(t-25)$$

$$\text{satd. KCl calomel electrode: } +0{\cdot}244-0{\cdot}0007\,(t-25)$$

Other sub-standard reference electrodes sometimes used are the silver–silver chloride electrode (section 13C), the quinhydrone—'standard acid'—cell (section 13F) and mercury–mercuric oxide cell.

It is relevant to mention here that discrepancies in the literature between values reported for these electrode potentials arise largely from the uncertainty of liquid junction potentials in the different cells used by various workers.

2. CONSTRUCTION OF CELLS: THE SALT BRIDGE

When both electrodes of a cell have the same electrolyte solution it is immaterial how the halves are connected; the two electrodes may be either in the same vessel or in separate compartments connected by a tube containing the electrolyte. With all other cells, however, a 'liquid junction' is inevitably present at some point, and the usual practice is to interpose a 'salt bridge' between the two solutions to minimize the liquid junction potential. The salt bridge usually consists of a saturated solution of KCl, but if chloride is objectionable, as with silver salts, ammonium nitrate or potassium nitrate is used.

Since the form of the junction can affect the junction potential, it is desirable to have a symmetrical plane of contact between the solutions, with diffusion as the only cause of mixing. It is usual to connect two half-cells, containing relatively dilute solutions, by dipping the connecting tubes of both electrode vessels into a small beaker containing the concentrated salt-bridge solution, as in Fig. 13B.2(a).

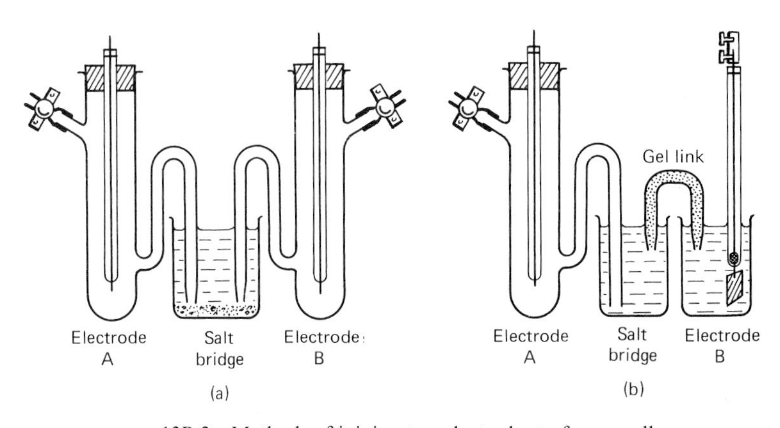

FIG. 13B.2 Methods of joining two electrodes to form a cell.

When this arrangement is not applicable because the electrode vessel has to be open to the air and syphoning would take place (as in potentiometric titrations), a *solidified* link consisting of a conducting jelly is used to make electrical contact, as in Fig. 13B.2(b). A suitable jelly is prepared by dissolving about 3 g of KCl and 0·3 g of agar powder in 10 cm^3 of water; the constituents are heated in a small beaker on a steam-bath, and when a clear solution is obtained it is sucked or poured while hot into U-tubes made from narrow quill tubing. Great care must be taken to avoid air bubbles in the bridge which will break the circuit. For the same reason the gel should protrude slightly from the ends of the capillary. The mixture sets to a gel on cooling. The tubes can be stood in saturated KCl solution until required. After use they must be renewed if there is any possibility of contamination having occurred. Agar gels with KNO_3 or NH_4NO_3 in place of KCl can be used where necessary.

BIBLIOGRAPHY 13B: Reference Electrodes and the salt bridge.
See bibliography 13A.
Ives and Janz (eds), *Reference Electrodes: Theory and Practice*, 1962 (Elsevier, Amsterdam).

[1] Hitchcock and Taylor, *J. Amer. Chem. Soc.*, 1937, **59**, 1812; MacInnes, Belcher and Shedlovsky, *ibid.*, 1938, **60**, 1094.

13C Metal electrodes

1. THE SILVER ELECTRODE

The Ag|Ag^+ electrode is one of the most convenient and reproducible metal electrodes, because silver is readily obtainable in a state of high purity, is little affected by oxygen and, being soft, is not much influenced by strain and other physical factors such as grain size.

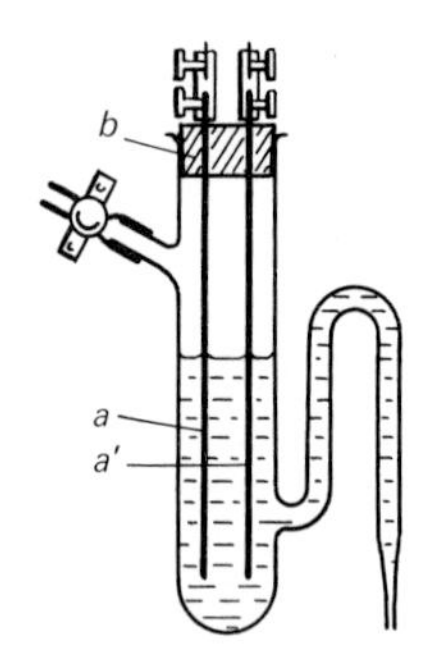

FIG. 13C.1 Construction of duplicate silver electrode unit.

A simply constructed silver electrode unit is shown in Fig. 13C.1 in which *a* and *a'* are duplicate rods of pure silver inserted through the rubber bung *b*. The rods are first cleaned for a few moments in dilute nitric acid, and then electroplated in a 1% solution of $AgNO_3$ in 90% methanol–water, using as anode a platinum wire mounted in a glass tube. A 2-volt accumulator provides sufficient current to produce a thin coating of small crystals of silver on the wires in the course of two or three hours. The electrodes are rinsed with the solution that is to be used in the cell, and are then mounted in position in the vessel containing the solution. A check on the quality of the electrodes is obtained by comparing the e.m.f. of the duplicates. The terminals on the two silver wires are connected to the potentiometer and the e.m.f. determined. Clearly, if the duplicates do not give almost exactly the same e.m.f. (so that when in opposition they give zero volts) they cannot be considered satisfactory and must be electroplated again.

EXPERIMENT

Determine the standard electrode potential of the silver electrode at room temperature, and examine the applicability of the theory of concentration cells.

Procedure. Prepare two 1 000 mol m^{-3} calomel electrodes (section 13B(1)) and *two* silver electrode units (Fig. 13C.1). Fill one of the latter with exactly 100 mol m^{-3} and the other with 10 mol m^{-3} silver nitrate solution. Arrange the two calomel vessels and two silver electrode vessels around a small beaker containing saturated potassium nitrate solution into which the tips of the electrode vessels should just dip.

Connect up a potentiometer circuit, with accumulator, and after 10 min standardize the potentiometer as described in section 4C(2). In all work with a potentiometer the *key should be closed only momentarily* to avoid polarizing the standard cell or the cell on test. Alternatively a digital voltmeter may be used.

The e.m.f. of the following electrode pairs should be determined to an accuracy of 0·1 mV. Only the connecting leads need to be moved to obtain all the combinations of half-cells. *The polarity of every cell should be noted.*

(a) Calomel against calomel.
(b) Ag | 100 mol m^{-3} $AgNO_3$ duplicate electrodes against one another.
(c) Ag | 10 mol m^{-3} $AgNO_3$ duplicate electrodes against one another.

Any of these duplicates that differ by more than, say, 1 mV should be dismantled and prepared afresh. If the duplicates agree, then the following cells should be measured:

(d) Ag | 100 mol m^{-3} $AgNO_3$ ‖ satd. KNO_3 ‖ 100 mol m^{-3} KCl, Hg_2Cl_2 | Hg

(e) Ag | 10 mol m^{-3} $AgNO_3$ ‖ satd. KNO_3 ‖ 100 mol m^{-3} KCl,
Hg_2Cl_2 | Hg

(f) Ag | 100 mol m^{-3} $AgNO_3$ ‖ satd. KNO_3 ‖ 10 mol m^{-3}
$AgNO_3$ | Ag

The effect of dispensing with the salt bridge can be tried by replacing the potassium nitrate in the beaker by 100 mol m^{-3} $AgNO_3$ into which the two silver electrode vessels are dipped. The liquid in the connecting tubes is first flushed clear of any KNO_3 which may have diffused in. Thus the e.m.f. of the following cell can be obtained.

(g) Ag | 100 mol m^{-3} $AgNO_3$ ‖ 10 mol m^{-3} $AgNO_3$ | Ag

The temperature of the cells should be recorded at the time of measuring their e.m.f.

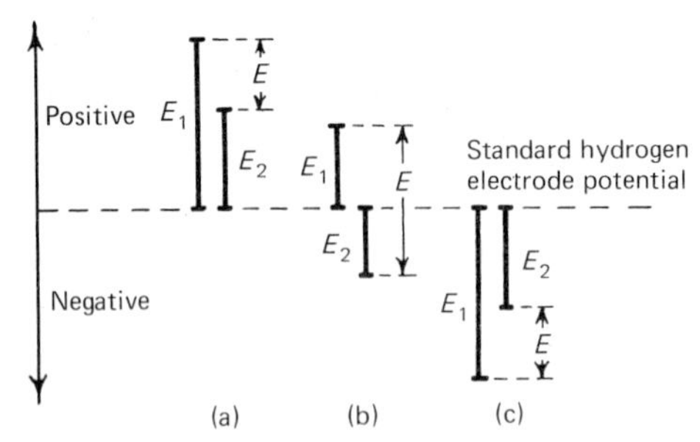

FIG. 13C.2 Diagrammatic representation of e.m.f. of a cell as the algebraic difference between separate electrode potentials.

The first step is to decide the *sign* and value of the separate electrode potentials (see section 13A(3)). A diagram such as that in Fig. 13C.2 helps to avoid confusion. The single electrode potentials are represented by vertical lines which begin at a level corresponding to the potential of the standard hydrogen electrode. Electrodes which are positive to the hydrogen electrode rise above the zero level, and those which are negative fall below, the length of the lines being proportional to the e.m.f. of the electrodes. The measured e.m.f. of the cell is the difference of potential, E, between the outer ends of the lines for the pair of electrodes. The positive pole of the cell is clearly the electrode rising the higher on the diagram.

Three cases are possible: (a) both electrodes positive with respect to the hydrogen electrode, (b) one positive, one negative, (c) both negative. Since one of the half-cells will generally be a reference electrode such as a calomel half-cell, its line on the diagram will be known in sign and magnitude. The potential of the normal calomel electrode, for example, is 0·281 V (25°C) *positive*. If the other electrode is positive with respect to the calomel electrode, it must have a potential more positive than 0·281 V as in case (a) ($E_1 > E_2$). If the other electrode is negative with respect to calomel, the cell could be the opposite instance of case (a), the calomel being the higher, in which case E ($=E_1 - E_2$) will be less than 0·281 V, or of case (b), if E is more than 0·281 V.

In short, consideration of the polarity of the cell, its e.m.f. and the sign and e.m.f. of the reference electrode will always give a check on the sign of the unknown electrode potential. The correct result can, of course, be obtained by rigid application of the algebraic difference formula (sections 13A(3), 13A(7)).

In the present experiment, the e.m.f. of the calomel electrode at the temperature of the experiment should be calculated from the equations of section 13B(1); hence the e.m.f. of the silver electrode in cell (d) can be calculated (E_{Ag}).

The e.m.f. of a silver electrode is given by

$$E_{Ag} = E^\circ_{Ag} + \frac{\boldsymbol{RT}}{\boldsymbol{F}} \log_e a_{Ag^+} = E^\circ_{Ag} + \frac{2{\cdot}303\boldsymbol{RT}}{\boldsymbol{F}} \log_{10} a_{Ag^+}$$

where E°_{Ag} is the standard electrode potential of the Ag | NO_3^- electrode, and a_{Ag^+} is the activity of silver ions in the solution.

*The value of the numerical factor 2·303**RT**/**F**, which occurs in all calculations on cells, is given for various temperatures in the table below.* (For intermediate temperatures it may be calculated from the approximate formula 0·05420 + 0·0001985t(°C).)

Value of the factor 2·303 ***RT***/***F*** for different temperatures

t (°C)		t (°C)	
0	0·05420	22	0·05857
10	0·05619	24	0·05897
12	0·05658	25	0·05916
14	0·05698	26	0·05936
16	0·05738	28	0·05976
18	0·05777	30	0·06016
20	0·05817	40	0·06214

In order to obtain E°_{Ag} one needs to know a_{Ag^+}; this is taken to be equal to the mean ion activity of the silver and nitrate ions in the solution, and is therefore obtained by multiplying the molality of the

$AgNO_3$ by the appropriate activity coefficient, values of which are given in Table A10. Thus, for 100 mol m^{-3} $AgNO_3$,

$$m_{Ag^+} \approx c_{Ag^+}/\text{density} \approx 100/1\,000 = 0{\cdot}1$$
$$a_{Ag^+} \approx 0{\cdot}1 \times 0{\cdot}733 \text{ mol (kg H}_2\text{O)}^{-1}$$

Thus, a value for E°_{Ag}, the standard electrode potential of the silver electrode can be obtained from cell (d). Another value results similarly from cell (e).

The e.m.f. of cell (f) should be exactly equal, of course, to the difference of (d) and (e), and should be given by the formula for a concentration cell without liquid junction potential ('concentration cell without transport'), namely:

$$E = \frac{RT}{F} \log_e (a_{Ag^+})_1/(a_{Ag^+})_2$$

Cell (g) differs from cell (f) in having a direct liquid junction between 100 mol m^{-3} $AgNO_3$ and 10 mol m^{-3} $AgNO_3$, and therefore the difference between their e.m.f. values gives an indication of the liquid junction potential E_j (assuming that in (f) E_j is entirely eliminated by the KNO_3 salt bridge). The experimental value of E_j may be compared with that predicted by the theoretical equation of Henderson which, for $AgNO_3$ solutions, gives

$$E_j = (2t_{NO_3^-} - 1)\cdot\frac{RT}{F} \log_e (a_{Ag^+})_1/(a_{Ag^+})_2$$

where $t_{NO_3^-}$ is the *transport number* of the Ag^+ ion, and may be taken as 0·53. The total e.m.f. of cell (g) should therefore be given by

$$2t_{NO_3^-}\cdot\frac{RT}{F} \log_e (a_{Ag^+})_1/(a_{Ag^+})_2$$

2. EXPERIMENTS WITH OTHER METAL ELECTRODES

Similar experiments may be performed with electrodes of cadmium, copper, lead, mercury, thallium, tin or zinc in 10 mol m^{-3} and 100 mol m^{-3} solutions of their salts. The metals should be as pure as possible. Cadmium and zinc electrodes may be amalgamated in order to improve their stability. This is done by placing them in dilute sulphuric acid and rubbing mercury over them with a mop of cotton-wool. Copper electrodes may be scraped, cleaned in dilute nitric acid and electroplated with copper from a solution containing 15 g $CuSO_4$, 5 g H_2SO_4, 5 g ethanol, 100 g water, using a pure Cu anode and passing a current of only 5 mA/cm^{-2}.

Results. Experimental results for standard electrode potentials should be compared with the following values.[1]

Metal electrode	Ag	Hg	Tl	Cu	Zn	Cd	Sn	Pb
Standard potential at 25°C (V)	0·799	0·798	−0·335	0·34	−0·763	−0·402	−0·140	−0·126

3. APPLICATIONS OF METAL ELECTRODES

3.1. *Determination of activity coefficients.* It is an obvious extension of the previous work to use a metal electrode to determine the activity of metal ions in other solutions. Either the standard electrode potential may be taken as known, or a solution of a salt of the metal of known activity may be used as one side of a concentration cell similar to cell (f) above.

EXPERIMENT

Determine the activity coefficient of a solution of a silver salt, e.g. silver acetate up to 50 mol m^{-3}, or silver fluoride up to 1 000 mol m^{-3}. Alternatively, the effect of KNO_3 *(up to 1 000 mol m^{-3}) on 10 mol m^{-3}* $AgNO_3$ *may be studied.*

Procedure. Set up pairs of silver electrodes in duplicate as described previously, and measure the e.m.f. of appropriate concentration cells, taking 100 mol m^{-3} $AgNO_3$ as a solution of known activity and using satd. KNO_3 as salt bridge.

Compare the results with the predictions of the Debye–Hückel theory and the Güntelberg–Guggenheim equation (section 13A(5)).

3.2. *Solubility of sparingly soluble salts.* When a salt M^+X^- is sparingly soluble it exhibits the 'common ion effect', the amount of M^+ ion solution being related to the concentration of X^- ion by the solubility product $K_s = [M^+][X^-]$ or, strictly, $K_s = a_{M^+} \times a_{X^-}$. K_s can be found by measuring a_{M^+} with an electrode of M, in a solution of definite a_{X^-} saturated with M^+X^-. In the ordinary saturated solution of M^+X^- in pure water $[M^+] = [X^-]$ and activity coefficients may be taken as unity if the solution is very dilute. Hence, the solubility $= \sqrt{K_s}$.

EXPERIMENT
Determine the solubility of silver chloride in water.

Set up the cell

Ag|AgCl, 10 mol m^{-3} KCl‖satd. KNO_3‖10 mol m^{-3} $AgNO_3$|Ag

The left-hand electrolyte is obtained by adding one drop of silver nitrate solution to 10 mol m^{-3} KCl. Determine the e.m.f. as before and note the temperature.

Calculation. Treat the cell as a concentration cell without liquid junction. Take the activity of Ag^+ in 10 mol m^{-3} ($m \approx 0{\cdot}01$) $AgNO_3$ as $0{\cdot}01 \times 0{\cdot}892$ mol (kg H_2O)$^{-1}$ and hence calculate a_{Ag^+} in 10 mol m^{-3} KCl. Take the activity of chloride ion in this solution as $0{\cdot}01 \times 0{\cdot}902$, and hence calculate K_s $(=a_{Ag^+} \times a_{Cl^-})$ and the solubility in mol m^{-3} $(=\sqrt{K_s})$ and finally in kg m^{-3}. The value[2] is about $1{\cdot}9 \times 10^{-3}$ kg m^{-3} at 25°C.

3.3. *Anion-reversible electrodes.* Given the result of the preceding experiment, one can employ an electrode consisting of Ag + AgCl to determine chloride ion activities, since the e.m.f. of the electrode is given by $E = E^\circ_{Ag} + (\boldsymbol{RT/F}) \log_e a_{Ag^+} = E^\circ_{Ag} + (\boldsymbol{RT/F}) \log_e (K_s/a_{Cl^-}) = E^\circ_{Ag/AgCl} - (\boldsymbol{RT/F}) \log_e a_{Cl^-}$. Here $E^\circ_{Ag/AgCl}$ is the standard electrode potential of the composite Ag/AgCl electrode, which is readily calculable from the above results.

The silver–silver chloride electrode is often used as a reference half-cell since it is reproducible and stable, and can be inserted into cells containing chlorides without introducing liquid junctions. The standard e.m.f. is given[3] by

$$E_{Ag/AgCl} = 0{\cdot}222_4 - 0{\cdot}0006\,(t-25)$$

If used as a reference electrode in place of a calomel, the electrode may be placed in 100 mol m^{-3} KCl ($a_{Cl^-} = 0{\cdot}1 \times 0{\cdot}771$ mol (kg H_2O)$^{-1}$). It is usual to ensure that the solution in immediate contact with the silver surface is properly saturated by providing a porous coating of AgCl on the Ag, for example, by electrolysing a solution of 100 mol m^{-3} HCl at very low current density for half an hour using an ordinary silver electrode (section 13C(1)) as anode.[4]

3.4. *Equilibrium constants of complex ions.* This application is an obvious extension of the first. For example, if the activity of silver in a very dilute solution containing a known amount of ammonia is measured, the extent to which the complex ion $Ag(NH_3)_2^+$ is formed can be estimated.

EXPERIMENT

Determine the e.m.f. of the following cell:

Ag | 25 mol m^{-3} $AgNO_3$ in 1 000 mol m^{-3} NH_3 solution ‖ satd soln KNO_3 ‖ 10 mol m^{-3} $AgNO_3$ | Ag.

Calculation. The reaction is $Ag(NH_3)_2^+ \rightleftharpoons Ag^+ + 2NH_3$ and hence the thermodynamic equilibrium constant is given by

$$K = \frac{a_{Ag^+} \times a_{NH_3}^2}{a_{Ag(NH_3)_2^+}}.$$

The e.m.f. determination gives a_{Ag^+} directly. The activity coefficient for univalent ions in a solution of ionic strength $I_m = 0{\cdot}025$ mol (kg H_2O)$^{-1}$ may be estimated by comparison with silver nitrate as about 0·9. Hence $[Ag^+]$ can be calculated, and then $[Ag(NH_3)_2^+]$ and $[NH_3]$. $a_{(AgNH_3)_2^+}$ can be taken as $0{\cdot}9 \times [Ag(NH_3)_2^+]$ and a_{NH_3} as $[NH_3]$, since this is non-ionic. Thus K can be calculated.

3.5. *Analytical applications—potentiometric titrations.* See section 13G.

BIBLIOGRAPHY 13C: Metal electrodes

See bibliography 13A.

[1] Bockris and Herringshaw, *Faraday Soc. Disc.*, 1947, **1**, 328.
[2] Brown and MacInnes, *J. Amer. Chem. Soc.*, 1935, **57**, 4459.
[3] Harned and Ehlers, *ibid.*, 1932, **54**, 1350.
[4] Brown, *ibid.*, 1934, **56**, 646.

13D The hydrogen electrode

(For theory, see section 13A(6); applications to pH, section 13F.)

Preparation of the electrodes. The metallic part of the electrode consists of a small rectangle of platinum foil welded to a piece of platinum wire which is then sealed into a glass tube (Fig. 13D.1). Connection is made by a drop of mercury inside the tube into which a copper lead is dipped. Alternatively, the copper lead can be 'soldered' to the platinum inside the tube by means of a small piece of Wood's

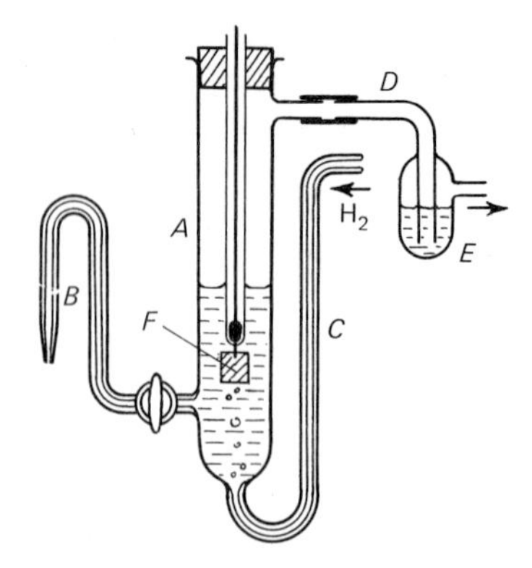

FIG. 13D.1 Hydrogen electrode vessel.

metal (m.p. 65·5°C). It is important that the platinum–glass seal should be sound, and free from cracks.

The platinum electrode must be platinized before it will function reversibly. The electrode is first cleaned in chromic acid and then well washed with water and finally electroplated in a solution of chloroplatinic acid (see section 11A(8) for details), using another platinum foil electrode as anode. *Only a very thin, translucent coating of platinum black is wanted*: less than 5 min of electrolysis is necessary and there is no need to use current reversal.

After the electrolysis, the electrode is washed and then made the cathode in dilute sulphuric acid in order to remove occluded chlorine by the stream of electrolytically-generated hydrogen. After another washing, the electrode is ready for use. It will normally remain active for days if kept under water and protected from 'poisoning' (see below).

The electrode vessel. Since the electrolyte has to be saturated with hydrogen and protected from oxygen, it must be contained in an enclosed vessel through which a stream of hydrogen can be passed, and there must be provision for connecting the electrolyte to the other half-cell. A suitable form of apparatus is shown in Fig. 13D.1. The tube *A* and connecting arm *B* contain the solution. A stream of hydrogen passes in through the tube *C*, and out via *D* and the small bubbler *E* which contains water and serves to keep back air. It is best to have the electrode tube in a ground-glass joint or polythene stopper; if a rubber stopper is used, it should first be boiled in concentrated NaOH solution and then in water.

The hydrogen gas for hydrogen electrodes must be as pure as possible. That supplied commercially in cylinders is generally satisfactory, but it should be passed through wash-bottles containing (a) a solution of alkaline pyrogallol (2 g pyrogallol in about 35 cm^3 4 000 mol m^{-3} sodium hydroxide), (b) very dilute sulphuric acid, (c) water, and finally (d) a sample of the solution used in the cell. Wash-bottle (d) is to ensure that the hydrogen stream does not alter the concentration of the solution in the cell. It is advisable to avoid the use of rubber in connecting the gas train owing to its liability to contain sulphur compounds which may 'poison' the electrode; polyvinyl chloride tubing is very satisfactory, and polythene can also be used. If rubber must be used, it should first be boiled in concentrated NaOH solution and then washed. The apparatus must be air-tight.

Better removal of oxygen can be achieved by passing the hydrogen through a palladium membrane (B.O.C. Ltd), or over palladized asbestos at 400°C or reduced copper wire at 600°C.

Operation of hydrogen electrodes. A steady stream of hydrogen should be bubbled through the cell until the electrode assumes a constant e.m.f.; this may take 10–20 min, and thereafter the e.m.f. should be independent of the rate of bubbling. The correct value is not reached until oxygen has been removed.

Unfortunately, platinized platinum electrodes are susceptible to 'poisoning'; that is to say, they sooner or later lose their catalytic activity, often as a result of adsorption of traces of substances such as arsenic, sulphur and mercury compounds on the active surface. The symptoms of poisoning are (i) irreproducibility, (ii) sluggishness in attaining a steady e.m.f., and (iii) dependence of the e.m.f. on the rate of bubbling of H_2. If the platinum electrode becomes poisoned it must be cleaned and platinized again as already described.

Limitations. Inorganic oxidizing agents (Fe^{+++}, $Cr_2O_7^{--}$, MnO_4^-, etc.) and various organic substances (e.g. alkaloids, per-acids, dyestuffs) are reduced by the hydrogen electrode; similarly, strongly reducing substances like stannous chloride, and metallic salts of Cu, Ag, Bi, Hg, etc., interfere. Some colloidal materials inactivate the platinum. It is seen, therefore, that these restrictions, together with the liability to poisoning and the precautions needed to obtain reliable results, tend to limit the practical usefulness of the hydrogen electrode. It is employed chiefly in accurate research work, while the more convenient glass electrodes are used elsewhere.

EXPERIMENT

Determine the e.m.f. (25°C) of the cell H_2(Pt) | HCl, *molality m* | AgCl | Ag *with various concentrations of* HCl *from about 100 mol* m^{-3} *to about 0·1 mol* m^{-3}*, and hence obtain the activity coefficients of* HCl.

Procedure. Prepare two (duplicate) hydrogen electrode half-cells (Fig. 13D.1) and two silver–silver chloride electrodes (section 13C(3)). Fill all four electrode vessels with the first solution of HCl. Since the above cell is one without liquid junction, no salt bridge is needed, and the two half-cells can be coupled together by a short length of polyvinyl chloride tubing containing the same HCl solution. Set up the cells in a thermostat at 25·0° C, and check the e.m.f. of the duplicates against one another. If they agree, measure the e.m.f. of the above cell, making sure that a steady value has been reached. Repeat with other concentrations of HCl. To expedite the work, the HCl solutions in bottles may be placed in readiness in the thermostat and a stream of H_2 passed through in order to remove most of the dissolved air. One cell can be prepared while the other is coming to equilibrium. The barometric pressure should be noted.

Calculation. The e.m.f. of the hydrogen electrode is given (section 13A(6)) by $E_H = (\boldsymbol{RT}/\boldsymbol{F}) \log_e (a_{H^+}/p_{H_2}^{\frac{1}{2}})$ and that of the silver–silver chloride electrode by $E_{Ag/AgCl} = E^\circ_{Ag/AgCl} - (\boldsymbol{RT}/\boldsymbol{F}) \log_e a_{Cl^-}$. The net e.m.f. of the cell is

$$E_H - E_{Ag/AgCl} = -E^\circ_{Ag/AgCl} + (2{\cdot}303\boldsymbol{RT}/\boldsymbol{F}) \log_{10} a_{H^+} \times a_{Cl^-}$$

If only a few measurements can be made, one can assume the value of $E^\circ_{Ag/AgCl}$ (0·222 V) and calculate $a_{H^+} \times a_{Cl^-}$ which is equal to $m_+ f_+ m_- f_- = m^2 f_\pm^2$ (where m is molality, f activity coefficient) and hence obtain $f_\pm$ for the various solutions. Alternatively, $f_\pm$ for one of the solutions can be taken from other data.

If a number of accurate measurements can be made down to very low concentrations, the following treatment[1] can be applied, thus dispensing with auxiliary data. It is known that activity coefficients of 1:1 electrolytes follow a law of the form $\log_{10} f_\pm = -0{\cdot}51 \sqrt{m} + Bm$. The e.m.f. of the cell (E) is equal to $-E^\circ_{Ag/AgCl} + (2\boldsymbol{RT}/\boldsymbol{F}) \log_e mf_\pm$. Substituting and rearranging,

$$E - 0{\cdot}118 \log_{10} m + 0{\cdot}0603\sqrt{m} = -E^\circ_{Ag/AgCl} + 0{\cdot}118\, Bm$$

A graph of the left-hand side of this equation against m should therefore be a straight line which can be confidently extrapolated to $m=0$, giving $E^{\circ}_{Ag/AgCl}$, and the slope of the line gives the constant B, from which the $f_{\pm}$ values of the electrolyte can then be calculated, according to the equation given above.

Other experiments. The hydrogen electrode may be used to determine the pH of buffer mixtures (section 13F). The thermodynamic dissociation constants of weak acids may be determined by extrapolation to zero ionic strength of results for buffer mixtures having different salt concentrations. For potentiometric titrations with the hydrogen electrode see section 13G.

BIBLIOGRAPHY 13D: The hydrogen electrode

See monographs cited under Bibliography 13F.

[1] Hitchcock, *J. Amer. Chem. Soc.*, 1928, **50**, 2076.

13E Oxidation–reduction ('redox') potentials

1. When a platinum electrode is placed in a solution containing ferric and ferrous ions, the potential of the electrode relative to a standard hydrogen electrode is given by

$$E = E^{\circ} + \frac{2{\cdot}303\boldsymbol{RT}}{n\boldsymbol{F}} \log_{10} \frac{a_{Fe^{+++}}}{a_{Fe^{++}}}$$

where n is the difference in valency of the ions and $a_{Fe^{+++}}$ and $a_{Fe^{++}}$ are the activities of the ferric and ferrous ions respectively. Since $n=1$, at 25°C:

$$E = E^{\circ} + 0{\cdot}0591 \log_{10}(a_{Fe^{+++}}/a_{Fe^{++}})$$

E° is the standard potential of the couple $Fe^{+++} + e \rightleftharpoons Fe^{++}$, and is a measure of the tendency of the ion to pass from the higher to the lower state of oxidation: the more positive its value the more effective is the ion in the higher state of oxidation as an oxidizing agent relative to the ion in the lower state of oxidation as a reducing agent, and vice versa.

In many oxidation–reduction systems, hydrogen ions take part in the electrode reaction, and consequently the 'redox' potential set up is dependent on the activity of hydrogen ions in the solution (see example of the quinhydrone electrode, section 13F), and the standard redox potential is that for a solution in which $a_{H^{+}} = 1\,\mathrm{mol\,(kg\,H_2O)^{-1}}$.

It generally happens that oxidation–reduction processes are studied in the presence of acids or other electrolytes, i.e. in solutions of high ionic strength. Consequently, the ionic activities, especially of polyvalent ions such as Fe^{+++}, Ce^{++++}, etc., may differ considerably from concentrations. Unfortunately there is no satisfactory method for calculating activity coefficients of polyvalent ions in solutions of high ionic strength. Standard redox potentials have sometimes been determined by extrapolating observed redox potentials to zero ionic strength. These true standard potentials are of little practical significance, however, and it is more useful to report the so-called *formal oxidation–reduction potential*, obtained by taking concentrations instead of activities in the above equation. Thus, for the $Fe^{+++}-Fe^{++}$ system,

$$E = E^{\circ}_{\text{formal}} + (\boldsymbol{RT/F}) \log_e [\mathrm{Fe}^{+++}]/[\mathrm{Fe}^{++}]$$

The formal redox potential is not expected to be a true constant, but it is effectively constant for solutions of the particular ionic strength employed in, say, a redox titration.

2. REDOX INDICATORS

Certain organic dyes which can undergo reversible oxidation–reduction exhibit different colours in the oxidized and reduced states. In many cases the reduced form ('leuco' compound) is colourless. Generally the reduction involves one or more hydrogen atoms and consequently the colour change of these dyes occurs at different redox potentials according to the pH of the solution. For example, the redox potential at which methylene blue exists half in the blue form and half in the reduced (colourless leuco) form is +0·101 V at pH 5, +0·011 V at pH 7, and −0·050 V at pH 9.

Redox indicators with different E° values (at a given pH) are available, and are used in biochemistry to determine the approximate oxidation–reduction conditions prevailing in biological systems. For example, at a redox potential of 0·2 V and pH 7, phenol-*m*-sulphonate indo-2:6-dibromo phenol is colourless while thionine is violet; at −0·05 V both these indicators are colourless, but indigo di-sulphonate is coloured. For biological purposes indicators are used having E° at pH 7 ranging from +0·35 to −0·35 V.

Examples of important oxidation–reduction systems which occur in living systems are: $2RSH \rightleftarrows RSSR + H_2$, e.g. the cysteine–cystine system: various haemoglobin systems: biological oxidations involving 'respiratory pigments', vitamin C, adrenaline, luciferin, and many pigments.[1]

Redox indicators are also employed in inorganic volumetric analysis.[2]

EXPERIMENT

Determine the formal redox potential of the ferrous–ferric system and of o-phenanthroline, and ascertain whether this indicator is suitable for use in the volumetric determination of ferrous salts.

Procedure. Prepare (a) 100 mol m^{-3} Fe^{++} solution and (b) 100 mol m^{-3} Fe^{+++} solution by weighing out ferrous ammonium sulphate and ferric alum respectively, both solutions to be made up in approximately 1 000 mol m^{-3} sulphuric acid solution. Pipette 25 cm^3 of ferrous solution into a 100 cm^3 beaker which is placed on a magnetic stirrer: the stirrer bar and a platinum electrode are placed in the solution, also the salt bridge leading to a calomel electrode (Fig. 13B.1(b)).

The platinum electrode consists of a square of platinum foil welded to a platinum wire which is mounted in a glass tube in the usual manner. It is cleaned in concentrated nitric acid, washed and then heated to redness in an alcohol flame. The platinum electrode assumes the redox potential of the solution as soon as it is immersed in it. The calomel electrode acts as reference electrode.

Titrate the ferrous solution with the dichromate solution, taking e.m.f. readings every 1 cm^3, and more frequently as the e.m.f. changes rapidly near the end point. Continue well past the end point.

Convert the e.m.f.s to the corresponding value E on the hydrogen scale using the relevant value of the calomel electrode potential (section 12B(1)). Plot the values of E against the titre V cm^3, and also the differential $\Delta E/\Delta V$ against V. Compare the endpoint titre (V_e) (i) obtained visually, (ii) from the plot of E against V, (iii) from the differential plot and (iv) expected from the known concentrations. Since $[Fe^{+++}]/[Fe^{++}] = V/(V_e - V)$, a plot of E against $\log_{10} (V/(V_e - V))$ should be a straight line of slope $2{\cdot}303\ \boldsymbol{RT}/2\boldsymbol{F}$. Also at $[Fe^{+++}] = [Fe^{++}]$, i.e. $\log_{10} (V/(V_e - V)) = 0$, $E = E^{\circ}$, the formal redox potential for the $Fe^{+++} - Fe^{++}$ system. Find E_0 and compare with the expected value.

Ortho-phenanthroline indicator solution is made by dissolving 0·23 g *o*-phenanthroline + 0·12 g ferrous sulphate in 100 cm^3 of water. The 'end-point' for an indicator can be considered to be the solution in which the colour change from one extreme to the other has proceeded half-way, i.e. [oxidized form] = [reduced form]. This solution corresponds to the formal redox potential of the indicator. The potential may be found approximately as follows. Add a little solution to some approx. 1 000 mol m^{-3} H_2SO_4 in a beaker, and then add two drops of the indicator solution. Titrate the solution with an oxidizing agent (potassium dichromate or ceric sulphate) also dissolved in 1 000 mol m^{-3} H_2SO_4 until the indicator begins to change colour. By trial obtain about half the full intensity of colour. Then, using this solution, determine the redox potential as described.

Repeat the ferrous–ferric titration with the addition of a few drops of this indicator, noting the colour changes and the e.m.f.s near the end point. Compare the colorimetric and the potentiometric end-points. Are the e.m.f.s altered by the addition of this indicator? Comment on the accuracy of the titration using the two methods.

BIBLIOGRAPHY 13E: Oxidation–reduction potentials

Latimer, *The Oxidation States of the Elements and their potentials in Aqueous Solution*, 2nd edn, 1952 (Prentice-Hall Inc., New York).

[1] Green, *Mechanisms of Biological Oxidations*, 1940.

[2] Tomicek (trans. Weir), *Chemical Indicators*, chapter 6, 1951 (Butterworth's Scientific Publ., London); Vogel, *Textbook of Quantitative Inorganic Analysis*, 3rd edn, 1961 (Longmans, London).

13F pH Determination

1. THE pH SCALE

The *hydrogen ion* plays a particularly important part in many processes and reactions—chemical, biological and industrial. For example, extremely minute changes in the concentration of hydrogen ions may determine whether a metal will dissolve, or whether a metallic salt or a protein will precipitate. The determination and control of the concentration of hydrogen ions is therefore of the greatest importance in chemistry. The range of concentration which may be encountered is very great, extending from, say, a solution of 1 000 mol m^{-3} (molality $m_{H^+} \approx 1$ mol (kg $H_2O)^{-1}$) down to pure water in which $[H^+] \approx 10^{-4}$ mol m^{-3} ($m_{H^+} \approx 10^{-7}$) and as far beyond into alkaline solutions. In a 1 000 mol m^{-3} NaOH solution $m_{OH} \approx 1$, and since the ionic product of water, $m_{H^+} \times m_{OH^-} \approx 10^{-14}$ is maintained, $m_{H^+} \approx 10^{-14}$. Sörensen (1909) introduced the convenient pH *scale of hydrogen ion concentration*; on this scale, a solution having $m_{H^+} = 10^{-x}$ is said to have a pH of x. Clearly, $\mathrm{pH} = -\log_{10} m_{H^+} = \log_{10}(1/m_{H^+})$. x is usually a positive number ranging from about 0 (for a 1 000 mol m^{-3}) acid through 7 (for a neutral solution), to about 14 (for a 1 000 mol m^{-3} alkali).

The pH scale was originally defined as $\mathrm{pH} = \log_{10}[H^+]$ where the concentration $[H^+]$ was in units of moles per litre. For dilute aqueous solutions c in these units is approximately equal to the molality in mol (kg $H_2O)^{-1}$.

It is obvious that these infinitesimal (but vitally important) hydrogen ion concentrations cannot be determined by ordinary analytical means. In fact, all methods of determining pH are based directly or indirectly on the original method introduced by Sörensen, namely, e.m.f. measurements with the hydrogen electrode. Using Nernst's theory of electrode potential, Sörensen took the e.m.f. of a hydrogen electrode having $p_{H_2} = 1$ atm. to be given by

$$E_H = (\boldsymbol{RT/F}) \log_e c_{H^+} = -0{\cdot}0571 + \text{pH (at 15°C)}$$

Therefore the e.m.f. of a hydrogen electrode is directly proportional to the pH of the solution. In practice, various other methods are often more convenient to use than the hydrogen electrode for pH determination, but they all relate back to the above definition and basic method of determination.

2. THE SIGNIFICANCE OF pH

The pH scale suffers from a certain lack of precision which nevertheless scarcely detracts from its value in practical chemistry. There are actually two sources of uncertainty. In the first place, Sörensen originally intended pH to be related to hydrogen ion *concentration*, but his fundamental method of measurement—the hydrogen electrode—is now known to depend on thermodynamic *activities* rather than concentrations, i.e. $E_H \propto \log a_{H^+}$ (*not* $\log m_{H^+}$) and $a_{H^+} = m_{H^+} f_{H^+}$. In very dilute solutions—say, those having an ionic strength of less than 1 mol (kg $H_2O)^{-1}$—f_{H^+} is near enough to unity for a_{H^+} to be taken as m_{H^+}, but this is certainly not true in concentrated solutions. Now the great value of pH comes in connection with *equilibria* involving the hydrogen ion, e.g. dissociation of weak acids or bases, and here again a_{H^+} is the quantity involved, rather than m_{H^+}. pH as measured by the hydrogen electrode therefore gives the quantity required. The original quantity defined by $-\log_{10} c_{H^+}$ is not only of little theoretical significance, but, in fact cannot be measured directly. It has therefore come to be accepted that

$$\text{pH} \equiv -\log_{10} a_{H^+}$$

Curiously enough, this definition is still lacking in precision. Strictly, single electrode potentials and single ion activities cannot be determined unambiguously. Thus, a cell without liquid junction such as that discussed in section 13D(1) measures in fact $a_{\pm}$, not a_+ and a_- separately. (For example, it is impossible to state unambiguously the concentration of a solution of HCl in which $a_{H^+} = 1$; one can but suppose that $f_{H^+} \approx f_{Cl^-} \approx f_{\pm}$ and hence $a_{H^+} \approx mf_{\pm}$, which would mean that m would have to be about 1·2 mol (kg $H_2O)^{-1}$.) If, on the other hand, one attempts to get a_{H^+} from single electrode potentials derived from cells having liquid junctions, the uncertainty as to the value of the liquid junction potential limits the precision of the result. Most pH measurements are based on cells of the type

$$H_2(Pt) \mid \text{solution containing } H^+ \underset{(a)}{\parallel} \text{satd} \quad KCl \underset{(b)}{\parallel} \text{calomel cell}$$

There are theoretical reasons (section 13A.8) for believing that the liquid junction (b) has negligible junction potential (say $<0{\cdot}1$ mV), but there is no such assurance for junction (a) since such junctions are liable to contain significant concentrations of the highly mobile H^+ or OH^- ions. The degree to which saturated KCl does or does not 'eliminate' such liquid junction potentials is uncertain. Consequently, the significance of pH values is uncertain to the same degree. Nevertheless, the value of the pH scale is not depreciated by this realization. Its practical usefulness is firmly established in many fields of applied chemistry, and fortunately, a high degree of precision is rarely, if ever, required.

At the present time the pH scale is accepted on its merits because it is empirically indispensable. pH is deemed to be defined by the method of measurement. The pH value may be taken to be *very roughly* $= -\log_{10} m_{H^+}$; the higher the ionic strength, the greater the error in this assumption. Rather more accurately, pH $= -\log_{10} m_{H^+} f$, where f can be considered as the activity coefficient of a typical uni-univalent electrolyte in the appropriate solution. This interpretation is valid within $\pm 0{\cdot}02$ pH units for the range pH 2 to pH 12 in solutions of ionic strength not greater than 0·1 mol (kg $H_2O)^{-1}$ (i.e. f values between 1 and about 0·75). Outside these limits the precision of interpretation of pH decreases.

3. BUFFER SOLUTIONS

The pH of water and of very dilute solutions of salts is extremely sensitive to traces of impurities, such as dissolved CO_2, alkali derived from glass surfaces, etc. When a solution of definite pH is required, it is rarely satisfactory merely to add acid or alkali until the desired pH is produced, because the pH of the resulting solution is unstable. Instead, 'buffer mixtures' are used. These are solutions of considerable total concentration in which the pH is controlled by a chemical equilibrium. The constituents are weak acids (or bases) and their salts. When, for example, one adds a small quantity of hydrochloric acid to a mixture of acetic acid and sodium acetate, the hydrogen ions unite to a large extent with acetate ions to form un-ionized acetic acid, and the concentration of hydrogen ion is thereby prevented from increasing to any considerable extent. Similarly, when a small quantity of alkali is added to the mixture, the hydroxide ions react with the acetic acid to form acetate ions and water, and again the hydrogen ion concentration remains practically unchanged.

Numerous buffer mixtures have been recommended, including mixtures based on acetic, boric, citric, phosphoric, malonic, phthalic and oxalic acids and on ammonium salts, and any pH from 1 to 10 can be obtained by making up a suitable buffer solution. The following are examples of mixtures giving integral pH values

pH Value	*Composition of mixture*
1	47·5 cm^3 200 mol m^{-3} HCl + 25 cm^3 200 mol m^{-3} KCl; mixture dil. to 100 cm^3.
2	5·3 cm^3 200 mol m^{-3} HCl + 25 cm^3 200 mol m^{-3} KCl; mixture dil. to 100 cm^3.
3	20·55 cm^3 200 mol m^{-3} Na_2HPO_4 + 79·45 cm^3 100 mol m^{-3} citric acid.
4	41·0 cm^3 200 mol m^{-3} acetic acid + 9·0 cm^3 200 mol m^{-3} sodium acetate.
5	14·75 cm^3 200 mol m^{-3} acetic acid + 35·25 cm^3 200 mol m^{-3} sodium acetate.
6	9·0 cm^3 200 mol m^{-3} acetic acid + 191·0 cm^3 200 mol m^{-3} sodium acetate.
7	12·0 cm^3 50 mol m^{-3} borax + 188·0 cm^3 of a solution containing 12·40 g H_3BO_3 and 2·93 g NaCl in 1 dm^3.
8	11·0 cm^3 50 mol m^{-3} borax + 29·0 cm^3 of a solution containing 12·40 g H_3BO_3 and 2·93 g NaCl in 1 dm^3.
9	40·0 cm^3 50 mol m^{-3} borax + 10·0 cm^3 of a solution containing 12·40 g H_3BO_3 and 2·93 g NaCl in 1 dm^3.
10	Equal parts of 25 mol m^{-3} $NaHCO_3$ and 25 mol m^{-3} Na_2CO_3.

Four other solutions of accurately known pH, which can be made up readily from single substances, are given in section 13F(7).

Since the dissociation constants of ammonium hydroxide and acetic acid are identical, solutions of ammonium acetate provide valuable neutral buffer solutions with pH = 7·0, and the pH remains constant over a wide range of concentrations. The solutions should not be prepared from the salt but by mixing together equal volumes of accurately prepared 2 000 mol m^{-3} solutions of ammonium hydroxide and acetic acid. Dilution of this neutral solution does not alter the pH.

4. DETERMINATION OF pH WITH THE HYDROGEN ELECTRODE

The cell usually employed is of the type

$$H_2(Pt)|\text{solution of unknown pH}\,\|\,\text{satd. KCl}, Hg_2Cl_2|Hg$$

i.e. a hydrogen electrode operating in the solution is connected to a saturated KCl calomel electrode, either by dipping the calomel electrode vessel directly into the solution or via a saturated KCl bridge (Fig. 13B.1). In place of a saturated KCl calomel electrode one may use a 1 000 mol m^{-3} KCl calomel or a silver–silver chloride electrode (section 13C(3)). The hydrogen electrode is prepared as previously described (section 13D(1)), and should be contained in a vessel such as that shown in Fig. 13D.1. If the e.m.f. of the calomel or other reference electrode is $E_{ref.}$ on the standard scale, the e.m.f. of the hydrogen electrode, E_H, is (e.m.f. of cell $-E_{ref.}$), and the pH can be calculated from the equation

$$E_H = -2{\cdot}303(\boldsymbol{RT}/\boldsymbol{F}) \times \text{pH}$$

Thus, at 18°C, using a saturated calomel electrode ($E_{ref.} = 0{\cdot}249$ V), the pH of the solution may be computed from the formula

$$\text{pH} = \frac{\text{e.m.f. (measured)} - 0{\cdot}249}{0{\cdot}0578}$$

The hydrogen electrode, which has an accuracy, at constant temperature, of 0·02 pH (section 13D(1)) should not be used in the presence of oxidizing agents, of metals which lie below hydrogen in the electromotive series (e.g. copper, silver), or of organic compounds which undergo hydrogenation. It is also unsuitable for use in presence of colloids.

EXPERIMENT
Determine the pH of a given solution.

The pH of the buffer solutions given above may be tested, or the pH of mixtures of 50 mol m^{-3} acetic acid and 50 mol m^{-3} sodium acetate solutions in different proportions may be measured. In the latter case, it may be shown that the pH is approximately linear with $\log_{10}$ [sodium acetate]/[acetic acid] in the mixtures (section 13G).

5. THE QUINHYDRONE ELECTRODE

Theory. Biilmann (1921) introduced the use of the quinhydrone electrode for pH determinations. This electrode is a redox electrode at which the reversible reaction

$$\underset{\text{quinone (Q)}}{C_6H_4O_2 + 2H^+} = \underset{\text{hydroquinone (QH}_2\text{)}}{C_6H_4(OH)_2 - 2\,e}$$

takes place. The electrode potential is given by

$$E = E_Q^\circ + \frac{\boldsymbol{RT}}{2\boldsymbol{F}} \log_e \frac{a_Q \times a_{H^+}^2}{a_{QH_2}}$$

where a_Q and a_{QH_2} are the activities of quinone and hydroquinone respectively in the solution, and E_Q° is the standard electrode potential of the quinhydrone electrode. In practice the ratio a_Q/a_{QH_2} is maintained constant at unity by saturating the solution with the substance 'quinhydrone', which is a 1:1 molecular compound of these substances. The potential of such a quinhydrone electrode is then

$$E = E_Q^\circ + (\boldsymbol{RT}/\boldsymbol{F}) \log_e a_{H^+} = E_Q^\circ - (2{\cdot}303\boldsymbol{RT}/\boldsymbol{F})\text{pH}$$

The electrode can therefore be used to measure pH in the same way as the hydrogen electrode. The value of E_Q° is given by the expression $0{\cdot}699 - 0{\cdot}0007_4\,(t-25)$ V (t in °C).

Advantages and limitations. The quinhydrone electrode is extremely simple to set up, and it gives its true reversible potential immediately. It is applicable to many types of solution and gives pH values as accurately as the hydrogen electrode. It can sometimes be used in solutions containing reducible substances (e.g. dilute HNO_3, NO_3^-, Cu^{++}, Pb^{++}, Cd^{++}, unsaturated acids, alkaloids) where the hydrogen electrode is not applicable. Very small quantities of solution suffice for a measurement, and air need not be removed. On the other hand, the electrode cannot be used with solutions containing any substances which might react with quinone or hydroquinone (e.g. ferrous salts, manganese dioxide, aniline, boric acid, some proteins). Also, inaccurate results are obtained in solutions more alkaline than pH 8·5 because the hydroquinone begins to ionize as a weak acid. Further, in alkaline solutions quinhydrone slowly decomposes and undergoes oxidation from the air, resulting in a rising pH.

The quinhydrone electrode is subject to 'salt error' when the concentration of electrolyte is greater than about 100 mol m^{-3}. This defect can be overcome by saturating the solution not only with quinhydrone but also with quinone or hydroquinone. In the former case, $E^{\circ} = 0{\cdot}75$ V and, in the latter case, 0·619 V, at 18°C. It is advisable to have a small well at the bottom of the electrode vessel into which the solids settle and surround the platinum electrode.

EXPERIMENT

Determine the pH of a number of buffer solutions (as section 13F(3)) or carry out a potentiometric titration (section 13G).

Practical details. The cell employed takes the form

Pt | quinhydrone (satd.) in solution of unknown pH ||
satd. KCl salt bridge || reference electrode

The platinum electrode is a bright foil (section 13E(2)). The solution under test is contained in a small beaker, and pure quinhydrone is added in sufficient quantity to leave some in excess (1–2 g per 100 cm^3 of solution). The salt bridge is conveniently an inverted U-tube containing KCl—agar jelly (section 13B(2)). The reference electrode may be a saturated KCl calomel half-cell or a 1 000 or 100 mol m^{-3} KCl calomel or Ag/AgCl electrode. If the e.m.f. of the reference electrode is $E_{\text{ref.}}$, then $E_{\text{observed}} = E_{\text{quin.}} - E_{\text{ref.}} = E^{\circ}_{\text{Q}} - (2{\cdot}303\boldsymbol{RT}/\boldsymbol{F})\text{pH} - E_{\text{ref.}}$. Hence,

$$\text{pH} = \frac{E^{\circ}_{\text{Q}} - E_{\text{ref.}} - E_{\text{observed}}}{2{\cdot}303\boldsymbol{RT}/\boldsymbol{F}}$$

For example, with a saturated calomel electrode and at a temperature of 18°C,

$$\text{pH} = \frac{0{\cdot}455 - E_{\text{observed}}}{0{\cdot}0578}$$

6. THE GLASS ELECTRODE

Haber and Klemensiewicz (1909) showed that if a thin membrane of glass is used to separate two solutions, it develops a membrane potential which is dependent chiefly on the pH of solutions. This principle forms the basis of a valuable method of measuring pH. The cell employed is of the form

Reference electrode (1) in solution (1)	Glass membrane	solution (2), pH $= x$ ‖	satd KCl salt bridge	reference electrode (2).

The precise e.m.f. of the cell is not of particular interest because it includes an indefinite 'asymmetry potential' (perhaps due to strain) which generally exists across the glass membrane even if the two sides of the cell are of identical composition. However, this asymmetry potential is practically constant. If, therefore, measurements are made with all parts of the cell unchanged except solution (2), it is found that the total e.m.f. varies in the manner expected for a similar cell containing a hydrogen electrode, i.e.

$$E = \text{const} + (2{\cdot}303\boldsymbol{RT}/\boldsymbol{F}) \log_{10} a_{H^+} = \text{const} - (2{\cdot}303\boldsymbol{RT}/\boldsymbol{F})\,\text{pH}$$

In practice, the glass electrode assembly is *calibrated* by measuring first the e.m.f. of the cell with solution (2) of known pH. This solution is then replaced by the solution of unknown pH. The *change of e.m.f.* is given by

$$\Delta E = (2{\cdot}303\boldsymbol{RT}/\boldsymbol{F})(\text{pH}_1 - \text{pH}_2)$$

It is clear that any types of reference electrode can be used on the two sides of the cell; their e.m.f., since they do not enter into the final expression, need not be known, but must be constant from one reading to another. Usually electrode (1) is a silver–silver chloride electrode in dilute HCl and electrode (2) is a satd. KCl calomel half-cell.

Various forms of glass electrode have been used. That most commonly employed consists of a thin-walled soda glass bulb which contains a small silver–silver chloride wire electrode in dilute acid, the tube being sealed at the top. New glass electrodes should be soaked for a period in dilute HCl and thereafter always stored in distilled water. The bulb is simply suspended in a small beaker of the solution to be tested (Fig. 13F.1(a)), but it is better for routine work to use an assembly which ensures adequate protection to the electrode. Glass electrodes must never be cleaned with chromic acid.

The calomel electrode may be of any type. A compact form suitable for dipping direct into the test solution is often used. It is permissible to use a very small calomel electrode in work with the glass electrode because only a minute current is drawn from the cell during the measurements (see below) and consequently the danger of polarizing

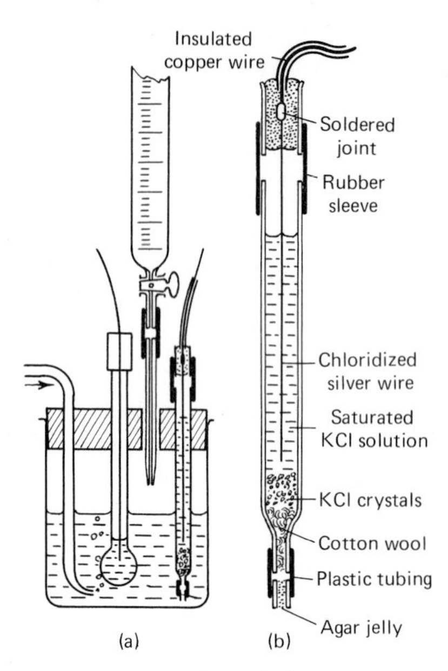

FIG. 13F.1 Arrangement for potentiometric titration with a glass electrode.

the electrode is negligible. Although it is traditional to use a calomel electrode for reference electrode (2), a simpler and more compact dipping electrode is a silver–silver chloride wire (similar to that inside the glass electrode bulb) in a tube containing saturated KCl (Fig. 13F.1(b)). Compact inexpensive versions are available commercially.

7. OTHER SELECTIVE ELECTRODES

The realization that the ordinary glass electrode operates by an ion exchange between the solution and the glass has led to a considerable extension of potentiometric methods. By altering the composition of the glass the electrode can be made sensitive to other monovalent cations (e.g. Na^+, K^+). If the glass membrane is replaced by other ion exchange membranes (liquids or solids such as silver halides or lanthanum fluoride), the electrode become sensitive to many divalent cations and some anions. All these are sensitive not only to the ion for which they are designed but also to some extent to other ions in solution, unlike the ordinary glass electrode for H^+. However membranes of biological material are highly specific, and can, for example, be used to measure the concentration of a small amount of K^+ ions in the presence of a large concentration of Na^+.

8. THE pH METER

Glass electrodes generally have a resistance of the order of 10^7–$10^8\ \Omega$. Since it is necessary to measure the electrode potential to better than 1 mV for an accuracy of 0·02 pH units, the instrument for detecting the balance point of the potentiometer must be sensitive to less than 10^{-10}–10^{-11} A. This is too small a current to be detected by ordinary galvanometers. Early workers used quadrant electrometers, but nowadays transistor or thermionic valve circuits are always employed.

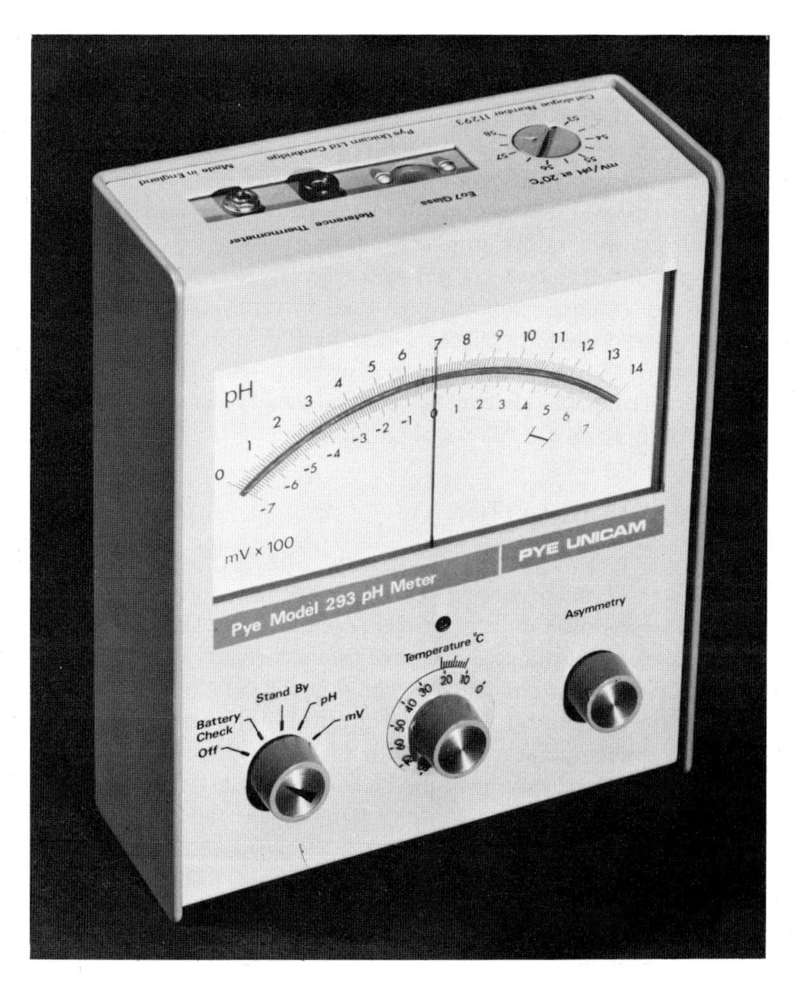

FIG. 13F.2 pH meter, direct reading. (Courtesy Pye Unicam Ltd, Cambridge.)

A valve electrometer can be used to adapt an ordinary potentiometer (section 4C(1)) for work with glass electrodes. More often, however, a potentiometer and electronic voltmeter are combined into a single instrument, the dials of which are graduated in terms of pH instead of mV, and a rheostat is provided for balancing out the asymmetry potential of the glass electrode. Such instruments are known as *pH meters*.

Two types of pH meter are in common use. The most accurate consists of a potentiometer (calibrated in pH units) whose voltage is adjusted until it balances that of the cell; the out-of-balance signal is detected by an electrometer circuit and then amplified. In the second type the cell voltage is amplified and displayed, either on a scale calibrated in pH units (Fig. 13F.2), or using a digital voltmeter (section 4C(3)) to display the pH in digits.

All pH meters have a dial which is set to the temperature of the solution, and which compensates for the variation of e.m.f. of the cell with temperature (section 13F(6)). Most pH meters can also be used as electronic millivoltmeters.

It is important to connect pH-meter circuits to earth at one point and to use screened leads between the electrodes and the meter in order to avoid capacity effects. Difficulty is also sometimes experienced, particularly in humid weather, owing to leakage currents; the trouble can often be removed by standing the pH meter and the electrode assembly on flat blocks of paraffin wax.

9. STANDARDIZATION OF GLASS ELECTRODES

All pH meters have provisions for standardizing the glass electrode in a buffer solution of known pH. This is necessary because different electrodes have different asymmetry potentials. Once the adjustment has been made, so that the meter registers correctly the known pH of the buffer solution, the instrument gives the pH of other solutions without any calculation. The temperature of the buffer solution must be set on the temperature dial of the pH meter before standardization. The standardization should be checked occasionally.

The buffer solutions shown on the next page have been recommended for standardizing glass electrodes.

If the electrode is standardized with solution *A*, it can be tested for satisfactory functioning by measuring the pH of solution *B*. If very acid or very alkaline solutions are to be studied, it is better to standardize with *C* or *D* respectively in order to minimize the errors to which the glass electrode is liable at extreme pH values (see below).

For pH range		*pH Standards for glass electrodes (25°C)**		
		Buffer solution	*kg m^{-3}*	*pH*
3–10	*A*	50 mol m^{-3} potassium hydrogen phthalate	10·21	4·00$_5$
	B	50 mol m^{-3} borax	19·07	9·18
<3	*C*	10 mol m^{-3} potassium tetroxalate ($2H_2O$)	2·54	2·15
>10	*D*	10 mol m^{-3} trisodium phosphate ($12H_2O$)	3·80	11·72

* Only slightly dependent on temperature; solution *A* becomes 4·01 at 38°C and 4·10 at 60°C.

After an electrode has been standardized, it is rinsed with water and then with the solution to be studied, the pH of which is read at once. If the electrode is then washed again and returned to the first buffer it should, of course, give the correct pH again. If it does not, a perforation of the electrode should be suspected. Solutions *A* and *B* should then be used for checking the reproducibility of the electrode.

Limitations of the glass electrode. Ordinary glass electrodes function satisfactorily from pH 0 to pH 9, and can be used even in unbuffered solutions in the range 6–8 provided the solution is agitated with a stream of nitrogen to remove CO_2 and to condition the electrode surface. As the pH is increased above 9 they begin to give results which are increasingly too high compared with those given by a hydrogen electrode in the same solutions. The error at high pHs is found to depend particularly on the *sodium* ion concentration; for example, at pH 12 the electrode might read 12·8 in a solution which is 10 mol m^{-3} in respect to Na^+. The 'alkali error' depends on the temperature, type of glass, composition of solution, etc. The standardization of the asymmetry potential of the electrode may also be disturbed by strong alkali. In recent years a special glass electrode has been introduced for the range pH 9–14, but high accuracy cannot be expected with it.

Glass electrodes are also subject to errors in highly acid solutions, at high temperatures, in concentrated alcohol solutions, and in the presence of ionized colloids.

EXPERIMENTS

(1) *Test the functioning of a glass electrode.* Measure the pH of a number of buffer solutions of pH ranging from 1 to 11·7 (section 13F(3)), noting particularly the reproducibility of the results on transfering the electrode from buffers of high pH to those of lower pH.

(2) *Investigate the 'salt effect' on pH.* Determine the effect of KCl on the pH of solutions of HCl. Prepare 10 mol m^{-3} HCl and add to portions of it various amounts of KCl, by weight, up to 100 mol m^{-3}. Determine the pH with a glass electrode calibrated on buffer *C*. Since $pH \approx -\log_{10} m_H f_\pm$, the effect of ionic strength I on $f_\pm$ can be obtained approximately, taking $f_\pm$ for pure 10 mol m^{-3} HCl as 0·906. Plot $\log f_\pm$ against $\sqrt{I}$; see sections 13A(4) and 13A(5).

(3) *Determine the degree of hydrolysis of aniline hydrochloride.* Prepare 40, 20, 10 and 5 mol m^{-3} solutions of aniline hydrochloride, and determine their pH values with a glass electrode.

Preparation of aniline hydrochloride. Pass dry HCl gas (made by dropping conc. HCl into conc. H_2SO_4 *in a fume cupboard*) through a 10% solution of aniline in dry ether. Filter off the precipitated hydrochloride, wash with dry ether, and store in a desiccator over KOH.

Calculation. The hydrolysis reaction is

$$C_6H_5NH_3^+Cl^- = C_6H_5NH_2 + H^+Cl^-$$

If x is the degree of hydrolysis, the classical hydrolysis constant K_h is given by $x^2m/(1-x)$, where m is the overall molality of aniline hydrochloride. The thermodynamic hydrolysis constant would be $a_{C_6H_5NH_2} \times a_{H^+}/a_{C_6H_5NH_3^+}$ which is practically the same as the classical expression since $f_{C_6H_5NH_2} \approx 1$ and $f_{H^+} \approx f_{C_6H_5NH_3^+}$, and the only correction needed is to allow for the fact that $pH \approx m_{H^+} f_\pm$. The value of $f_\pm$ can be estimated approximately (section 13A(5)).

The dissociation constant of aniline as a base,

$$C_6H_5NH_2 + H_2O = C_6H_5NH_3^+ + OH^-$$

is

$$K_b = m_{AH^+}\, m_{OH^-}/m_A \quad (A = C_6H_5NH_2)$$

and hence $K_h = K_w/K_b$, where K_w is the ionic product of water ($=0·6 \times 10^{-14}$ mol^2 (kg $H_2O)^{-2}$ at 18°C). Hence K_b can be calculated approximately.

(4) *Other experiments:* as for hydrogen and quinhydrone electrodes (sections 13F(4) and 13F(5)). For potentiometric titrations with a glass electrode, see section 13G.

10. pH DETERMINATION WITH INDICATORS

It is convenient to include in this section on pH measurements a short account of the use of *indicators* for approximate pH determination. pH indicators are weak acids or bases which exhibit different colours in their ionized and unionized states. For example,

$$\underset{\text{(red)}}{\text{methyl orange acid}} \rightleftarrows \underset{\text{(yellow)}}{\text{methyl orange anion}} + H^+$$

The dissociation constant K_a is given by $K_a = [\text{anion}][H^+]/[\text{acid}]$ and hence $\text{pH} = \text{p}K_a + \log_{10} [\text{anion}]/[\text{acid}]$, where $\text{p}K_a = -\log_{10} K_a$. If p$K_a$ (which is a constant at a given temperature) is known, the pH of a solution can be determined by finding an indicator which is at its end point for the solution, i.e. for which both coloured forms are present when the indicator comes to equilibrium with the solution. The pK_a values for a number of common indicators may be read from the table below.

Conversely the table indicates which indicator should be chosen for acid–base titrations, i.e. the indicator whose pK_a value most closely matches the pH of the solution at the endpoint.

Table showing pK_a of common indicators

Bromphenol blue	*Methyl red*	*Bromcresol purple*	*Bromthymol blue*	*Phenol red*	*Cresol red*	*Thymol blue*
4·1	5·0	6·3	7·1	7·7	8·1	8·8

BIBLIOGRAPHY 13F: pH Determination

Bates, *Determination of* pH, *Theory and Practice*, 2nd edn, 1964 (John Wiley, New York).
Kolthoff and Laitenin, pH *and Electrotitrations*, 2nd edn, 1944 (John Wiley, New York).
Gold, pH *Measurements*, 1956 (Methuen, London).
Dorst (ed.), *Ion Selective Electrodes*, 1969 (Nat. Bureau Stand. Publn 314).

13G Potentiometric titrations

1. THEORY

Since the potential of an electrode dipping in the solution of an electrolyte depends on the concentration of the ions with which the electrode is in equilibrium, it is possible to use determinations of the potential as an 'indicator' in volumetric analysis. The electrode potential depends on the *logarithm* of the concentration of ions, and is therefore not suitable for obtaining concentration directly with any accuracy, but the change of potential with concentration *during a titration* provides an accurate indication of the equivalence point. Thus, the cell

$$H_2(Pt)\,|\,\text{acid solution}\,\|\,\text{KCl } aq.\,\|\,\text{calomel electrode}$$

will have a certain e.m.f. depending on the pH of the solution. On adding small portions of a standard solution of alkali to the acid, the e.m.f. of the cell will alter slowly at first, because the change in the electrode potential depends on the *fraction* of hydrogen ion removed. As the amount of alkali added approaches equivalence to the amount of hydrogen ion in the solution, the fraction of the hydrogen ion concentration removed by each drop of alkali solution rapidly increases, and there is a correspondingly rapid change in the e.m.f. Later, as excess of alkali is added, the e.m.f. again shows a slow change. Consequently, when the e.m.f. of the cell, E, is plotted against the volume v of standard alkali added, a curve of the form shown in Fig. 13G.1 is obtained. The end-point of the titration is given by the

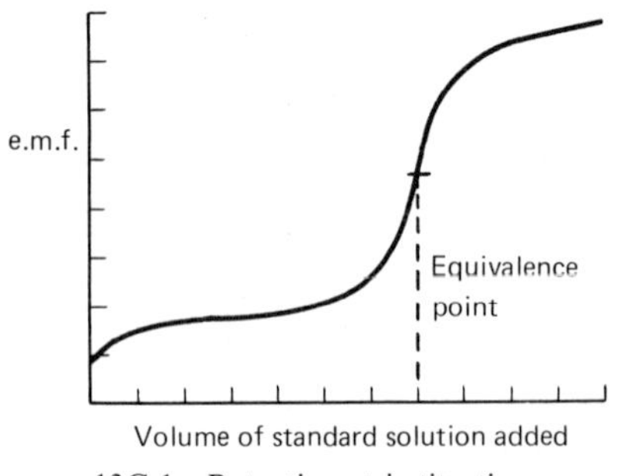

FIG. 13G.1 Potentiometric titration curve.

point of inflection on the curve. The point of inflection can be most easily found by plotting the values of $\Delta E/\Delta v$ against volume of titrating solution. At the point of inflection the value of $\Delta E/\Delta v$ is a maximum.

In carrying out the titrations, the titrating liquid is run, in small quantities at a time, from a burette into the solution to be titrated, the solution being kept well mixed by means of a stirrer. As the titration approaches the equivalence point, the titrating liquid is added in smaller and smaller amounts so that the graph in the neighbourhood of the equivalence point is obtained with precision.

Advantages. (1) Potentiometric titrations are applicable to any reactions for which an appropriate electrode is available; for example, sulphides can be titrated with lead salts, using a lead electrode. Many reactions for which no colour indicator is available can be employed in the potentiometric method.

(2) The determination is very reliable, since the result depends on a number of independent readings, not on one subjective judgment of an 'end-point' (which may be rather ill-defined in indicator titrations). This is particularly important in research work when the quantity of material available is limited.

(3) An accuracy comparable with that of the best gravimetric analysis can be achieved if weight-burettes are employed instead of volumetric burettes, and the potentiometric titration is more rapid.

(4) Potentiometric titrations can be carried out on a microscale with little difficulty and can often be extended to extremely dilute solutions which would be beyond the sensitivity of conventional volumetric analysis.

2. TECHNIQUE: SIMPLIFIED POTENTIOMETRY

The obvious method of carrying out potentiometric titrations is to set up a cell, one electrolyte of which is to be titrated, and to determine its e.m.f. accurately in the usual way with a potentiometer after each addition of titrant. While this is most satisfactory, it is possible to dispense with a precision potentiometer since the highest accuracy

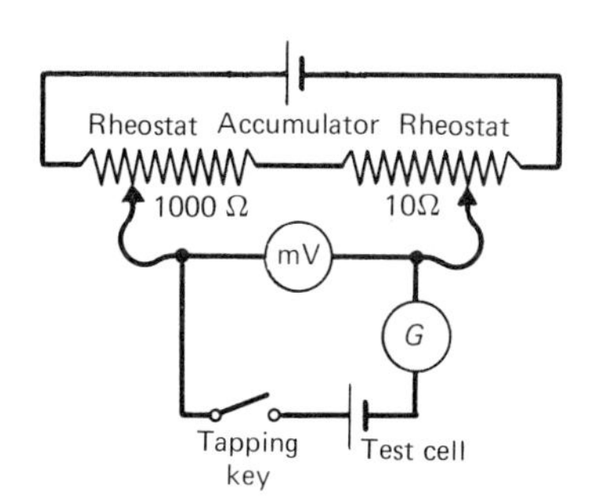

FIG. 13G.2 Simple substitute for a laboratory potentiometer.

of e.m.f. measurement is not needed. Figure 13G.2 shows a simple arrangement requiring only two wire-wound rheostats (coarse and fine), a millivoltmeter (shunted if too sensitive), a galvanometer and a tapping key. The rheostats are adjusted until, on closing the key, no deflection is produced on the galvanometer. The millivoltmeter then reads the e.m.f. of the cell directly. The two rheostats can be replaced by a single 10 turn 'Helipot'; the millivoltmeter can then be omitted, and the helipot dial setting at balance used as a measure of e.m.f.

Another method, and one particularly suitable for potentiometric titrations since no balancing is needed at each reading, is the use of a direct reading instrument: either a pH meter used directly for acid–base titrations or as an electronic millivoltmeter, or some other form of electronic millivoltmeter or a digital voltmeter (section 4C(3)). The high input impedance of these instruments ensures that negligible current is drawn from the cell and consequently the electrodes can be quite small. This in turn means that small samples can be titrated.

3. DIFFERENTIAL POTENTIOMETRIC TITRATIONS

It has been mentioned that the point of inflection on a potentiometric titration (E against v) curve, which gives the equivalence point, can be most accurately located by plotting a graph of v against $\Delta E/\Delta v$ where ΔE and Δv are the increments of e.m.f. and of volume respectively at each addition during the titration. The graph takes the form shown in 13G.3. A sharp, and more or less symmetrical, peak is obtained at the equivalence point. Instead of recording E and v and later computing $\Delta E/\Delta v$ values, one can obtain the differential curve directly, by a device such as that shown in Fig. 13G.4 by means of which the difference between successive E values is read with any type of potentiometer. A and B are duplicate electrodes, reversible with respect to the substance in solution which is to be titrated (e.g. Ag electrodes in $AgNO_3$ solution, to be titrated with KCl solution). B is enclosed in a tube which temporarily keeps back a small portion of the solution. Initially, of course, the e.m.f. between A and B is zero, but after a portion of titrant has been added, the e.m.f. of A is

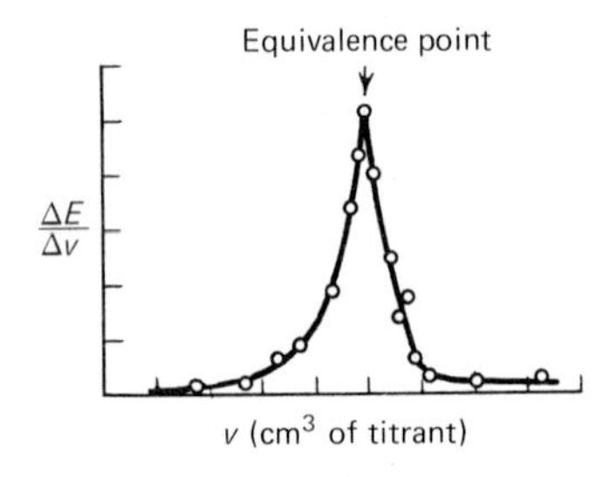

FIG. 13G.3 Differential potentiometric titration graph.

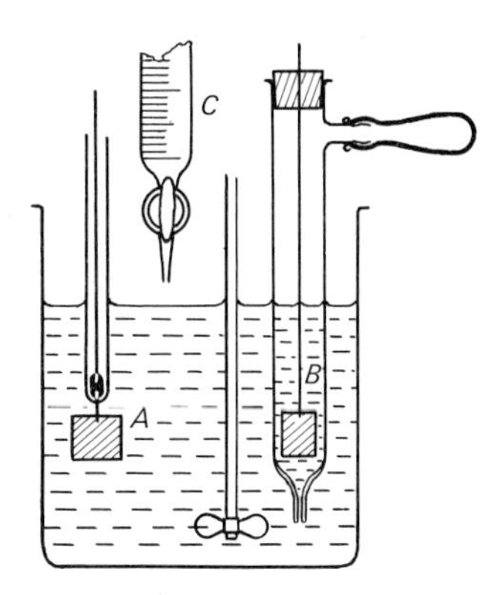

FIG. 13G.4 Method of carrying out a differential potentiometric titration, using a pair of similar electrodes.

thereby changed whereas that of *B* is unaffected because the solution around *B* is not mixed with the main bulk. The small e.m.f., ΔE, between *A* and *B* is noted, and then the solution is expelled from the enclosure (for example, by compressing a rubber teat attached at the top of the tube). When the compartment refills with solution, the e.m.f., of course, returns to zero and the next addition of titrant can be made. This technique is widely applicable and has the advantages that full sensitivity of the potentiometer can be used, and the early stages of the titration can be accomplished rapidly because the approach of the end-point is apparent from the increasing ΔE values.

Some modern pH meters have provision for a higher sensitivity over a limited range of potential: this can be used to give increased precision near the end point. If the normal burette is replaced by a mechanically operated piston type, and the potentiometer by a potentiometric recorder (section 4C(4)) the titration curve can be recorded automatically. Alternatively the titration can be halted automatically when the potential reaches a predetermined value corresponding to the end point.

4. EXAMPLES OF POTENTIOMETRIC TITRATIONS

The following experiments illustrate some of the principal types of titration that can be conveniently carried out potentiometrically. Any of the methods of e.m.f. measument described above can be used if only the equivalence points are required, but a potentiometer accurate to 1 mV should be used if the theory of the titration curves is to be studied in the manner suggested in the text.

5. ACID–ALKALI TITRATIONS

The pH can be determined in the titration with a hydrogen electrode (section 13D(1)), quinhydrone electrode (section 13F(5)), or glass electrode (section 13F(6)). Some suitable cell assemblies are shown in

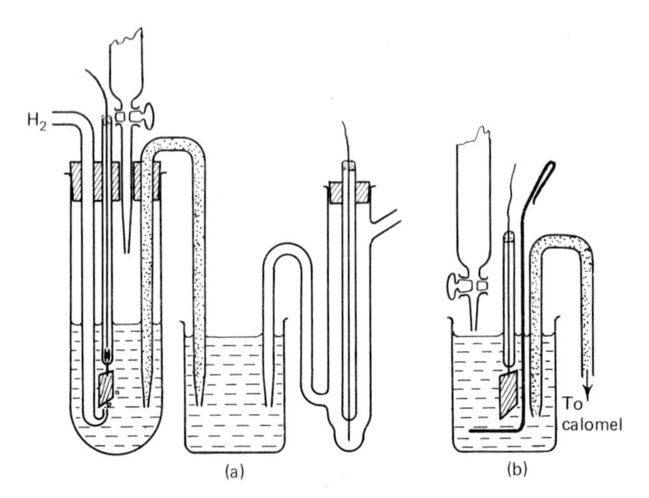

FIG. 13G.5 Arrangements for potentiometric titrations with (a) a hydrogen electrode, (b) quinhydrone or other redox electrode.

Figs. 13F.1 and 13G.5. Many variations are possible; for example a dipping type of reference electrode with a narrow capillary tip can be used with advantage provided an instrument with high input impedance of the type described above is used for the e.m.f. measurements.

EXPERIMENTS

Use the hydrogen electrode for the following titrations. (Quinhydrone or glass electrodes are unsuitable because of the extreme pH values involved.) Titrate (a) 100 mol m^{-3} HCl *with 100 mol m*$^{-3}$ NaOH, (*b*) *100 mol m*$^{-3}$ *acetic acid with 100 mol m*$^{-3}$ NaOH, (*c*) *50 mol m*$^{-3}$ H_2CrO_4 *or* H_3PO_4 *with 100 mol m*$^{-3}$ NaOH, (*d*) *aniline with acetic acid.*

Results. Plot pH against volume of titrant. Note the forms of the curves and the pH at the equivalence points and at the half-neutralization stages. The experiments illustrate the following types of titration curve: (a) strong acid–strong base, (b) weak acid–

strong base, (c) dibasic or tribasic acid–strong base (the first stage being a strong acid, the second a weak one and the third, in phosphoric acid, so weak that no inflection point in the curve is seen for Na_3PO_4—i.e. it is largely hydrolysed into HPO_4^{--} and OH^-), (d) weak base–weak acid (salt almost entirely hydrolysed).

One may study also the titration of mixtures of weak and strong acids (or bases), and dibasic acids having dissociation constants which do not differ greatly and hence show 'overlap' between the first and second stages of neutralization (e.g. oxalic, tartaric, succinic, phthalic acids).

An amphoteric substance like glycine may be titrated with acid and with alkali. At the extremes of pH a correction for the OH^- or H^+ ions in the solution should be applied; a 'blank' with an equal volume of pure water is run, and hence the acid or alkali actually *combined* by the substance can be plotted against pH.

EXPERIMENT

Carry out an accurate electrometric titration of a weak acid with a strong alkali, using a quinhydrone or glass electrode, and determine from the results the dissociation constant(s) of the acid.

Suitable acids: monobasic–lactic, benzoic or acetic; dibasic, with groups separated–malonic; dibasic, groups overlapping–tartaric, succinic.

Procedure. Set up a cell as in Fig. 13G.5(b) or 13F.1(a) with a platinum quinhydrone electrode or glass electrode to measure the pH and a calomel or Ag/AgCl electrode of known e.m.f. (sections 13B(1), 13C(3.3)) as reference electrode. In the case of the quinhydrone electrode a preliminary check may be made by measuring the e.m.f. of the cell with 50 mol m^{-3} potassium hydrogen phthalate in the cell; this should give a pH of 4·00 at 15°C. To carry out the titrations, place 100 cm^3 of 10 mol m^{-3} acid in the titration vessel and titrate it with 200 mol m^{-3} carbonate free NaOH from a 10 cm^3 microburette. The concentration of either the acid or the alkali must be known accurately. The titration should be carried a little past the equivalence point, notwithstanding the reduced accuracy of the electrodes at high pH; the precise pH values are not needed for the calculation, but the inflection point of the titration curve can be judged better if the complete curve is drawn. The complete titration curve should be repeated with a fresh portion of acid as a check on the accuracy.

Calculation

(a) *Monobasic acids.* First locate the equivalence point as accurately as possible by a differential graph of $\Delta E/\Delta v$ or $\Delta pH/\Delta v$ against v (Fig. 13G.3). Then compute for a series of points along the titration curve the following quantities: pH, concentration of salt Na^+A^- formed (=concentration of added base=b), concentration of free acid HA remaining (=$(a-b)$ where a is the initial concentration of acid). According to the elementary theory (neglecting activity coefficients) the dissociation constant of the acid is given by $K_a = m_{H^+}m_{A^-}/m_{HA}m_{H^+}$ obtained from the measured pH. It derives from the dissociation of the free acid, and therefore $m_{HA} = a-b-m_{H^+}$. Also, the concentration of anions, m_{A^-}, derives largely from the acid neutralized (=b) but partly from dissociation of the acid; hence $m_{A^+} = b+m_{H^+}$. Therefore

$$K_a = m_{H^+}(b+m_{H^+})/(a-b-m_{H^+})$$

This expression may be used to calculate values of K_a for several positions along the titration curve. It can be further simplified if the acid is weak. Then m_{H^+} is negligible compared with $(a-b)$ or b, and the expression can easily be reduced to

$$\begin{aligned} \text{pH} &= \text{p}K_a + \log_{10}[b/(a-b)] \\ &= \text{p}K_a + \log[(\text{cm}^3 \text{ of titration})/(\text{cm}^3 \text{ still needed to reach equivalence point})]. \end{aligned}$$

A graph of pH against $\log[b/(a-b)]$ should give a straight line of unit slope, and the pH at the point where $(a-b)=b$ (half-neutralization point) gives the pK_a of the acid ($= -\log_{10}K_a$). As a and b are only required as a ratio, they can be in any convenient units.

Activity correction. The molality terms in the above equations ought to be multiplied by their proper activity coefficients. Remembering that pH can be taken as $-\log a_{H^+}$ (section 13F(1)), the final equation becomes

$$\text{pH} = \text{p}K_a' + \log(m_{a^-}/m_{HA}) + \log f_A - \log f_{HA}.$$

Provided the solutions are dilute, f_{HA} can be neglected (HA being an uncharged species) and the only correction needed is f_{A^-}. This can be estimated roughly from the simple Debye–Hückel theory as $-\log_{10} f_{A^-} = 0{\cdot}51\sqrt{I}$ where I is the ionic strength of the solution ($=\sum \frac{1}{2}m_i z_i^2$, m_i=molality of ion of valency z). Alternatively, the Güntelberg equation (section 13A(5)) can be used to get a better estimate of f_{A^-}. For the present purpose, since only univalent ions are concerned, $I = m_{Na^+A^-}$ = (moles of base added)/(mass of water in kg) ≈ (moles added)/(volume in $m^3 \times 0{\cdot}001$). The activity correction need be applied only to the data at the half-neutralization point, as correction of the other points would not alter the result obtained for K_a.

(b) *Dibasic acids:* $K_1 > 100\,K_2$. For the first dissociation of a dibasic acid $H_2A \rightleftharpoons HA^- + H^+$, $K_1 = m_{H^+} m_{HA^-}/m_{H_2A}$. The theory is identical with that given for a monobasic acid, and pK_1 = pH of the solution when half an equivalent of alkali has been added. For the second dissociation, $HA^- \rightleftharpoons A^{--} + H^+$, and $K_2 = m_{A^-} m_{H^+}/m_{HA^-}$. If $K_1 > 100\,K_2$, the first stage is practically complete before the second begins, and the form of the curve between the first and the second equivalents follows the same type of equation as for another monobasic acid. Thus pK_2 can be obtained as the pH of the solution when $1\frac{1}{2}$ equivalents of alkali have been added (i.e. $m_{A^-} = m_{HA^-}$). In this case the dibasic acid shows two points of inflection, the cm^3 of alkali being in the ratio 1:2, corresponding to the formation of NaHA and Na_2A.

(c) *Dibasic acids:* $K_1 < 100\,K_2$. In this case the titration curve may show no signs of two inflections, only the final Na_2A stage giving a step. This is because the dissociation reactions 'overlap', the second starting before the first is nearly complete. If d is the total molality of acid to the solution and b the molality of added NaOH, it can be shown by approximate theory similar to that employed above that for any point on the titration curve the following relationship holds:

$$\frac{b + m_{H^+}}{m_{H^+} + 2K_2} = \frac{dK_1}{m_{H^+}^2 + K_1 + K_1 K_2}$$

In order to get K_1 and K_2 it is necessary to choose two points on the titration curve (say, after $\frac{1}{2}$ and $1\frac{1}{2}$ equivalents of alkali have been added) and solve the resulting pair of simultaneous equations. The approximation m_x = (moles of x)/(volume of soln in $m^3 \times 0{\cdot}001$) can again be used for dilute solutions.

It should be added that none of the above equations can be used for strong or for extremely weak acids. For further study of titration curves the reader is referred to monographs on pH.[1]

6. PRECIPITATION TITRATIONS

Any precipitation reaction can form the basis of a potentiometric titration provided one of the ions can participate in a reversible electrode reaction. However, the reaction may not necessarily provide an accurate analytical method; in some reactions the end-point may be obscured by partial solubility (e.g. $PbCl_2$) or adsorption.

EXPERIMENT

Titrate approximately 10 mol m^{-3} $AgNO_3$ *with approximately 10 mol m*$^{-3}$ KCl.

The differential method illustrated in Fig. 13G.4 is very convenient for such a titration. The electrodes are simply two silver wires, freshly cleaned in nitric acid. The $AgNO_3$ solution is pipetted into the beaker, and the KCl solution run in from the burette. The theoretical titration curve for this reaction can readily be predicted from the theory of the silver and silver–silver chloride electrodes (section 13C(3.3)). The titration could also be reversed, and thus chlorides, bromides, iodides or thiocyanates, could be determined.

Other examples: titration of Ag, Hg, Pb, Cd, Zn, with sodium sulphide, zinc with potassium ferrocyanide.

7. 'REDOX' TITRATIONS

Any reversible redox reaction can be used for a potentiometric titration, with great ease. For example, a Pt electrode in a solution of a ferrous salt takes up a potential dependent on $\log_e(m_{Fe^{++}}/m_{Fe^{+++}})$. $m_{Fe^{+++}}$ is initially infinitesimal but if the solution is titrated with an oxidizing agent the ratio $(m_{Fe^{++}}/m_{Fe^{+++}})$ changes rapidly. There is a 'step' in the E–v curve at the equivalence point and beyond that the e.m.f. is determined by the redox behaviour of the excess oxidizing agent. In a potentiometric titration it is best to choose oxidizing or reducing agents with widely different redox potentials in order to make the 'step' at the equivalence point very large and steep. Such an example is the titration of a ferrous salt with a ceric salt; the equivalence point can be located extremely accurately.

EXPERIMENT

Titrate ferrous sulphate potentiometrically with ceric sulphate (or potassium dichromate), and determine the formal redox potential of the ferrous–ferric system (and of the cerous–ceric system).

Procedure. Prepare approx. 100 mol m^{-3} solutions of ferrous ammonium sulphate and of ceric sulphate, both solutions being made up in approx. 2 000 mol m^{-3} sulphuric acid to prevent hydrolysis. Pipette out 25 cm^3 of the ferrous solution into the titration vessel (similar to that shown in Fig. 13G.5(b)) and dip into it a bright platinum wire electrode (section 13E(2)). Connect the solution to a calomel electrode via a narrow U-tube containing agar jelly (section 13B(2)); alternatively, the reference electrode can be a glass electrode immersed in the solution, since this electrode is independent of redox reactions.

Carry out the titration, measuring the cell e.m.f. with any type of potentiometer or pH meter. Take numerous readings in the neighbourhood of the equivalence point, where the e.m.f. changes rapidly, and carry the titration well beyond the equivalence point.

Results. Plot the e.m.f. against cm^3 of ceric sulphate (v), and $\Delta E/\Delta v$ against v. Hence deduce the equivalence point v_0 and calculate the formal redox potential from the e.m.f. of the cell at the point where $v_0/2$ cm^3 had been added. A graph of e.m.f. versus $\log_e (m_{Fe^{+++}}/m_{Fe^{++}})$ should give a straight line of slope $2{\cdot}303\boldsymbol{RT}/\boldsymbol{F}$. Similarly, one can study the cerous–ceric potential in exactly the same way, using the e.m.f. values obtained in the presence of excess ceric sulphate.

BIBLIOGRAPHY 13G: Potentiometric titrations

Weissberger, Vol. I, Pt 4, chapter XLIV.

Charlot *et al*, *Electrochemical Reactions: the Electrochemical Methods of Analysis*, 1962 (Elsevier, Amsterdam).

Kolthoff and Laitenin, *pH and Electrotitrations*, 2nd edn, 1944 (John Wiley, New York).

Wilson, *Ann. Reports Chem. Soc.*, 1952.

[1] See Bibliography 13F.

13H Determination of thermodynamic quantities from e.m.f. measurements

1. The e.m.f. of a reversible cell is a direct measure of the maximum available energy ($w'_{max.}$) released by the cell reaction, and this gives at once the Gibbs free energy change (ΔG) of the reaction, i.e. $z\boldsymbol{EF} = w'_{max.} = -\Delta G$.

A knowledge of ΔG for the cell reaction makes it possible to calculate the equilibrium constant K for the reaction as it would proceed 'unharnessed'—for example, if the cell were short-circuited. The relationship is $\Delta G^\circ = -\boldsymbol{R}T \log_e K$; here ΔG° is the *standard* free energy of the reaction, i.e. it is the value of ΔG for the reaction conducted with reactants and products in their standard states. If the cell reaction involves pure solids or pure gases at 1 atm. only, $z\boldsymbol{EF}$ gives ΔG° directly. If solutions are involved in the cell reaction, then ΔG° must be calculated from the ΔG value by means of the van't Hoff reaction isotherm, namely, $\Delta G = \Delta G^\circ + \boldsymbol{R}T \log_e X$, where X stands for the mass law product for resultants and reactants at the activities appropriate to the cell. An example of a reaction for which K can be obtained both potentiometrically and by analysis is TlCl (s)+ KSCN (aq) ⇄ TlSCN (s)+KCl (aq) (see below).

The *temperature coefficient* of the e.m.f. of a cell provides a method of obtaining the entropy change ΔS and heat of reaction. Since $(\partial \Delta G/\partial T)_p = -\Delta S$ and $\Delta G = -zEF$, hence $\Delta S = z\boldsymbol{F}(\mathrm{d}E/\mathrm{d}T)$. Also, from $\Delta G = \Delta H - T\Delta S$, it follows that $\Delta H = -zEF + Tz\boldsymbol{F}(\mathrm{d}E/\mathrm{d}T)$. $-\Delta H$ is equal to the heat evolved when the reaction is allowed to proceed unharnessed (e.g. cell short-circuited). $T\Delta S$ is equal to the heat absorbed when the reaction is conducted reversibly, as in the balanced cell. As $\mathrm{d}E/\mathrm{d}T$ is usually small, it is necessary to measure E very accurately (say, to $\pm 0{\cdot}01$ mV) if ΔS and ΔH are to be obtained reliably: consequently, the method is restricted to cell reactions involving the more satisfactory types of electrode.

EXPERIMENT

Determine the equilibrium constant of the reaction

$$\mathrm{TlCl}\,(s) + \mathrm{KSCN}\,(aq) = \mathrm{TlSCN}\,(s) + \mathrm{KCl}\,(aq)$$

by the e.m.f. method and also by analysis. Determine the ΔS and ΔH values of the reaction.

Method. Set up the following cell.

Tl-amalgam | TlCl (satd), 80 mol m^{-3} KCl ‖ satd
NH_4NO_3 ‖ 100 mol m^{-3} KSCN, TlSCN (satd) | Tl-amalgam

Two half-cells similar to those used for calomel electrodes can be employed, the junction being made by dipping the delivery tubes of both electrode vessels in a tube of satd NH_4NO_3 solution. All three vessels must be supported in a thermostat. The thallium amalgam, which should be of the same composition in both electrodes, should contain 1–2% thallium. It is best prepared by electrolysing a solution of thallium sulphate containing the calculated amount of thallium, with a known weight of mercury as cathode. The e.m.f. of the cell is small; it is determined with an accurate potentiometer, used on its sensitive range, in combination with a sensitive galvanometer or an electronic null-point detector. The temperature of the thermostat is changed in steps of 5° from 30 to 40°C. Sufficient time must be allowed for the cells to reach equilibrium at each temperature.

Equilibrium by chemical analysis. The chemical equilibrium may be determined by shaking, say, 200 cm^3 of 100 mol m^{-3} KSCN with 10 g of solid thallium chloride, or 200 cm^3 of 100 mol m^{-3} KCl with 10 g of solid thallium thiocyanate. To a portion of the equilibrium solution potassium hydrogen iodate and sulphuric acid are added in order to oxidize the thiocyanic acid to hydrocyanic acid, and the hydrocyanic acid and iodine liberated are driven off by boiling (in a fume cupboard). The chloride remaining is then titrated with silver nitrate.

Another equal portion of the equilibrium solution is titrated with silver nitrate so as to give the total chloride plus thiocyanate. By subtracting from this total the amount of chloride present, one obtains the amount of thiocyanate.

Calculation. When the reaction is allowed to proceed to equilibrium (not harnessed in a cell):

$$\frac{a_{\mathrm{TlSCN}} \times a_{\mathrm{KCl}\ aq}}{a_{\mathrm{TlCl}} \times a_{\mathrm{KSCN}\ aq}} = K$$

Since TlSCN and TlCl are both present as pure solids in excess, their activities are taken as unity, so

$$a_{\mathrm{KCl}\ aq}/a_{\mathrm{KSCN}\ aq} = K$$

Now $a_{\mathrm{KCl}} = a_{\mathrm{K^+}} \times a_{\mathrm{Cl^-}}$ and $a_{\mathrm{KSCN}} = a_{\mathrm{K^+}} \times a_{\mathrm{SCN^-}}$, and $a_{\mathrm{K^+}}$ is the same for both in an equilibrium mixture: hence $K = a_{\mathrm{Cl^-}}/a_{\mathrm{SCN^-}}$. The activity coefficients of these ions may be taken as equal, and thus $K = m_{\mathrm{Cl^-}}/m_{\mathrm{SCN^-}} = c_{\mathrm{Cl^-}}/c_{\mathrm{SCN^-}}$ (at equilibrium). (K is also equal to the ratio of the solubility products of TlCl and TlSCN.)

If a cell of the kind used above were constructed with KCl and KSCN concentrations such that $a_{\mathrm{Cl^-}} = K a_{\mathrm{SCN^-}}$, it would show zero e.m.f. If $a_{\mathrm{Cl^-}}/a_{\mathrm{SCN^-}} < K$, positive current will flow in the cell from right to left, thereby causing an increase in the concentration of the chloride ion and a decrease in the concentration of the thiocyanate ion. The 'driving force' of the cell is then given by

$$\Delta G (= -EF) = -RT \log_e K - RT \log_e (a_{\mathrm{Cl^-}}/a_{\mathrm{SCN^-}})$$

Hence, if E is measured, and a_{Cl^-}/a_{SCN^-} is taken as c_{Cl^-}/c_{SCN^-} (used in the cell), then K can be calculated. The value is to be compared with that found by analysis of the equilibrium mixture.

ΔH and ΔS are obtained from dE/dT as explained above.

2. OTHER REACTIONS SUITABLE FOR STUDY BY THE E.M.F. METHOD

2.1. The equilibrium constant of the reaction, $Fe^{+++} + Ag \rightleftarrows Fe^{++} + Ag^+$, may be calculated from the e.m.f. of the cell:

Pt	25 mol m^{-3} $Fe(NO_3)_3$ 75 mol m^{-3} $Fe(NO_3)_2$ 50 mol m^{-3} HNO_3	50 mol m^{-3} HNO_3	50 mol m^{-3} $AgNO_3$ 50 mol m^{-3} HNO_3	Ag

The ferrous nitrate is best obtained by mixing equivalent quantities of $Ba(NO_3)_2$ and $FeSO_4$.

2.2. The oxidation of hydroquinone by silver ions leading to the equilibrium, $C_6H_4(OH)_2 + 2Ag^+ \rightleftarrows C_6H_4O_2 + 2Ag + 2H^+$, may be studied by means of the cell,

$$\text{Pt} \mid \text{quinhydrone, 100 mol m}^{-3}\ HNO_3 \parallel \text{100 mol m}^{-3}\ HNO_3\text{, 1 mol m}^{-3}\ AgNO_3 \mid \text{Ag.}$$

2.3. For the calculation of heat of reaction from e.m.f., one may study the cell formed by a cadmium (or cadmium amalgam) electrode and a silver–silver chloride electrode immersed in a saturated solution of $CdCl_2$, $2{\cdot}5H_2O$. The reaction involved is:

$$Cd + 2AgCl + 2{\cdot}5H_2O = CdCl_2,\ 2{\cdot}5H_2O + 2Ag$$

It should, however, be remembered that such measurements are of little value unless the e.m.f. can be determined with an accuracy of about 0·01 mV.

2.4. The free energy, entropy and heat of the reaction

$$Hg_2Cl_2\ (s) + 2KOH\ (aq) = 2KCl\ (aq) + H_2O + HgO\ (s) + Hg\ (l)$$

are obtained from measurements of the following cell:

$$\text{Hg} \mid \text{HgO (satd.), 1 000 mol m}^{-3}\ \text{KOH} \parallel \text{KCl (satd.)} \parallel \text{1 000 mol m}^{-3}\ \text{KCl, } Hg_2Cl_2 \mid \text{Hg}$$

If the e.m.f. of the 1 000 mol m^{-3} KCl calomel electrode is assumed (section 13B(1)), the measurements give the free energy, etc., of the reaction

$$Hg + 2OH^- = HgO + H_2O + 2\,e$$

3. THE THERMODYNAMIC PROPERTIES OF ALLOYS

The thermodynamic properties of an alloy can often be readily obtained by using it as an electrode in, for example,[1] the cell

$$Cd\ (s)\,|\,CdSO_4\ (100\ mol\ m^{-3})\,|\,Cd(Hg)$$

If this cell is reversibly discharged until 2 faradays of electricity have passed, the electrode and cell reactions are:

at the left-hand electrode : $Cd(s) \rightarrow Cd^{++} + 2e^-$
at the right-hand electrode : $Cd^{++} + 2e^- \rightarrow Cd(Hg)$
∴ net cell reaction : $Cd(s) \rightarrow Cd(Hg)$

The e.m.f. E of the cell is therefore given by the equation

$$-2FE = \Delta G = \mu_{Cd(Hg)} - \mu_{Cd(s)} = RT \log_e a_{Cd(Hg)} \qquad (1)$$

where μ is the chemical potential. ΔG is the increase in Gibbs free energy attending the formation of one mole of cadmium in the amalgam from pure cadmium at T K. The entropy and enthalpy changes for this reaction are

$$\Delta S = -\left(\frac{\partial\,\Delta G}{\partial T}\right)_p = 2F\left(\frac{\partial E}{\partial T}\right)_p \qquad (2)$$

$$\Delta H = \Delta G + T\,\Delta S = -2FE + 2FT\left(\frac{\partial E}{\partial T}\right)_p \qquad (3)$$

It is clear from Eq. (1) that the activity of cadmium in the amalgam has been based on pure solid cadmium as the standard state. If x is the mole fraction of cadmium in the amalgam, its activity coefficient f is defined by the relation

$$a = xf \qquad (4)$$

f is unity for an *ideal* solution of cadmium in mercury, and in that case it follows from Eqs (1) and (3) that $\Delta H = 0$.

EXPERIMENT

Determine ΔG, ΔH and ΔS for the formation of 1 mole of cadmium in a 1 wt% amalgam at 25°C, and the activity coefficient of the solution.

The cell is illustrated in Fig. 13H.1. Both electrodes (*A* and *B*) are first prepared by plating. The only solution required is 100 mol m^{-3} cadmium sulphate.

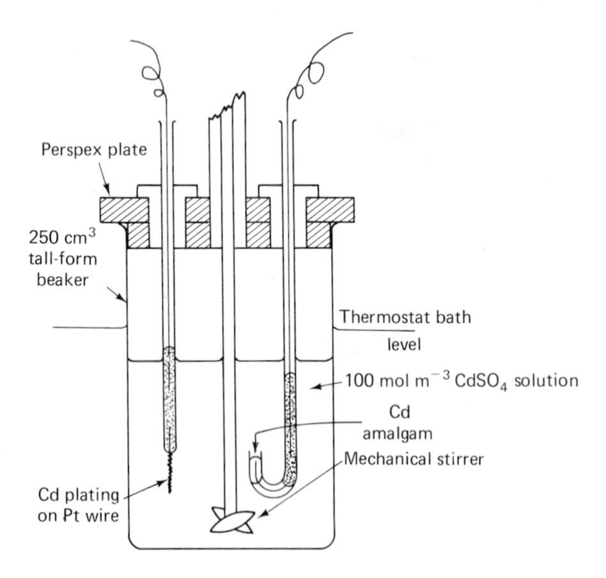

FIG. 13H.1 Cadmium amalgam cell.

The electrodes should always be suspended in a beaker from Perspex plates: they should never be put on a dirty bench top or allowed to rest under their own weight on the bottom of a beaker as this cracks the platinum–glass seals. Once the two electrodes have been prepared, they should be immersed in distilled water or in $CdSO_4$ solution at all times.

The cadmium electrode (*A*) is prepared by plating from the $CdSO_4$ solution onto the cleaned platinum wire protruding from the straight electrode. Electrode *A* is the cathode (−) and a cadmium rod serves as anode (+). The 1 mA current used is conveniently derived from a 110 V d.c. supply (Fig. 13H.2). Current should be passed until a fairly heavy deposit of cadmium is visible. A higher plating current gives a fuzzy deposit that falls off easily. When the cadmium electrode is transferred to the cell, care should be taken to avoid flaking off any of the deposit.

Amalgam electrode (*B*). Cadmium amalgams whose cadmium content exceeds 5 wt % are solid and it is believed that Cd_2Hg_7 is formed.

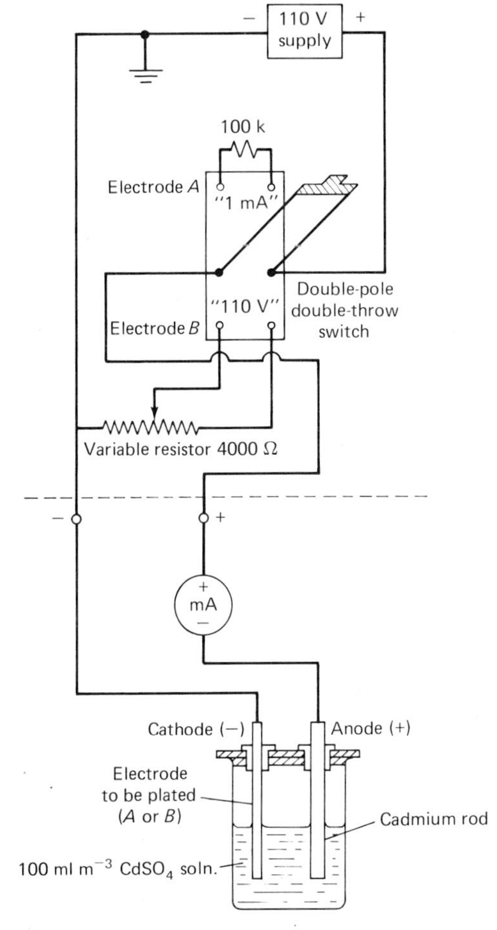

FIG. 13H.2 Plating circuit for cadmium electrodes.

One of the electrodes in the Weston standard cell consists of 12·5 wt % of cadmium amalgam. Amalgams containing less than 5 wt % cadmium are liquid, and in this experiment a 1·00 wt % amalgam is used. This is prepared by electrolysis. The cup of the J-shaped electrode (Fig. 13H.1) is cleaned with nitric acid, rinsed well and dried thoroughly. An accurately weighed drop of mercury is placed in the cup: about 0·1 cm^3 should completely cover the platinum wire.

The method of electrically depositing cadmium in the mercury is similar to the cadmium plating of electrode (*A*) described above, but the electrolysis is now quantitative. The number of coulombs needed to make a 1·00 wt % amalgam are calculated and the current set so that electrolysis takes about 45 min. Electrode B is the cathode (−) and the cadmium rod is the anode (+). If the voltage is applied with the wrong polarity even momentarily, the electrode is ruined; the preparation must be repeated with fresh mercury after thorough cleaning. 110 V d.c. is used together with a variable resistor to provide the appropriate current (Fig. 13H.2). The resistor is adjusted from time to time to keep the current constant at the selected value. Care must be taken that the cadmium plated out dissolves in the mercury and does not float off into the solution; continuous stirring of the amalgam with a finely drawn-out glass rod is advisable. On completion, the amalgam electrode (homogenized by stirring) and the cadmium electrode should be transferred to the cell.

The cell is suspended in the thermostat by means of the support. The e.m.f. is measured at 5 min intervals until it is steady at four or five different temperatures in the range 0–40°C. For the lower temperatures, a beaker filled with cold water and ice makes a fair thermostat.

The cadmium in the amalgam is slowly attacked by oxygen dissolved in the solution. E.m.f. measurements should therefore be carried out on the same day as the preparation of the amalgam.

Calculations. The e.m.f. values are plotted against temperature. From the best smooth curve through the experimental points, the value of E at 25°C and the slope at this temperature are found.

The activity and activity coefficient of cadmium in the 1·00 wt % amalgam at 25°C and the values of ΔG, ΔS and ΔH for the formation of one mole of cadmium in this amalgam are obtained using Eqs (1)–(4) of the previous section (13H(3)). Is the alloy an ideal solution?

BIBLIOGRAPHY 13H: Determination of thermodynamic quantities from e.m.f. measurements

See monographs cited under 13A and 13E.

[1] Parks and LaMer, *J. Amer. Chem. Soc.*, 1934, **56**, 90.

13I Determination of transport numbers from e.m.f. measurements

In certain cases cells may be used to obtain transport numbers. This application is based on a comparison of the e.m.f. of concentration cells with and without liquid junctions ('transport'), (section 13A(7)).

Instead of eliminating the junction potential by means of a salt bridge, one can interpose between the two half-cells a metal electrode reversible to the ion of opposite sign to that concerned at the primary electrodes. For example, the cell

$$H_2(Pt) \mid HCl(a_\pm)_1 \parallel HCl(a_\pm)_2 \mid H_2(Pt)$$

has an e.m.f. given by $E_1 = 2t_-(\boldsymbol{RT}/\boldsymbol{F}) \log_e (a_\pm)_1/(a_\pm)_2$ where t_- is the transport number of the anion. The double cell

$$H_2(Pt) \mid HCl(a_1), AgCl \mid Ag \mid AgCl, HCl(a_2) \mid H_2(Pt)$$

gives

$$E_2 = [(\boldsymbol{RT}/\boldsymbol{F}) \log_e (a_{H^+})_1 - E^\circ_{Ag/AgCl} + (\boldsymbol{RT}/\boldsymbol{F}) \log_e (a_{Cl^-})_1]$$

$$- [(\boldsymbol{RT}/\boldsymbol{F}) \log_e (a_{H^+})_2 - E^\circ_{Ag/AgCl} + (\boldsymbol{RT}/\boldsymbol{F}) \log_e (a_{Cl^-})_2]$$

$$= \frac{\boldsymbol{RT}}{\boldsymbol{F}} \log_e \frac{(a_{H^+})_1 \times (a_{Cl^-})_1}{(a_{H^+})_2 \times (a_{Cl^-})_2} = (2\boldsymbol{RT}/\boldsymbol{F}) \log_e (a_\pm)_1/(a_\pm)_2$$

Hence the ratio of $E_1/E_2 = t_-$, the transport number of the anion.[1]

EXPERIMENT
Determine the transport number of the lithium ion in lithium chloride.

The cells required are:

(1) (with transport) $Ag \mid AgCl, LiCl(a_1) \parallel LiCl(a_2), AgCl \mid Ag$,

and

(2) (without transport) $Ag \mid AgCl, LiCl(a_1) \mid$ Li-amalgam $\mid$ $LiCl(a_2), AgCl \mid Ag$

The ratio a_1/a_2 should be about 10 (see Appendix Table A10). The theory, by analogy with that given above, leads to $E_1/E_2 = t_{Li^+}$. The transport number varies a little with the concentrations employed.

BIBLIOGRAPHY 13I: Determination of transport numbers from e.m.f. measurements

See monographs cited under 13A.

[1] Mason and Mellon, *J. Chem. Ed.*, 1939, **16**, 512.

14
Chemical kinetics

14A Introduction

1. A knowledge of thermodynamic quantities such as heats of reaction, free energies, equilibrium constants, etc., enables one to predict the *direction* of a chemical reaction, but gives no information about the *velocity* with which the reaction will take place. Whereas some reactions are practically instantaneous (e.g. reactions between salts in solution), others—particularly in organic chemistry—are normally slow: some potential reactions do not take place at detectable rates unless catalysed or otherwise initiated (e.g. $2H_2+O_2=2H_2O$ at room temperature). A study of the *kinetics* of a reaction (i.e. its rate under various conditions of concentration, temperature, etc.) is usually made in order to elucidate the path or *mechanism* by which it takes place.

Most commonly reactions proceed by steps via intermediate ions, molecules or free radicals, the existence of which is rarely apparent from the stoichiometric equation of the reaction. For this reason the *kinetics* often bear no relation to what might be expected from the classic ideas of Guldberg and Waage. For example, the rate of the reaction $H_2+Br_2=2HBr$ is *not* proportional simply to $[H_2]\times[Br_2]$ but is found experimentally to be proportional to

$$\frac{[H_2]\times\sqrt{[Br_2]}}{\text{const}+[HBr]/[Br_2]}$$

This expression is consistent with a mechanism involving Br atoms. Similarly, although many unimolecular gas reactions of the type $A=$products are known, all proceed by one or more *bimolecular* steps despite the molecularity suggested by their equations.

It is important to distinguish the *stoichiometric* equation for a reaction, which describes the change in the number of molecules as the overall reaction proceeds, from the *reaction steps*, which describe the molecular collisions responsible. For example the high temperature dissociation of bromine can be represented by the stoichiometric equation $Br_2 = 2Br$; however, it takes place by the bimolecular reaction step $Br_2 + M \rightarrow 2Br + M$ where M = any molecule present. In this chapter we distinguish stoichiometric equations and reaction steps by the use of = and → respectively.

Reaction steps can be unimolecular, bimolecular or termolecular according to whether 1, 2 or 3 molecules (on the l.h.s. of the step) collide simultaneously. The term 'unimolecular reaction' is, however, also used to describe any reaction with a single molecule on the l.h.s. of the *stoichiometric* equation. As noted, the mechanism for these reactions may involve both bimolecular and unimolecular *steps*.

The first step in the empirical study of the kinetics of a reaction (after the stoichiometry and existence or absence of side reactions have been established), is to decide whether it is *homogeneous* or *heterogeneous*, and *catalysed* or *uncatalysed*. These factors must be considered in order to get reproducible reaction rates for further study.

The next step is to determine how the rate of the reaction depends on the concentration of reactants (and products).

If the rate of a reaction A + B = products is found to be proportional to [A] it is said to be *first order* with respect to A. If the rate is proportional to $[A]^x$, it is of xth order with respect to A. If the rate is proportional to $[A]^x[B]^y$, the reaction as a whole is of the $(x+y)$th order. If the mechanism is simple, x and y will be 0, 1, 2 or 3; non-integral values indicate a complex mechanism (cf. the formation of HBr). If $x = 0$ (*zero order*), the *rate-determining step* is independent of the amount of the substance present in the system. Most often $x > 0$, so the majority of reactions slow down as the reactants are used up; a few, however, are accelerated by the products of the reaction (autocatalytic). The net rate of reaction, of course, falls to zero as chemical equilibrium is attained.

Some reactions proceed by *chain mechanisms* in which reaction occurs by rapid steps in which *free radicals* react and are reformed. If more than one is formed for each consumed, the reaction is *branching*, leading to explosion; otherwise it is *non-branching* as in most catalysed polymerization reactions.

All *rate-processes* are accelerated by rise of temperature. The *velocity constant* k of the reaction—that is, the specific velocity of the reaction with unit concentration of all reactants—usually follows the Arrhenius equation, $k = A\,e^{-E/RT}$, where A and E are practically independent of temperature. E, the *energy of activation* of the process, is interpreted as the minimum thermal energy which the molecules must get before they can undergo reaction. The energy of activation is needed either to break a chemical bond, as in *free radical* gas reactions, or to permit an electronic rearrangement when the reacting molecules collide. In heterogeneous reactions a relatively small activation energy is often found; this generally indicates that the rate of the reaction is controlled by diffusion or other *transport* mechanism rather than by a slow chemical process.

The deduction of the precise molecular mechanisms of a reaction from its kinetics, i.e. the choice of the correct set of reaction steps, often proves a difficult and complex problem. The present chapter will be confined to experimental methods of studying rates of reaction in order to establish the form of their kinetics under changing conditions.

2. METHODS OF MEASURING RATES OF REACTION

The most convenient reactions to study are those requiring between 1 min and 1 day to go to completion. The conditions of concentration, temperature, catalysts, etc., must be carefully controlled and known so that a reproducible rate is obtained and the course of the reaction can be quantitatively interpreted.

Innumerable methods have been used for following the progress of reactions. The most obvious one is by chemical analysis; if the reaction can be stopped after a known time by cooling or by addition of some reagent one can analyse the mixture at leisure. It is usually more convenient, however, to use a physical method in order to measure the amount of reaction which has occurred without disturbing the system. Any physical property which changes in a definite manner as the reaction proceeds can be employed as a means of measurement. The following list of examples gives a few of the many properties which have been used; pressure (for gas reactions involving a change in the number of moles of gas); volume; electrical conductivity; optical rotation; thermal conductivity (gas mixtures); weight (e.g. oxidation of a metal); refractive index; pH; viscosity (especially convenient for polymerization studies). However, methods which follow the change in concentration of a single species, e.g. optical absorption, are in general to be preferred to measurements of bulk properties, unless both the stoichiometry and mechanism are known unambiguously.

3. FORM OF THE KINETICS

Since the velocity of a reaction depends on the concentration of reactants, the actual rate in terms of grams of substance reacted in unit time changes during the course of the reaction. The rate can be defined in any convenient way; usually the rate of change of concentration $[x]$ of one of the reactants or products will be chosen. Thus, the rate of a reaction $2A+B = C+D$ could be expressed as $-d[A]/dt$, $-d[B]/dt$, $d[C]/dt$ or $d[D]/dt$. Note that since $(-d[A]/dt)$ is twice $(-d[B]/dt)$ the reactant chosen must be specified. The concentration is normally expressed in mol m^{-3}, but for intermediate calculations it is often convenient to use other units such as partial pressure or cm^3 of titrant.

The experimental measurements will show how the concentration of the chosen substance, say x, changes with time, and the rate at any moment is therefore given by the tangent of the $[x]$–t curve.

The chief problem in experimental kinetics is to ascertain how the rate $d[x]/dt$ depends on the concentrations of the various substances concerned. One method is to vary the concentration of each substance in turn (keeping all other factors constant) and then measure the rate in each instance. Since, however, the rate varies even while the reaction is being studied, it is usually necessary to deduce the order of the reaction from the mathematical form of the $[x]$–t results. The shape of the $[x]$–t curve can readily be predicted for all the simplest types of theoretical reaction mechanism. It is usually convenient to measure *initial* rates of reaction to investigate mechanism, to avoid complications due to the back reaction and other factors.

4. KINETICS OF FIRST ORDER REACTIONS

By definition, the rate is proportional to the concentration, i.e. $-d[x]/dt \propto [x]$, or $-d[x]/dt = k[x]$, where k is the *velocity constant*; k has the dimensions of time^{-1}.

There are several ways of testing kinetic data for conformity to first-order kinetics.

4.1. *Tangent method.* Plot a graph of $[x]$ against t. Draw tangents at a number of points along the curve, measure their slopes, and see whether slope/$[x]$ = const ($=k$). This is not the most convenient method.

4.2. *Half-change method.* It is a characteristic of first order kinetics that the *time of half-change*, that is, the time needed for $[x]$ to fall to half its initial value, is independent of the initial concentration. (The same holds for any other fraction besides $\frac{1}{2}$.) This follows from integration of the defining equation. Separating the variables, one has $d[x]/[x] = -k dt$. Integrating between (x_1, t_1) and (x_2, t_2) gives $\log_e ([x]_2/[x]_1) = -k(t_2 - t_1)$. If $[x]_2 = \frac{1}{2}[x]_1$,

$$k = \frac{2{\cdot}303}{(t_2 - t_1)} \log_{10} 2.$$

Thus, the interval between t_2 and t_1 should be found constant whatever value of $[x]_1$ is selected on the curve.

4.3. *General integral method.* The general integration of the rate equation gives

$$\log_e [x] = -kt + \text{const}$$

A graph of $\log_{10} [x]$ against t should therefore give a straight line of slope $-k/2{\cdot}303$. This is the most commonly used method of calculating first order velocity constants, as it makes full use of all data. It can be applied to any part of the $[x]$–t curve. x can be replaced by any quantity *directly proportional* to the concentration, whose absolute value need not be known.

4.4. *Guggenheim's method.*[1] It sometimes happens that the readings in the middle section of a kinetic determination can be made more accurately than those near the beginning or the end. Method **4.3** of treating the results gives undue weight to the initial concentration, and this quantity is often not known with as high an absolute accuracy as are the relative differences in the middle of the curve. The following method is then appropriate.

A series of 'readings', V (e.g. titrations, value of a physical property, etc.), is made at times $t_1, \ldots, t_n$ spread over an interval two or three times the time of half change of the reaction. A second series of 'readings' V' is then made at times $(t_1 + t'), \ldots, (t_n + t')$, each exactly at constant interval t' after the corresponding 'reading' V. The interval t' should be equal to 2–3 times the time of half-reaction. If the values of $\log_{10} (V' - V)$ are then plotted as ordinates against the times t as abscissae, a straight line is obtained, the slope of which, $\log_{10} (V' - V)/t$, is $-k . \log_{10}$ e; or, $2{\cdot}303 \times$ slope $= -k$.

The advantage of this method is that the concentration of the reacting substance need not be known at all provided some property 'V', which is a *linear function* of the composition of the system, can be observed.

First order kinetics are encountered not only with unimolecular reactions, but also with bimolecular reactions in which the concentration of the second substance is relatively very great ('pseudo-first order reactions', see below) or which is maintained constant. The 'decay' of radioactive elements and of the glow of some phosphorescent solids is also first order.

5. KINETICS OF SECOND ORDER REACTIONS

If the reaction $n_A A + n_B B =$ products is first order with respect to both A and B, the overall kinetics will be second order. Let the initial concentrations of A and B be a and b respectively, and let the amount of reaction that has occurred after time t be x. Thus, the rate is given by $dx/dt = k(a - n_A x)(b - n_B x)$. Normally $n_A = n_B = 1$ so $dx/dt = k(a-x)(b-x)$. k is the second order rate constant with dimensions of concentration^{-1} time^{-1}. Integration of the above equation between 0 and t gives

$$k = \frac{1}{(a-b)t} \log_e \frac{(a-x)b}{(b-x)a} = \frac{2{\cdot}303}{(a-b)t} \log_{10} \frac{(a-x)b}{(b-x)a}$$

When $n_A = n_B$ and the initial concentrations a and b are the same, the corresponding equations are

$$\frac{dx}{dt} = k(a-x)^2, \quad \text{and} \quad k = \frac{1}{t} \cdot \frac{x}{a(a-x)}$$

This equation also applies to a second order reaction which is simply second order with respect to a single reactant, e.g. $2NO_2 = 2NO + O_2$ which is second order with respect to NO_2. Note that in these cases if c is the concentration of A at time t, i.e. $c = a - x$, then $1/c = 1/a + kt$; a plot of reciprocal concentration against time is linear with slope k.

If $b \gg a$, $(b - n_B x)$ does not alter much during the reaction, since it changes only from b to $(b - n_B a/n_A)$. Writing the reaction with $n_A = 1$, i.e. $A + n_B B =$ products,

$$dx/dt = k(a-x)(b - n_B x) \approx kb(a-x) = k'(a-x).$$

This is a 'pseudo-first order reaction', which appears to obey first order kinetics during the course of a single run, with a *pseudo-first order* rate constant $k' = bk$ s^{-1}.

Expressions for the velocity of reactions of a higher order, in which the concentrations of three or more molecular species undergo change, can be obtained in a similar manner. Such reactions are, however, few in number and will not be considered here.

Expressions can also be deduced for reversible reactions, and for free radical chain reactions.

BIBLIOGRAPHY 14A: Introduction to chemical kinetics

Weissberger, Vol. VIII, *Investigation of Rates and Mechanisms of Reactions*, 1953 (Interscience, New York).

Laidler, *Reaction Kinetics*, 1963 (Pergamon, Oxford).

Bamford and Tipper (ed.), *Comprehensive Chemical Kinetics*, Vols I–III, 1969 (Elsevier, Amsterdam).

Amdur and Hammes, *Chemical Kinetics*, 1966 (McGraw-Hill, New York).

Gardiner, *Rates and Mechanisms of Chemical Reactions*, 1969 (Benjamin, New York).

Benson, *Foundations of Chemical Kinetics*, 1960 (McGraw-Hill, New York).

Moelwyn-Hughes, *Chemical Statics and Kinetics of Solutions*, 1971 (Academic Press, London).

Emmett (ed.), *Catalysis*, Vols I–VI, 1954–8 (Reinhold, New York).

Frost and Pearson, *Kinetics and Mechanism*, 2nd edn, 1961 (John Wiley, New York).

[1] Guggenheim, *Phil. Mag.*, 1926, **2**, 538.

14B First order reactions

Numerous examples of reactions which follow first order kinetics are known.

1. HYDROLYSIS OF AN ESTER, CATALYSED BY ACID

When an ester, such as methyl acetate, reacts with water, it is partially converted into alcohol and acid, according to the equation

$$CH_3COOCH_3 + H_2O = CH_3COOH + CH_3OH$$

When the amount of water is relatively large, the reaction takes place practically completely as represented, from left to right, and the kinetics are first order with respect to the ester. The hydrolysis takes place slowly even with pure water and is catalysed by acids. The rate is approximately proportional to the concentration, or rather, activity of hydrogen ions, and the kinetics of the catalysed reactions are thus really pseudo-first order.

EXPERIMENT

Determine the velocity constant of the hydrolysis of methyl acetate at 25°C, catalysed by 500 mol m^{-3} HCl.

Procedure. Prepare a standard solution of baryta, approximately 50 mol m^{-3}, and determine its titre by means of pure succinic acid, using phenolphthalein as indicator. By means of this baryta solution, prepare also 500 mol m^{-3} hydrochloric acid. Clean, steam and dry a 250 cm^3 Pyrex conical flask. Pipette 100 cm^3 of 500 mol m^{-3} HCl into it, fit a rubber stopper, and suspend the flask in a thermostat at 25°C. Use a ring of lead to weight it down. Also suspend in the bath a small stoppered bottle of methyl acetate.

After, say 10 min, when the liquids will have assumed the temperature of the bath, pipette 5 cm^3 of methyl acetate into the flask of acid, shake well, and immediately withdraw 5 cm^3 of the solution. This is allowed to run into 25 cm^3 of ice-cold water in order to arrest the reaction, and the acid titrated as soon as possible by means of the baryta solution. Note the moment, to the nearest second, at which the solution is run into the water; this is taken as the starting-point of the reaction.

About 10 min after the first titration, again withdraw 5 cm^3 of the mixture from the flask, and determine the titre as before, noting the moment at which the reaction is arrested. Further titrations are made at successive intervals of 20, 30, 40, 60, 120 min. The remainder of the reaction mixture should be left (stoppered) for 48 h for the reaction to go to completion.

Calculations. The velocity coefficient k can be calculated by one or more of the methods given in section 14A(4)). The initial concentration of methyl acetate is proportional to $(T_\infty - T_0)$, where T_∞ is the final titration (HCl + acetic acid) and T_0 is the initial titration (HCl alone). The concentration of methyl acetate is $(T - T_t)$ where T_t is the titre of the sample taken at time t.

Plot a graph of [methyl acetate] (in units of the titre difference) against time, and also of log [methyl acetate] against time. Calculate k in s^{-1} (first order) and in m^3 $mole^{-1}$ s^{-1} (second order).

OTHER EXPERIMENTS

The variation of the first order constant with hydrogen ion concentration can be investigated. Since the rate of hydrolysis of methyl acetate is proportional to the activity of hydrogen ions, determinations of the velocity constant in the presence of different catalysts may also be used to study the degree of dissociation of weak acids, the degree of hydrolysis of salts of weak bases (e.g. urea hydrochloride), the influence of neutral salts on the activity of strong acids, and the general phenomena of acid–base catalysis.[1]

The activation energy may be determined by repeating the measurements at 30, 35 and 40°C. The experiment described in section 14C(2), is, however, more convenient for studying the influence of temperature on velocity of reaction.

2. VELOCITY OF INVERSION OF SUCROSE

Another acid-catalysed pseudo-first order reaction which has been much studied[2] is the hydrolysis of sucrose into glucose and fructose. The sugars are optically active. Sucrose is dextro-rotatory, while the product ('invert sugar') is laevo-rotatory; hence, during the reaction the optical rotation changes sign, and the reaction can be followed by means of the polarimeter.

EXPERIMENT

Determine the velocity coefficient of inversion of sucrose by 500 mol m^{-3} hydrochloric acid at 25°C.

Before commencing the experiment, read through the section on polarimetric measurements, section 9F.

Procedure. Prepare a solution of cane sugar by dissolving 20 g of pure cane sugar in water and making the volume up to 100 cm^3, and, if necessary, filter the solution so that it is quite clear. Add a small crystal of mercuric iodide as preservative. Prepare, also, a normal solution of hydrochloric acid. Place 25 cm^3 of the sugar solution and about 30 cm^3 of the acid in separate flasks which have previously been steamed out and dried, and stand them in a thermostat at 25°C.

After having set up the polarimeter and determined the zero, place a jacketed observation tube in the polarimeter, and cause water at 25°C to circulate through the mantle of the observation tube. The circulation of water through the tube must be so regulated that the temperature remains constant to within 0·1° during the whole of the experiment. This should first be tested with the observation tube full of water, in which the bulb of a thermometer is immersed.

The observation tube is dried and replaced in the polarimeter. When the temperature has again become constant, add 25 cm^3 of the acid to the sugar solution, mix quickly, and as soon as possible pour the mixture into the observation tube. Determine the angle of rotation, and note the time at which the reading is made. Preserve a quantity of the mixed sugar and acid solution in a stoppered flask in the thermostat for the final reading of the rotation.

As the angle of rotation alters rather quickly during the first few minutes, a series of five or six readings should be made, one after the other, and the time noted at which the first and last readings are made. The mean value of the angles read, and the middle point of the time period between the first and last readings, should be taken as the initial value of the rotation (A_0) and the starting-point of the reaction, respectively. Further readings, up to the number of eight or ten, of the angle of rotation should be made after periods gradually lengthening from 10 min to 2 h. The observation tube may then be cleaned out and the final rotation determined after at least 48 h with the solution which had been kept in the thermostat.

Calculation. If A_0 represents the initial angle, and A_∞ the final angle of rotation, after complete inversion has occurred, the initial amount of cane sugar will be proportional to the total change in rotation, i.e. to $A_0 - A_\infty$. Similarly, at time t_n, if the angle of rotation is A_n, the amount of cane sugar present will be represented by $A_n - A_\infty$. Hence, in accordance with the formula for a unimolecular reaction, one obtains the expression:

$$k = \frac{2{\cdot}303}{t_n}\left[\log_{10}(A_0 - A_\infty) - \log_{10}(A_n - A_\infty)\right]$$

The angles of rotation must be given their proper sign, rotations to the right being reckoned positive and those to the left negative.

3. DECOMPOSITION OF DIAZONIUM SALTS

In illustration of another method of following the course of a reaction, the decomposition of diazonium salts, e.g. benzene diazonium chloride[3] may be studied. When this salt is warmed with excess of water, it undergoes decomposition according to the equation

$$C_6H_5.N_2Cl + H_2O = C_6H_5OH + HCl + N_2$$

and the course of the decomposition can be followed by measuring from time to time the volume of nitrogen evolved.

Since the concentration of benzene diazonium chloride in the solution is proportional to the total volume of nitrogen which the solution is capable of yielding, the rate constant of the decomposition can be calculated from:

$$k = \frac{2 \cdot 303}{t} [\log_{10} V_{\infty} - \log_{10} (V_{\infty} - V_n)]$$

where V_{∞} is the total volume of gas obtained, and V_n the volume of gas collected up to time t_n.

EXPERIMENT

Determine the velocity constant of the decomposition of benzene diazonium chloride.

Procedure. Prepare a solution of benzene diazonium chloride as follows: dissolve 6·64 g of aniline in 21·4 cm^3 of hydrochloric acid (sp. gr. = 1·16); cool in ice-water, and add gradually from a dropping funnel a cold solution of 4·9 g of sodium nitrite in 75 cm^3 of water. After the addition of the sodium nitrite, make the solution up to 100 cm^3. Place 30–35 cm^3 of this solution in a tube of about 3 cm in diameter, into which a side tube, of about 1-mm bore is sealed. The tube, which should be well cleaned and dried before use, should be chosen of such a length that the air-space above the solution is small.

The tube is connected to a Henkel gas burette by rubber pressure tubing, immersed in the thermostat at 35°C and shaken vigorously by a motor-driven eccentric or by a laboratory shaking machine. A better method of stirring liquid inside a closed apparatus is with a magnetic stirrer: a powerful U-shaped magnet outside the flask is rotated by an electric motor, and causes a piece of iron enclosed inside glass or 'Perspex' to revolve inside the flask. See also section 8B(6).

At first the tube with the solution should be in open communication with the air, and the gas evolved allowed to escape. After 5–7 min, communication with the gas burette is effected while that with the outside air is stopped. The time at which this is done is noted, and this is taken as the starting-point of the reaction. Gas will now collect in the burette, and its volume should be read off at intervals of about 15 min. The temperature of the gas and the height of the barometer should also be noted.

The end point of the reaction is determined by repeatedly immersing the tube with the reaction mixture in a larger beaker of hot water, until, on cooling again to the temperature of the experiment, there is no further increase in the volume of gas evolved.

The velocity constant of the decomposition is then calculated by means of the equation given above.

4. HYDROLYSIS OF TERTIARY AMYL IODIDE

This reaction can be followed conveniently by measurements of electrical conductivity. The hydrolysis is not catalysed by hydrogen or hydroxyl ions. The rate-controlling process is probably slow ionization, $C_5H_{11}I \xrightarrow{\text{slow}} C_5H_{11}^+ + I^-$, followed by a rapid reaction of the cation with water, $C_5H_{11}^+ + HOH \xrightarrow{\text{fast}} C_5H_{11}OH + H^+$. The net chemical change is therefore $C_5H_{11}I + H_2O \rightarrow C_5H_{11}OH + H^+ + I^-$. The conductivity increases owing to the formation of the strong electrolyte, hydrogen iodide.[4]

EXPERIMENT

Determine the order, the rate constant and the activation energy of the hydrolysis of tertiary-amyl iodide.

Procedure. Place about 50 cm^3 of aqueous alcohol (80% ethanol by volume) in a large, well-corked, boiling tube in a thermostat at 25°C. A second tube containing a dipping-type conductivity cell is placed alongside. When both have had time to take the temperature of the thermostat, pipette about 0·3 cm^3 of tertiary amyl iodide into the solution, stir well with a glass rod, and insert the electrodes. A stop-watch is started and the conductivity measured on any available conductivity bridge. Readings are taken every 30 s for 5 min, then every minute for 10 min. The electrodes are then returned to the empty boiling tube, and the tube containing the reaction mixture is corked and placed in a beaker of water at about 60°C for about 5 min to complete the hydrolysis. The tube is then cooled, replaced in the thermostat, and the final conductivity is taken when the temperature is 25°C.

Other determinations should then be made in the same way, using 0·2 cm^3 and then 0·4 cm^3 of amyl iodide. To ascertain the temperature coefficient of the reaction, determinations can be made at 18, 30 and 35°C. More frequent readings of conductivity will of course be needed at the higher temperatures.

Results. The conductivity will be approximately proportional to the amount of hydrogen iodide produced. (Alternatively, a calibration curve can be constructed by successively diluting quantitatively the final reaction solution with the 80% ethanol-water solvent and measuring the conductivity at each dilution.) If the conductivity is λ_0 when the stopwatch is started ($t=0$), λ_t after time t, and λ_∞ at the end of the reaction, the velocity constant for a first order reaction would be:

$$k = \frac{1}{t}\log_e \frac{(\lambda_\infty - \lambda_0)}{(\lambda_\infty - \lambda_t)}$$

A graph of $\log_{10}[(\lambda_\infty - \lambda_0)/(\lambda_\infty - \lambda_t)]$ against t should therefore give a straight line of slope $k/2{\cdot}303$. The time of half-change should also be found to be independent of the initial amount of amyl iodide.

The value of k will be independent of the units in which the conductivity is measured, and therefore it is not necessary to know the cell constant of the conductivity cell. Moreover, the results are not affected by traces of electrolyte impurities, since the evaluation of k involves the ratio of two differences.

The energy of activation can be calculated from the values of k at different temperatures by means of the Arrhenius equation (section 13A(1)).

5. DECOMPOSITION OF DIACETONE ALCOHOL

Many reactions in solution can be followed conveniently by noting the small change of volume which occurs during the reaction. Provided the solution is dilute, the volume change is proportional to the amount of reaction that has occurred. The volume change is detected by enclosing a considerable volume of the solution in a vessel having a narrow capillary neck ('dilatometer'). The movement of the meniscus along the capillary can be followed, if necessary, with a cathetometer. The dilatometer method is applicable to many reactions: the chief limitation on its accuracy is set only by the accuracy with which the temperature can be controlled.

A reaction suitable for study by the dilatometer method is the decomposition of diacetone alcohol:

$$(CH_3)_2C(OH)CH_2COCH_3 = 2CH_3COCH_3$$

The forward reaction is catalysed by hydroxyl ions and goes practically to completion in dilute aqueous solution.

EXPERIMENT

Determine the influence of NaOH *on the rate of decomposition of diacetone alcohol at 25°C.*

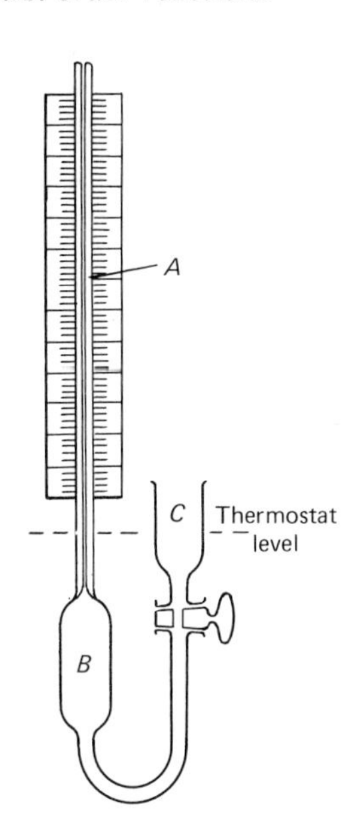

14B.1 Dilatometer for studying the rate of chemical reactions in solution.

Apparatus. The dilatometer (Fig. 14B.1) consists of a bulb B of capacity about 25 cm^3, a capillary stem A, a funnel C, and stopcock. The densities of diacetone alcohol and acetone are 931 and 788 kg m^{-3} respectively; consequently one may anticipate that 25 cm^3 of a 5% solution would expand by about 0·17 cm^3 during the reaction. This corresponds to a movement of about 20 cm in a capillary 1 mm bore, and can therefore be measured by means of a scale of glass or stainless steel laid along the capillary. Alternatively, of course, a cathetometer may be used. The large bulb connected to a capillary itself acts as a sensitive thermometer: a simple calculation shows that *it is essential to use a thermostat capable of controlling the temperature to within 0·01°C* (see section 3B).

Procedure. It is essential that the dilatometer be thoroughly clean to prevent air bubbles from sticking to the sides. Take the tap out and remove as much grease as possible from it and from the cell with tissue and a little carbon tetrachloride. Then clean the cell and the tap with chromic–sulphuric acid cleaning mixture to dissolve the last traces of grease, and rinse thoroughly several times with distilled water. Allow the water to drain and then rinse two or three times with small quantities of dry acetone. Dry the dilatometer by blowing warm air through it, being careful to avoid the introduction of dust. The tap should then be carefully lubricated with low melting high vacuum grease. The results will be meaningless if there is any leakage because the tap is insufficiently greased; on the other hand the cell will be difficult to fill if there is excess grease at the ends of the bore (this can be removed with a pipe-cleaner). An air bubble trapped under the tap during filling may be expelled by gentle suction or pressure at the top of the capillary. Fasten the dilatometer in a thermostat so that all the bulb is immersed. Pipette exactly 35 cm^3 of 100 mol m^{-3} NaOH into a small flask, and leave it in the thermostat for 10 min. Then add 2·0 cm^3 of diacetone alcohol from a pipette, mix well, and carefully pour the solution into the funnel C of the dilatometer. Open the tap and by means of suction at A adjust the level of the meniscus in the capillary until it stands near the bottom of the scale. Then close the tap. The dilatometer must, of course, be free from air bubbles.

This experiment is most conveniently treated by Guggenheim's method of calculation (section 14A(4)), as it is then unnecessary to know either the initial or final volumes. Accordingly, readings are taken of the position of the meniscus every 5 min for a period of about 45 min, and a second set of readings is taken (also at 5-min intervals) after the expiry of one hour from the start.

The experiment should be repeated with several different concentrations of NaOH, e.g. 50, 200, 500 mol m^{-3}. In these cases the time intervals between readings must be modified according to the velocity of reaction; but in every case the interval between the two sets of readings should be greater than $t_{\frac{1}{2}}$, the time of half-change of the reaction. As a rough guide it may be taken that

$$t_{\frac{1}{2}}\,(\text{min}) \times [\text{OH}^-](\text{mol m}^{-3}) \approx 5\,000$$

Calculation. Use Guggenheim's method (see section 14A(4)) to evaluate the first order velocity constant of the reaction at each concentration of NaOH. The change of volume may be assumed proportional to the amount of reaction. The results consist of a set of readings $r_1, r_2, r_3, \ldots$ taken at times $t_1, t_2, t_3, \ldots$, and a second set, $r'_1, r'_2, r'_3, \ldots$ taken at $(t_1+T), (t_2+T), (t_3+T), \ldots$ where T is the interval of time between the beginnings of the first and second sets of readings. For the nth readings of each set, the theory gives,

$$2{\cdot}303 \log_{10}(r'_n - r_n) = \text{const} - kt_n$$

A graph of $\log_{10}(r'_n - r_n)$ against t_n therefore gives a straight line of slope $-k/2{\cdot}303$.

Evaluate k for each concentration of NaOH and note the relation between these quantities.

BIBLIOGRAPHY 14B: First order reactions

See Bibliography 14A.

[1] Bell, *Acid-base Catalysis*, 1941 (Oxford University Press, London).
[2] Guggenheim and Wiseman, *Proc. Roy. Soc.*, 1950, **A 203**, 17.
[3] Cain and Nicoll, *J. Chem. Soc.*, 1902, **81**, 1412; 1903, **83**, 206, 470; Moelwyn-Hughes and Johnson, *Trans. Faraday Soc.*, 1940, **36**, 948.
[4] Hughes, *J. Chem. Soc.*, 1946, 968; Shorter and Hinshelwood, *ibid.*, 1949, 2412.

14C Second order reactions

Many gas reactions occur by bimolecular processes, and many reactions in solution follow second order kinetics, although their mechanisms may not necessarily be simple bimolecular ones.

1. HYDROLYSIS OF ESTERS BY ALKALIS

Many hydrolytic reactions are catalysed by alkalis as well as by acids, but the overall kinetics are now second order instead of first order because, with alkalis, the catalyst is consumed in the reaction:

$$R.COOR' + OH^- = R.COO^- + R'OH$$

The reaction can be followed by titration of the alkali remaining in solution, or, better, by measurements of the electrical conductivity. The net result of the reaction, as regards the conductivity, is that OH^- ions are progressively replaced by $R.COO^-$ ions. The latter have a lower electrical mobility (about one-fifth of that of OH^-); consequently, the conductivity falls. The change of conductivity is approximately proportional to the amount of reaction that has taken place.

EXPERIMENT
Determine the second order velocity constant of the hydrolysis of ethyl acetate by sodium hydroxide.

Method (a)—Titration. The technique is similar to that used for the acid hydrolysis of methyl acetate, but the reaction occurs more rapidly. Mix 50 cm^3 of 20 mol m^{-3} solution of ethyl acetate and 50 cm^3 of NaOH (carbonate-free) at 25°C. Take 10 cm^3 samples after 3, 5, 10, 20 min, etc., running them into 10 cm^3 of 50 mol m^{-3} HCl to stop the reaction. Titrate the excess acid with baryta. The final titration is made after 24 h. The rate constant is calculated by the formula given in section 14A(5).

The terms in the logarithm can be replaced by the number of cm^3 of acid required to neutralize the alkali in the reaction mixture at the beginning (T_0), after time t min (T_t), and at the end of the reaction (T_∞), but the term $(a-b)$ in front of the logarithm must be expressed in mol m^{-3}. Thus,

$$k = \frac{2{\cdot}303}{([\text{NaOH}]_0 - [\text{EtAc}]_0)t} \cdot \log_{10} \frac{T_t(T_0 - T_\infty)}{T_0(T_t - T_\infty)}$$

Method (b)—Conductivity. The following solutions are required: 100 cm^3 50 mol m^{-3} NaOH, 50 cm^3 200 mol m^{-3} ethyl acetate and 100 cm^3 sodium acetate, made by neutralizing 20 cm^3 of the 50 mol m^{-3} NaOH with acetic acid and making up to 100 cm^3. All the solutions should be prepared with CO_2 free water.

The reaction is carried out in a solution which is initially 10 mol m^{-3} with respect to both ethyl acetate and NaOH. The conductivity at the beginning will thus be that of 10 mol m^{-3} NaOH, while at the end it will be that of 10 mol m^{-3} sodium acetate. It is convenient to find these initial and final values of the conductivity before following the course of the reaction.

A conductivity cell having electrodes of about 1 cm^2, separated by 1–2 cm, is suitable. Any type of conductivity bridge can be used. (See chapter 11 for the technique of conductivity measurements.)

Procedure. First measure the cell resistance with (a) 10 mol m^{-3} NaOH ($=R_0$) and then (b) 10 mol m^{-3} CH_3COONa ($=R_\infty$) at 25°C.

Now suspend two flasks in the thermostat, one containing 20 cm^3 50 mol m^{-3} NaOH + 50 cm^3 CO_2-free water and the other 5 cm^3 200 mol m^{-3} CH_3COOEt + 25 cm^3 of CO_2-free water. While they are attaining the temperature of the bath (10 min) clean and dry the conductivity cell. Then mix the two solutions thoroughly and quickly, and start a clock. Fill the cell with the mixture and take readings of its resistance at frequent, timed intervals as long as the reaction is still occurring.

Calculation. The second order rate expression for a reaction in which the two reactants are at equal concentration a is

$$k = \frac{1}{t} \cdot \frac{x}{a(a-x)}$$

In this experiment $a = 10$ mol m^{-3}. The amount of reaction that has occurred, x, and the amount of reactants remaining, $(a-x)$, can be deduced from the conductances (C). These are calculated as reciprocals of the cell resistances; thus, initially $C_0 = 1/R_0$; after time t min $C_t = 1/R_t$, and finally $C_\infty = 1/R_\infty$. The total amount of reaction possible is proportional to the total change of conductivity, i.e. $a \propto (C_0 - C_\infty)$. Therefore the amount of reaction after time t s, namely x, is proportional to $(C_0 - C_t)$. Substituting in the above expression, one finds

$$k = \frac{1}{10t} \cdot \frac{C_0 - C_t}{C_t - C_\infty}$$

A graph of $(C_0 - C_t)/(C_t - C_\infty)$ against t should therefore give a straight line of slope $10k$. The units of k are mol^{-1} m^3 s^{-1}.

2. THE REACTION $H_2O_2 + 2HI = I_2 + 2H_2$

This reaction is kinetically of second order—not, as might be expected, third order. The mechanism is probably an initial, rate-determining step according to the equation $H_2O_2 + I^- \rightarrow H_2O + IO^-$ (slow), followed by a rapid reaction given by $IO^- + 2H^+ + I^- \rightarrow H_2O + I_2$ (fast). (The latter presumably occurs by consecutive steps, not by a simultaneous collision of four ions!)

The order of the reaction *with respect* to H_2O_2 can be studied conveniently by choosing conditions such that there is practically constant excess of HI. The kinetics then follow the first order law. Experimentally, this is achieved by continually adding small volumes of sodium thiosulphate solution to remove the iodine as soon as it is liberated and to regenerate iodide according to the reaction $2Na_2S_2O_3 + I_2 = Na_2S_4O_6 + 2NaI$. By using a large volume of solution and adding small amounts of *concentrated* thiosulphate solution from a microburette, one can neglect the small increase of volume of the solution and take the concentration of I^- ions as constant. The rate of the reaction then depends only on $[H_2O_2]$ and temperature. The course of the reaction can readily be followed by timing the appearance of iodine (indicated by starch solution) after a small, known volume of thiosulphate solution has been added. Another addition of thiosulphate is made immediately the blue colour reappears.

The order of the reaction with *respect to* HI can be determined by determining the first order velocity constant of the reaction with different concentrations of HI.

EXPERIMENT

Study the kinetics of the reaction between H_2O_2 *and* HI.

Procedure. Dissolve *exactly* 2 g of KI in 500 cm^3 of water in a flask (preferably the 'bolthead' type). Add 25 cm^3 of dilute H_2SO_4 (1 vol. conc. H_2SO_4 mixed with 2 vols of water) and 10 cm^3 of starch solution. Arrange the flask so that it is deeply immersed in a bath of ice and water. While the solution is cooling, prepare an approximately '2-volume' solution of H_2O_2 and standardize it by the following method:

Dilute 20 cm^3 of the sulphuric acid to 100 cm^3, add about 2 g of KI, warm to 30°C, and add 10 cm^3 of the H_2O_2 solution. Mix, and allow to stand for 10 min. Titrate the iodine liberated with 1 000 mol m^{-3} $Na_2S_2O_3$, using a microburette.

Arrange a microburette filled with 1 000 mol m^{-3} $Na_2S_2O_3$ over the flask in the bath so that it will deliver directly into the solution. It is convenient to attach to the bottom of the burette a bent tube with a narrow jet, so that the burette need not be vertically above the neck of the flask; this leaves room for a small electric motor and glass stirrer. The speed of the motor should be set at a suitable value by means of a rheostat. Rapid mixing is needed.

Set a stop-cock to zero. Take the temperature of the solution in the flask. Pipette in 25 cm^3 of the H_2O_2 solution, and start the clock when the pipette is half discharged into the flask. Immediately run in 0·3 cm^3 of the $Na_2S_2O_3$ solution from the burette so that the blue colour initially formed is discharged. Observe the solution closely and when the blue colour reappears note the time (without stopping the clock), and immediately run in 0·2 cm^3 of thiosulphate. Again time the reappearance of the blue colour.

Continue in this way until the reaction has reached about half its theoretical course. It is the reappearance of the blue colour that must be timed—not the addition of thiosulphate. The addition should, however, be made with as little delay as possible. At the end of the readings take the temperature of the solution again, and assume a mean value for the temperature of the experiment.

To determine the temperature coefficient of the reaction, repeat the experiment at 10 and 20°C. At 10°C the H_2O_2 solution should be cooled before adding it to the rest of the reaction mixture. Initial additions of 0·5 cm^3 and 1·0 cm^3 of thiosulphate solution, followed by 0·3 cm^3 and 0·5 cm^3 additions will be found suitable at the respective temperatures.

To determine the order of the reaction with respect to HI, the experiment may be repeated at any one temperature (say 10°C, for convenient rates) with half and double the amounts of KI and H_2SO_4 in the same total volume of reaction mixture.

Results. The experimental data obtained correspond to the amount of iodine which has been liberated by the reaction (measured in terms of equivalent cm^3 of $Na_2S_2O_3$ solution) at a series of times (as noted from the clock). The total amount of iodine which would be liberated in infinite time is also known—from the preliminary standardization. Since 1 molecule of H_2O_2 is destroyed for every molecule of I_2 liberated, the results provide values of the term $[H_2O_2]$ as a function of time and permit calculation of the first order velocity constant for the rate of reaction of H_2O_2 at constant concentration of iodide, i.e. $-d[H_2O_2]/dt = k_1[H_2O_2]$. The data obtained from the experiment at 0°C can be treated by methods **4.1**, **4.2** and **4.3** for first order reactions (section 14A(4)). Method **4.3** will be found the most convenient, and should be adopted for all the other experiments.

Also calculate the corresponding second order rate constants k_2, assuming the reaction is first order w.r.t. I^- ions (section 14A(5). In order to calculate the *energy of activation* of the reaction, the Arrhenius equation is put in the form $2{\cdot}303 \log_{10} k_1 = \log_e A - E/\boldsymbol{R}T$. Since $\log_e A$ is constant, a graph of values of $\log_{10} k_1$ against $1/T$(K) is therefore a straight line. The slope of the line is equal to $-E/(2{\cdot}303\boldsymbol{R})$. If $\boldsymbol{R}$ is taken as 8·31 J cal mol^{-1} K^{-1}, the units of E are J mol^{-1}.

The order of the reaction with respect to HI can be observed from the effect of [HI] on k_1 (at const temp.). If experiments are done with several values of [HI], the same value of k_2 should be obtained (at any one temperature).

3. HOMOGENEOUS CATALYSIS OF THE REACTION $S_2O_8^= + 3I^- = 2SO_4^= + I_3^-$

The stoichiometric equation for this reaction suggests it may obey fourth order kinetics (first order in persulphate and third order in iodide). It is in fact kinetically of the second order (first order in each reactant):

$$\frac{-\mathrm{d}[S_2O_8^{=}]}{\mathrm{d}t} = k'[S_2O_8^{=}][I^-]$$

The reason is that the reaction occurs in stages, with the first step

$$S_2O_8^{=} + I^- \rightarrow S_2O_8I^{3-}$$

much slower than the succeeding steps (*rate determining*).

When iodide is present in such excess that its concentration can be regarded as constant, the reaction becomes pseudo first order (section 14A(5)).

If the concentration of persulphate is initially a and becomes $a-x$ at time t, $k_1 t = \log_e [a/(a-x)]$; (section 14A(4)). The time $t_{1/n}$ for one nth of the reaction to occur (i.e. $x/a = 1/n$) is

$$k_1 t_{1/n} = \log_e [n/(n-1)]$$

If the same amount of reaction (one nth) is allowed to take place in solutions all containing the same iodide concentration, the pseudo-first order rate constants k_1 are inversely proportional to $t_{1/n}$. The time required to produce a fixed amount of I_3^- is measured by timing the appearance of a blue colour with starch after an initial addition of thiosulphate; this removes I_3^- until it is consumed (section 14D(2)).

When the reaction is carried out at different temperatures, the rate constant follows the Arrhenius equation

$$k_1 = \log_e \left(\frac{n}{n-1}\right) \Big/ t_{1/n} = A\,\mathrm{e}^{-E/\boldsymbol{R}T}$$

where E is the activation energy. Taking logarithms, we have

$$\log_{10}(t_{1/n}) = \frac{E}{2{\cdot}303\,\boldsymbol{R}T} + \text{constant}$$

The activation energy can therefore be found by plotting $\log_{10}(t_{1/n})$ against $1/T$.

Catalytic effect. At any given temperature, many reactions are speeded up by the addition of small quantities of catalytic substances. Oxidation–reduction processes such as the persulphate–iodide reaction are often catalysed by traces of transition metal ions such as Cu^{++}, Fe^{+++} or Ag^+ by virtue of the ability of these ions to change their valency state readily. The rate constant then generally increases linearly with the amount of catalyst added:

$$k_1 = k_0 + k_{cat}\,[\text{catalyst}]$$

The reciprocal of $t_{1/n}$ should therefore vary linearly with the catalyst concentration.

EXPERIMENT
Study the catalysed reaction $S_2O_8^{=} + 3I^- = 2SO_4^{=} + I_3^-$.

Make up 250 cm^3 of 500 mol m^{-3} potassium iodide solution, 250 cm^3 of 10 mol m^{-3} potassium persulphate solution and 250 cm^3 of 10 mol m^{-3} sodium thiosulphate solution. The persulphate solution should be prepared fresh each day, as it slowly decomposes. The thiosulphate solution is stable if it contains 0·1 g sodium carbonate per dm^3; the KI solution should be kept in the dark to prevent photochemical oxidation.

A large beaker can serve as a thermostat: it is filled with water and brought to the temperature required for an experiment either by heating or by adding ice. During a run the temperature is kept as constant as possible by further gentle heating or cooling. In a typical run, 20 cm^3 of 500 mol m^{-3} KI solution and 10 cm^3 of 10 mol m^{-3} $Na_2S_2O_3$ solution are placed in a boiling tube in the thermostat; 20 cm^3 of 10 mol m^{-3} $K_2S_2O_8$ solution with a few drops of starch solution are placed in another boiling tube and allowed to reach the thermostat temperature. The former mixture is then added rapidly with stirring to the latter, and the time measured for the appearance of the blue starch–iodine colour. This signifies that all the thiosulphate has been used up, so and the corresponding fraction of the persulphate–iodide reaction is complete.

Demonstrate that $t_{1/n}$ is independent of a at a fixed value of n and at room temperature, and hence that the reaction is first order in persulphate. Use two or more different persulphate concentrations not exceeding 4 mol m^{-3} in the mixture, and amounts of thiosulphate proportional to the amounts of persulphate.

It might seem that the first-order dependence on the iodide concentration could also be tested by varying the latter. However, this cannot be done without allowing for a 'salt effect'. The reaction involves the collision of two charged species and the rate is therefore sensitive to the total ionic strength whether produced by a reactant or an inert electrolyte. For this reason, the KI concentration is kept relatively large and constant throughout the whole experiment.

Temperature coefficient. The reaction with 4 mol m^{-3} persulphate should be carried out at not less than four temperatures between 5 and 35°C, one of which can be room temperature. The activation energy E is found as described above. The value of k at 25°C exactly should be obtained from the plot.

Catalytic effect. Add to the persulphate solution at a fixed low temperature several small amounts of copper sulphate, whose concentration in the reaction mixture should not exceed 0·1 mol m^{-3}, and measure $t_{1/n}$ as before. Allow for any dilution effect and for any 'blank' reaction between copper sulphate and potassium iodide alone. Test the equation $k_1 = k_0 + k_{cat}[Cu^{++}]$ and find k_{cat} from the graph in $mol^{-1}\ m^3\ s^{-1}$.

BIBLIOGRAPHY 14C: Second order reactions

See Bibliography 14A.

14D Heterogeneous reactions

1. All the reactions considered so far are homogeneous reactions in solution, and their velocities are generally governed by the rate of collision of the reacting species. Another class of reactions comprises those in which substances present in different phases participate—heterogeneous reactions.

In industrial chemistry heterogeneous reactions are more important than homogeneous reactions. Consider, for example, the large-scale reactions concerned in the extraction and refining of metals, the production of water-gas, the softening of water, dyeing of textiles, catalytic hydrogenation of oils, and catalytic synthesis of SO_3, NH_3, CH_3OH, etc. In such processes the rate is rarely controlled by slowness of chemical reaction; it is usually limited either by the rate at which the reacting substances can be brought together ('transport-controlled reactions')[1] or by factors concerned with adsorption or desorption of substances on a catalytic surface. These factors are illustrated by the following experiments.

2. REACTIONS OF ION-EXCHANGE RESINS

Modern ion-exchange resins (as used in water softening, etc.) consist of insoluble granules of a polymerized, cross-linked acid or base. An important type of 'strong-acid resin', for example, is composed of insoluble sulphonated polystyrene. All the sulphonic acid groups in the material are able to react with alkali to form stoichiometric salts, but in contrast to homogeneous ionic reactions, which are practically instantaneous, the rate of the reaction $H^+R^- + M^+OH^- \rightarrow M^+R^- + H_2O$ (where R^- stands for the resin substance and M^+ is a metal ion) is controlled by transport of MOH from the solution to the surface of the particle and into the interior of the particle.[2] The rate of reaction is therefore affected by the rate of stirring of the solution, the particle size of the resin grains, the concentration of the solution and by the temperature (in so far as it changes the rate of diffusion in the solution or inside the resin).

Under some conditions, the chief resistance to reaction lies in the transport of substance through the solution to the particles. Improved stirring then increases the rate by bringing fresh solute close to the surface. However, perfect transport cannot be achieved because the layer of liquid in contact with the solid surface is necessarily stationary. There is therefore a more-or-less stagnant film of liquid round the particles, and the only mechanism of transport through this film is *diffusion*, a relatively slow process. Nernst (1904) proposed that the kinetics of such heterogeneous processes could be represented by an expression of the type $dx/dt = AD.\Delta C/\delta$, where A is the area of the interface, D the diffusion coefficient of the substance diffusing, ΔC the difference between the concentration of the substance at the surface and in the bulk of the solution, and δ is the effective thickness of the unmixed layer of solution ('Nernst film'). The thickness δ depends only on hydrodynamic factors such as the stirring.

Under other conditions (efficient stirring, high concentration in solution, large ions) the rate of transport in the solution may become more rapid than transport within the particle; diffusion inside the resin then becomes the rate-determining factor.

EXPERIMENT

Study the kinetics of reactions with a sulphonic acid ion-exchange resin.

Materials. 'Zeo-Karb 225' (Permutit Co., Ltd, London) or 'Dowex 50' (Dow Chemical Co., Midland, U.S.A.) are suitable. These materials are supplied in the form of spherical grains, produced by suspension-polymerization. If possible, samples of two or more different grain-sizes should be used (e.g. coarse, 20–30 mesh, medium 50–100 mesh, fine 100–200 mesh sieve).

To prepare the resins in the acid form, they should first be treated in a column with portions of 2 000 mol m^{-3} HCl amounting in all to a considerable excess over the stoichiometric quantity. (The equivalent weight of the dried resin is about 200; the air-dried form has also about 40% water.) The resin is then washed thoroughly until the effluent is free from acidity; it is dried on filter-paper, and then overnight in the air. The resin should then be well mixed and stored in a stoppered bottle so that the moisture content remains constant.

Procedure. Weigh 0·5 g of resin into a beaker containing 100 cm^3 of water and three drops of phenolphthalein or bromo-cresol green indicator solution. Into a second beaker pipette 10 cm^3 of 100 mol m^{-3} KCl and 100 cm^3 of water. Arrange a stirrer driven by an electric motor to agitate the resin suspension. Vigorous and reproducible stirring during the runs is essential. The resin beads must move rapidly throughout the liquid, and none must lie stagnant on the bottom. The stirrer should be kept at the same height above the bottom of the beaker in all experiments, as this affects the degree of stirring and thus the thickness of the diffusion layer. For the same reason the total volume of solution should be the same in each run. Set up a burette containing 200 mol m^{-3} KOH solution to deliver into the beaker.

To start the exchange reaction, pour the contents of the second beaker into the first, start a clock, and quickly run in 1 cm^3 of KOH from the burette. When the indicator changes back again to the acid form, note the time, and add a further 1 cm^3 of KOH. Continue in this way, adding 1 cm^3 portions of KOH and noting the time at which it is neutralized, until the reaction becomes too slow for convenient observation. Further indicator can be added if the endpoint becomes faint. Stop the stirrer, allow the resin to settle, and tip a little solid KCl on to it in order to displace any remaining acidity still present in the resin. Allow the beaker and contents to stand for about a quarter of an hour, and then titrate the contents with the same burette of KOH until the equivalence point of the indicator is just reached. The final burette reading gives the total exchange capacity of the sample of resin (V_∞), while the intermediate volume readings (V_t) give the amount of exchange which has occurred at a series of times, t. It should be noticed that the concentration of KCl remains practically constant during the experiment. Note the temperature of each run.

The instructions given above apply to an experiment in which 20–30 mesh resin is used. With more finely-divided resin the reaction is faster, and it may be necessary to add larger portions of KOH or use smaller amounts of resin or greater volumes of more dilute KCl solution in order to obtain a convenient reaction velocity. A few preliminary experiments may be needed to find suitable conditions.

The determination should be repeated with several different quantities of KCl in the solution, using, of course, fresh portions of resin for each experiment. If the effect of particle size of the resin is to be studied, the mean surface area of each batch should be estimated from microscopical examination or sedimentation experiments.

Treatment of results. The form of the kinetics is seen by plotting a graph of V_t/V_∞ (=F) against t. It will be found that the kinetics do not fit to the first or second order equations. Instead, the kinetics are determined by diffusion of ions through the boundary film of liquid which surrounds the particles.[2]

Let δ be the effective thickness of the unmixed Nernst film; A the surface area of the resin; D_H and D_K the diffusion coefficients of H^+ and K^+ ions respectively through the solution; $[H^+]_s$ and $[K^+]_s$ the concentrations of these ions in the mixed solution; $[H^+]_r$ and $[K^+]_r$ their concentrations next to the resin surface; K the approximate mass action equilibrium constant for the ion exchange reaction, defined by equation (3) below. Q_t is the amount of exchange in g equiv. that has occurred in time t.

The following equations must hold according to Fick's law of diffusion: for diffusion of the K^+ ions from the solution towards the resin

$$\frac{dQ_t}{dt} = \frac{AD_K}{\delta}([K^+]_s - [K^+]_r) \tag{1}$$

For the diffusion of H^+ outwards from the resin through the Nernst film

$$\frac{dQ_t}{dt} = \frac{AD_H}{\delta}([H^+]_r - [H^+]_s) \tag{2}$$

As the solution is kept alkaline during the experiment, $[H^+]_s$ is negligible in this equation.

Thirdly, the assumption can be made that the electrolyte solution in immediate contact with the resin surface is in exchange equilibrium with the resin; consequently,

$$\frac{[H^+]_r \times [\text{K resin}]}{[K^+]_r \times [\text{H resin}]} = K \tag{3}$$

Putting

$$\frac{[\text{K resin}]}{[\text{H resin}]} = \frac{Q_t}{(Q_\infty - Q_t)}$$

it becomes possible to solve the three simultaneous equations, eliminating the unknown quantities $[H^+]_r$ and $[K^+]_r$. The resulting differential equation is

$$\frac{dQ_t}{dt}\left[1 + \frac{D_K}{KD_H}\left(\frac{Q_t}{Q_\infty - Q_t}\right)\right] = \frac{AD_K[K^+]_s}{\delta}$$

On integration, this finally gives a kinetic equation of the form

$$\frac{Q_\infty}{Q_t}\log_e\left(\frac{Q}{Q_\infty - Q_t}\right) = \frac{AKD_H[K^+]_s}{\delta}\left(\frac{t}{Q_t}\right) + \left(1 - \frac{KD_H}{D_K}\right)$$

or

$$-\frac{1}{F}\log_e(1-F) = \frac{AKD_H}{\delta}\cdot\frac{[K^+]_s}{Q_\infty}\left(\frac{t}{F}\right) + \left(1 - \frac{KD_H}{D_K}\right)$$

The experimental data should first be examined for conformity to simple first order kinetics by testing the constancy of the half reaction time (section 14A(4)).

Conformity to the mechanism is first tested by plotting $(1/F)\log(1-F)$ against (t/F) for each run. The plots should be straight lines with the same intercept on the ordinate axis. The slopes of the lines should be plotted against $[K^+]_s$, and a straight line passing through the origin should be obtained for runs with the same resin size. The variation with resin area can be checked by plotting the slopes of plots of $(1/F)\log(1-F)$ against (t/F) against area. Again a straight line through the origin should be obtained for runs with the same value of $[K^+]_s$. If samples of equal weight are used, the area will be inversely proportional to the average resin diameter; this can be roughly estimated from the mesh size.

The diffusion coefficient D_H of the hydrogen ion is five times that of the potassium ion, D_K; this can be used to estimate the equilibrium constant K from the intercept of the plots of $(1/F)\log(1-F)$ against (t/F).

When similar experiments are conducted with more concentrated solutions of KCl, it is found that the kinetics follow a different form. This is now consistent with a mechanism controlled by diffusion within the resin grains.[2]

3. DECOMPOSITION OF AMMONIA

Just as the reaction $N_2 + 3H_2 \rightarrow 2NH_3$ proceeds most readily in the presence of certain solid catalysts, so the decomposition of gaseous ammonia occurs more rapidly as a heterogeneous reaction on the surface of catalysts such as tungsten than as a homogeneous reaction.[3] Since the decomposition is accompanied by a doubling of the number of molecules, it can readily be followed by the increase of pressure in a constant volume apparatus. The increase of pressure is then proportional to the amount of ammonia decomposed.

The heterogeneous reaction can be studied conveniently by employing an electrically heated tungsten wire as the catalyst. The temperature of the tungsten can be controlled by adjusting the current supplied to the wire. The temperature-dependence of the resistivity R, in Ω m of tungsten is given below.[4]

Temp. (*K*)	900	1 000	1 100	1 200	1 300	1 400	1 500	1 600	1 700
$R(\times 10^{-8})$ (Ω m)	21·94	24·90	28·10	31·06	34·10	37·18	40·38	43·50	46·78

A given temperature can therefore be obtained by adjusting the current to produce the required resistance.

The rate of a surface reaction may be dependent on (a) the rate at which molecules of reactant are adsorbed on the surface, (b) the rate at which reaction of the adsorbed molecules takes place, or (c) the rate at which the products are desorbed, leaving vacant sites on the surface for further reaction. In the present reaction it is found that the reaction is of zero-order, i.e. the rate is independent of the ammonia pressure, except at low pressures (where the rate begins to fall off). This is taken to indicate that the surface is practically covered with adsorbed ammonia at all ordinary pressures. The rate-controlling process is thought to be the desorption of N_2 molecules from the surface. The evidence for this theory lies largely in the fact that on several catalysts the activation energy for ammonia decomposition is of the same order as that for the desorption of chemisorbed nitrogen.

EXPERIMENT

Determine the order of reaction and the activation energy of the heterogeneous decomposition of ammonia.

Apparatus. The glass and electrical apparatus is shown in Fig. 14D.1.

Ammonia gas is produced by dropping fresh 0·880-ammonia solution on to solid NaOH in a flask *A*; it is dried by passage over more NaOH in a U-tube *B* and is then admitted to the reaction bulb *C* and manometer *D*. This part of the apparatus can be isolated by stopcocks *E* and *F*; it is preferably made of Pyrex glass. The U-tube *J* contains calcium chloride and glass wool to protect the rotary oil vacuum pump from moisture and dust.

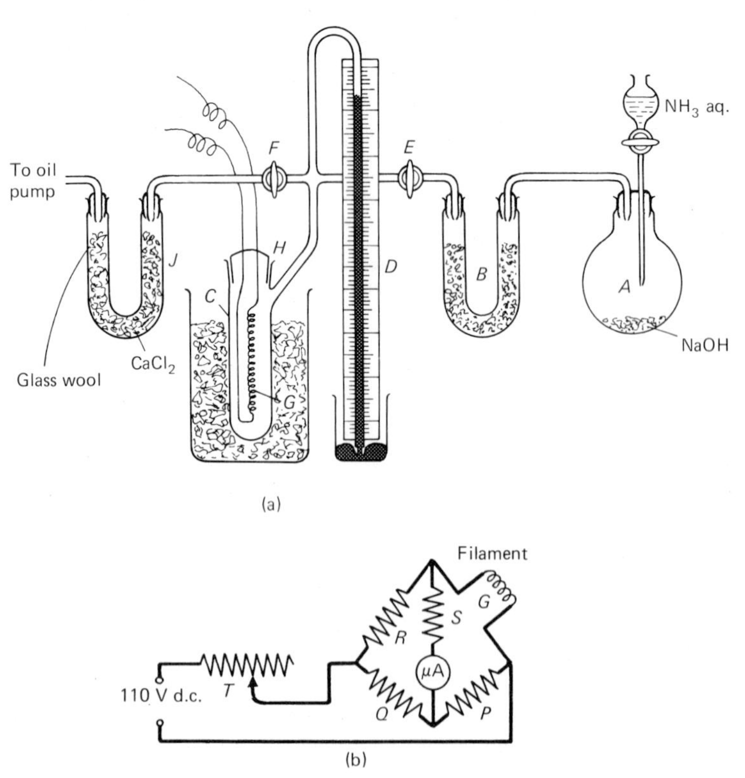

FIG. 14D.1 Apparatus for determining the rate of decomposition of ammonia gas by a tungsten filament. (a) Vacuum apparatus. (b) Electrical circuit.

The tungsten filament *G* consists of 30·0 cm of 0·1 mm diameter wire, and is attached to stout tungsten leads which are sealed through the removable Pyrex stopper *H*. In order to control the temperature of the filament, the current passed through it can be adjusted by a rheostat *T*, in series with a 110 V d.c. supply. Owing to the production of hydrogen during the reaction, the thermal conductivity of the gas will rise and the temperature will tend to fall. This must be counteracted by increasing the current so as to keep the *resistance* of the filament constant. The filament G is therefore connected into a Wheatstone bridge circuit $GPQR$ (Fig. 14D.1(b)), the balance of which is indicated on a centre-zero needle galvanometer or micro-ammeter protected by a resistance S of say 10 000–50 000 Ω. Resistance R is 15·3 Ω, made from nichrome wire. Arms P and Q of the bridge are provided by a Post Office box. Q is set at 1 000 Ω, and several values of P are employed in successive experiments to obtain different filament temperatures: 750, 800, 850 and 900 Ω are suitable values of P. *The current must not be switched on while the apparatus contains air, as the filament will burn out immediately. Also, the reaction vessel must be kept cool by an ice bath or the heat developed will crack the ground glass joint.*

Procedure. Place about 5 g of sodium hydroxide pellets in the 250 cm^3 round-bottomed flask A, and about 10 cm^3 of *fresh* 0·880-ammonia in the tap-funnel. Start the pump with all taps closed and then carefully open first tap F and then tap E. Then close taps E and F and see if the apparatus is vacuum-tight by watching the manometer for a few minutes. If there is no leakage, run in ammonia solution *drop by drop* from the tap funnel. When a pressure of about 40 cm is shown by the manometer close tap E and pump out the gas by opening F. After evacuation close tap F and admit 30 cm of ammonia by carefully opening tap E. A beaker of ice-water is placed around the reaction bulb and well stirred. After about 10 min the manometer is read, giving the initial pressure of gas at 0°C.

For the first determination set resistance P at 850 Ω. Immediately the current is switched on, the rheostat T must be quickly adjusted so that the galvanometer needle returns to zero, and throughout the experiment T must be continually adjusted to maintain zero deflection. At the beginning of each rate experiment the rheostat T will not be properly adjusted and a large error can arise if the tungsten filament is at too high a temperature during the few seconds needed for the balancing of the Wheatstone bridge. A preliminary experiment should therefore be carried out to note the first two or three rheostat settings needed, the cell pumped out again, and a fresh 30 cm of ammonia admitted for the experiment proper. A stop-watch is started immediately the current is switched on. After about 30 s, during which the rheostat is constantly adjusted, the current is switched off. The ice-bath is stirred, and the manometer tapped until the mercury level reaches a constant position. The pressure reading is then taken, the gas being at 0°C. The current is switched on again and the procedure repeated at convenient time intervals. After sufficient readings have been obtained, the remaining ammonia is decomposed by heating the wire to *dull* red heat (not hotter) for a minute or two. This should be repeated to make sure that decomposition is complete.

The whole experiment should be repeated once with the same resistances and same initial pressure of ammonia and then with 750, 800 and 900 Ω for P in place of 850 Ω.

At the highest temperatures readings should be taken every 15 s rather than every 30 s. The experiment may be repeated with differing pressures of ammonia at the same filament temperature. The increase in pressure, Δp, should be plotted against the total time t for which the filament is heated: the plots should be straight lines initially, the rate falling off as the reaction nears completion. If the plots are not straight lines, set P equal to 1 300 Ω. With about 40 cm of ammonia in the reaction vessel, flash the tungsten filament for about 30 s. See whether the results have improved and if not, repeat the treatment with the heating time increased to about 10 min. A new filament is required if the results are still not satisfactory or if the coils are touching each other and so shorting electrically.

Calculations. If the reaction is zero order, the slopes of the ΔP–t plots should be independent of the pressure of ammonia at one temperature. For each series of measurements calculate the temperature of the filament from its dimensions, resistance and resistivity. For each temperature deduce the initial rate of reaction at 20 cm pressure. The activation energy for the heterogeneous reaction can be calculated by applying the Arrhenius equation in the normal way, i.e. a graph of $\log_e$ (rate) against $1/T$ should give a straight line, the slope of which is equal to $-E/\boldsymbol{R}$.

BIBLIOGRAPHY 14D: Heterogeneous reactions

[1] Bircumshaw and Riddiford, *Quart. Rev.*, 1952, **6**, 157.
[2] Boyd, Adamson and Myers, *J. Amer. Chem. Soc.*, 1947, **69**, 2836; Kressman and Kitchener, *Faraday Soc. Disc.*, 1949, **7**, 90; Hale and Reichenberg, *ibid.*, p. 79; Grossman and Adamson, *J. Phys. Chem.*, 1952, **56**, 97.
[3] Hinshelwood and Burk, *J. Chem. Soc.*, 1925, 1105; Kunsman, *J. Amer. Chem. Soc.*, 1928, **50**, 2100; Kunsman *et al.*, *Phil. Mag.*, 1930, **10**, 1015; Hailes, *Trans. Faraday Soc.*, 1931, **27**, 601; Barrer, *ibid.*, 1936, **32**, 490.
[4] Jones, *Phys. Rev.*, 1926, **28**, 203.

14E Photochemical reactions

1. Many reactions, some of which do not proceed under ordinary conditions, take place if the reactants are irradiated with visible or ultra-violet light. The light may serve merely as catalyst to initiate a spontaneous reaction (e.g. $H_2+Cl_2=2HCl$) or it may provide the necessary free energy for a reaction which is otherwise thermodynamically impossible (e.g. CO_2+H_2O=carbohydrates, in photosynthesis). In either case the light must be *absorbed* to have any effect. This produces *electronic excitation* in the absorbing atoms or molecules. Since light consists of quanta, the extra energy thus introduced into individual molecules is given by $h\nu$ (h=Planck's constant, ν= frequency of absorbed light), and calculation shows that $h\nu$ is large compared with thermal energies and often great enough to break chemical bonds. Many photochemical gas reactions do, in fact, proceed by way of dissociation of molecules into atoms (e.g. $Cl_2 \xrightarrow{h\nu} 2Cl$). In other cases excitation provides energy of activation for a reaction, without intermediate dissociation (e.g. in the photo-dimerization of anthracene). The shorter the wavelength, the more powerful the light.

Absorption of light does not necessarily lead to reaction. The excited molecule may re-emit light energy (*fluorescence*), or the energy may be degraded to heat, or it may release electrons (*photo-emission effect* and *photoconductivity*). Consequently, although the *primary process* (absorption of light) in photochemistry is always at the rate of 1 molecule (or atom) excited per quantum absorbed, the *secondary process* (i.e. subsequent reactions, etc.) may not show any such equivalence. The *quantum yield* is the number of molecules reacting per quantum of light absorbed; it may be less than unity (for reasons indicated) or greater—namely, when the light starts a spontaneous chain reaction.

In some cases reactions can be made to proceed with light of a wavelength which is not absorbed by any of the reactants; the absorption can be effected by another substance which can transmit the absorbed energy to the reactants and itself remain unchanged. This is *photosensitization* (e.g. photosensitization of the decomposition of oxalic acid in the near ultra-violet by addition of uranyl salts). In some systems thermal reactions proceed simultaneously with photochemical ones, and may enhance or reverse the effect of light. In the latter case a *photo-stationary state* may be set up, as, for example, in the experiment described in section 14E(2).

The technique of research in photochemistry is somewhat specialized. Powerful mercury arc lamps are the favourite source of illumination since one can obtain a high intensity of monochromatic light of various wavelengths by use of filters or monochromator (chapter 9). Determination of the absorption spectra is essential, since one must identify the initial step in the process. The stoichiometric equation of the reaction must be established. Some physical or chemical means must be found for determining the amount of reaction which takes place during irradiation. The other principal problem is measurement of the amount of light which has been supplied, so that the quantum efficiency of the reaction can be evaluated. Light intensities are generally measured in photochemistry by means of multiple-junction thermopiles which are previously calibrated in absolute units by means of standard tungsten filament lamps. Thermopiles measure total radiant energy and are therefore not selective, whereas photocells have high spectral selectivity (section 4A(4)). Quartz apparatus must be used for work in the ultra-violet since light of wavelength less than about 350 nm is strongly absorbed by ordinary glass.

Another method of measuring high light intensities for photochemistry in the ultra-violet is by means of a chemical 'actinometer' reaction, the quantum efficiency of which has been previously determined by the absolute physical method. The decomposition of a uranyl oxalate solution is the best established method of actinometry.[1]

Photochemistry has been revolutionized by the development of *flash photolysis* by Norrish and Porter. The photolytic illumination is obtained from an electronic flash lamp, i.e. by a sudden high voltage discharge of a condenser through a gas. The short intense pulse of light produces photolysis—typically generating free radicals—within a few us. The spectrum of the transient species is obtained by passing the light from a second flash fired after a short delay through the reaction vessel into a spectrograph. By varying the delay the concentration of the transient species can be followed as a function of time. This technique has produced much detailed information about the spectra and hence structure of free radials, about their chemical kinetics, and about energy transfer processes in gases. Experiments using this technique are beyond the scope of this book; a suitable inexpensive apparatus is made by Messrs Applied Photophysics Ltd (Royal Institution, 20 Albemarle St, London W1X 2HA); a manual of experiments is available.[2]

2. THE PHOTO-REACTION OF THIONINE WITH FERROUS SALTS[3]

The dye thionine ('Lauth's violet') can be reduced reversibly by mild reducing agents to a colourless 'leuco' form. The leuco form reacts readily with ferric salts and gives thionine again. The redox potential (section 13E) of the reaction thionine + $e \rightarrow$ leuco-thionine is such that the redox equilibrium with the ferrous–ferric system lies to the left-hand side in the equation thionine + $Fe^{++} \rightleftarrows$ leuco-thionine + Fe^{+++}. However, by supplying energy in the form of light (absorbed by the thionine) one can cause the reaction to proceed to the right. As the concentration of products increases, the spontaneous reversal of the reaction ('dark reaction') becomes more rapid, and, clearly, with a given intensity of light a *photo-stationary equilibrium* is eventually reached. (Incidentally, the redox potential of the solution will be altered by the photochemical reaction; this provides one type of photo-voltaic effect.) The rate of the dark reaction increases with rise of temperature whereas the photochemical bleaching reaction is substantially independent of temperature.

Thionine solutions show a faint red fluorescence; this can best be seen by examining the solution at right angles to a powerful narrow beam of white light in a dark room. The primary process is therefore excitation of the thionine. The excited molecule can either return to its ground state with the emission of fluorescence, or it may react with a ferrous ion, capturing an electron from it. Consequently, addition of ferrous salt 'quenches' the fluorescence.

EXPERIMENT

Investigate the photochemical reaction between thionine and ferrous sulphate.

Procedure

Determination of the absorption spectrum. See section 9C for methods of spectrophotometry. Thionine is a dye which does not obey Beer's Law accurately; with increasing concentration of dye, a dimer (or aggregate ?) is formed which has a slightly different absorption spectrum.[2] The absorption spectrum should therefore be determined over a range of concentrations, using cells of different optical length (e.g. 0·1, 1 and 10 cm with solutions containing 1, 0·1 and 0·01 $kg\,m^{-3}$). A stock solution (e.g. 100 cm^3) of thionine containing 1 $kg\,m^{-3}$ should first be made up and carefully *filtered*. This stock is used for preparing other solutions as required.

The absorption spectrum of thionine has been reported by Rabinowitch and Epstein.[3]

Absorption and fluorescence. Set up a sodium lamp (section 8A(1)) or an intense tungsten filament lamp with a filter transmitting in the yellow-green region, and, by means of a lens and diaphragms, produce a strong, narrow beam of yellow light. Direct the beam horizontally through a cell about 10 cm long containing successively thionine solutions of 1, 0·1 and 0·01 kg m^{-3}, and examine the solution *at right angles* to the light beam. A red fluorescence can be seen, and from the distribution of the fluorescence it can be seen where the light is being absorbed. A darkened room is desirable for these observations. Confirm by use of other filters or by using a wavelength spectrometer as monochromator (section 8B(1)) that light of spectral regions not strongly absorbed by the dye does not cause fluorescence. Note that the fluorescent light is of longer wavelength than the absorbed light, i.e. of smaller energy.

Investigate qualitatively the effect of additions of ferrous sulphate to the dye solution on the intensity of fluorescence.

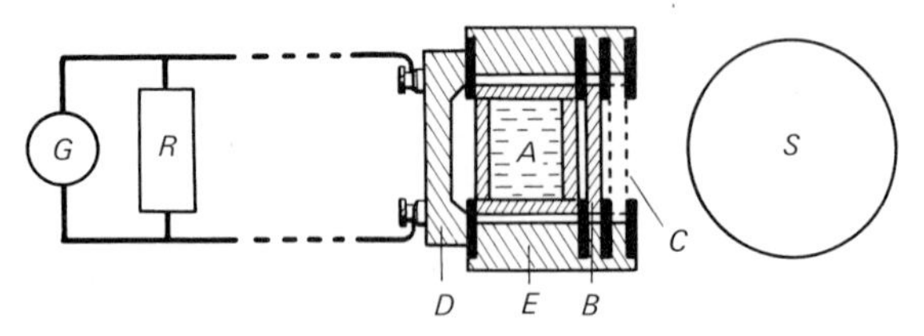

FIG. 14E.1 Apparatus for determining the rate of bleaching of a solution of thionine.

Kinetics of the bleaching reaction. The apparatus shown in Fig. 14E.1 is required. The essential parts are: *S*, a sodium lamp or 150 W pearl tungsten lamp with a yellow-green filter: *A*, an optical cell, e.g. 1 cm thick, 1·25 cm wide and 5 cm high: *B*, a piece of heat-absorbing glass: *C*, a sliding metal plate to act as shutter: *D*, a rectangular barrier-layer photocell: *G*, a galvanometer with variable shunt *R*. *G* is preferably of the type shown in Fig. 4B.1. Suitable characteristics are an internal resistance of about 100 Ω and a maximum sensitivity of about 200 mm/μA. It must be possible to reduce the sensitivity by factors of 10 and 100 by means of the shunt. *E* is a wooden mount for the cells. The whole of the solution in the cell *A* must be illuminated, and the stops must allow only light which has passed through the solution to fall on the photocell.

Solutions. A suitable reaction mixture contains about 1 mg of thionine, 0·1 g phosphoric acid and 0·1 g of ferrous sulphate. The phosphoric acid is added to repress hydrolysis of the iron salt and to reduce the effective ferric ion concentration by complex formation, thus slowing the 'dark', reversal reaction. Note that *the above reaction mixture cannot be kept for more than a quarter of an hour* as the ferrous salt is rather quickly oxidized by dissolved air, and the photo-reaction is then greatly inhibited. It is best, therefore, to make 500 cm^3 of a stock solution containing 50 mg of thionine and 3 cm^3 of syrupy phosphoric acid; then take 10 cm^3 portions of this stock, add 50 cm^3 of de-aerated water and 0·1 g of powdered ferrous sulphate crystals. Dissolve the crystals quickly and perform the measurements at once.

Measurements. Clean the cell and fill it with distilled water. Arrange the sodium lamp close to the cell-holder so as to obtain a full-scale deflection of the galvanometer with the shunt set for lowest sensitivity (i.e. about 1/100th of full sensitivity). Allow the lamp to run for $\frac{1}{2}$ h to become steady, and then read the galvanometer zero (shutter closed) and full deflection (shutter open) several times. Close the shutter, empty the cell, and refill it with the standard reaction mixture, freshly made up from the stock solution as described above. Open the shutter, start a clock, and read the galvanometer deflection every 15 s at first, and then less frequently as the reaction slows down. At first, the deflection should be only a few per cent of full scale, but it should gradually increase as the dye bleaches, and eventually come to a constant value. This indicates the photo-stationary equilibrium where the rate of bleaching by light is equal to the rate of thermal back-reaction. Then close the shutter, check the zero, and read again the deflection with pure water to see whether the lamp intensity has altered during the experiment.

Repeat the experiment with a fresh sample of mixture to ascertain whether the kinetics are reproducible. Then repeat the readings with two or three lower light intensities. For this purpose, the lamp should be moved away from the cell a little so that the deflection with water is 75, 50 and 25% of that employed in the first experiment. (When the intensity has been adjusted and measured in this way, one can alter the shunt, if desired, to produce a conveniently large deflection for taking the kinetic readings.)

Carry out two or three more kinetic determinations at a fixed light intensity but with different amounts of added ferrous sulphate.

Kinetics of the thermal ('dark') back-reaction. The reversal of the photo-reaction can be followed by using a very low light intensity. Move the lamp away from the cell until full-scale deflection of the galvanometer at maximum sensitivity is obtained with water in the cell. Then empty the cell and fill it with fresh reaction mixture. Bleach the mixture as fully as possible by exposing it near a tungsten filament lamp, interposing a cell of water to protect the solution from the heat of the lamp. When the solution has bleached, transfer the cell *as quickly as possible* to the photometric apparatus and take galvanometer readings at once and then every $\frac{1}{4}$ min, etc. Continue the readings until a steady value is reached. If the reaction is completely reversible, the colour should return fully, and the light transmission should then equal that of a fresh, unbleached solution.

Repeat the measurements with (i) a solution which has been cooled somewhat below room temperature (but not to the dew-point), and (ii) another which has been warmed to about 30°C.

Also make two or three dark-reaction measurements with solutions to which small, known amounts of ferric salt have been added.

Results. In each series of galvanometer–time readings the lamp intensity may have changed a little from beginning to end of the readings. A correction can be applied by assuming that the drift was linear with time. Thus one can calculate the fraction of light transmitted by the solution, I_t/I_0, for each reading. For the present purpose the small deviations from Beer's law are negligible, and hence the optical density, $D = \log_e (I_0/I_t)$, is a linear measure of the concentration (section 9D). Plot all the results in the form of graphs of D against t to show how the concentration of dye decreases with illumination or increases in the dark (or, rather, under very feeble illumination).

The form of the bleaching curves is mathematically complicated, especially as the solution has not been kept stirred, but the general form of the kinetics can be elucidated by the following approximate analysis. In the early stages of the bleaching (e.g. for the first 20% of change) the back-reaction is negligible and the slope of the concentration–time line, which is approximately linear, can be taken as a measure of the rate of the photochemical reaction. It should be noted that the first few points on the graph are liable to be less accurate than later ones as they represent small deflections; hence, the best mean slope over the first 20% of the graph should be taken. Plot graphs to show the influence of (i) light intensity, and (ii) ferrous ion concentration on the rate of bleaching.

Note also the effect of these two factors on the concentration of thionine present in the photo-stationary mixtures. Explain the observations qualitatively.

Examine the kinetics of the dark reaction for conformity to

first or second order kinetics in the usual way. Since the reactions involved are thionine + Fe^{++} $\rightleftarrows$ leuco-base + Fe^{+++} one might expect the reverse reaction (in the absence of added ferric salt) to be of second order. At any stage the concentration of leuco-base and also that of Fe^{+++} should be equal to (original dye concentration *minus* observed dye concentration) and this is the quantity to be used when examining the kinetics of the back-reaction. On the other hand, if an excess of Fe^{+++} is present, either as a deliberate addition or as a result of oxidation of ferrous salt by dissolved air, then the kinetics of the dark reaction would be expected to follow the first order law.

Experiments at different temperatures should show that the photochemical (forward) reaction is practically independent of temperature, whereas the dark reaction accelerates with temperature in the normal way.

The detailed *mechanism* of this photochemical reaction is much more complicated than might be supposed from the elementary discussion given above. This is frequently the case in such studies; the overall kinetics may not lead directly to all the details of the mechanism.

BIBLIOGRAPHY 14E: Photochemical reactions

Calvert and Pitts, *Photochemistry*, 1966 (John Wiley, New York).
Cundall and Gilbert, *Photochemistry*, 1970 (Nelson, London).

[1] Leighton and Forbes, *J. Amer. Chem. Soc.*, 1930, **52**, 3139; Forbes and Heidt, *ibid.*, 1934, **56**, 2363.

[2] Goodall, *Flash Photolysis—An Experimental Manual*, 1971 (Dept of Chemistry, University of York).

[3] Weiss, *Nature*, 1935, **136**, 794; Rabinowitch and Epstein, *J. Amer. Chem. Soc.*, 1941, **63**, 69; Rabinowitch, *J. Chem. Phys.*, 1940, **8**, 551, 560.

15
Surface chemistry and colloids

Introduction

Surface chemistry is a special branch of Physical Chemistry; at a boundary between two phases there is necessarily a rather abrupt change in the nature or intensity of the inter-molecular forces, and this fact gives rise to a number of phenomena which have no counterpart in homogeneous systems. The chief of these are *adsorption*, *molecular orientation* in monolayers, and the formation of *electrical double layers*; other phenomena of surface chemistry may be considered as secondary effects.

Surface phenomena are rarely important, however, unless the system has a relatively large amount of interfacial area per unit mass. Consequently, surface effects are generally noticed most clearly in connection with very thin films, highly porous solids or finely-divided dispersions of one phase in another. The latter category includes particularly *colloids*.

Colloid chemistry is best defined as the study of submicroscopic dispersions, but the term 'colloid' is often used to include coarse suspensions, emulsions, films, plastics, etc.—in fact, any material which is not obviously crystalline. In this sense, colloids include the great majority of materials used in industrial chemistry, and consequently colloid chemistry has great technological importance. However, much of the content of classical colloid chemistry is descriptive and qualitative, and it is only with the development of surface chemistry and of high polymer chemistry that it has become possible to find theoretical interpretations of some of the numerous phenomena that are known.

Major developments in modern colloid chemistry are concerned with the configuration of linear molecules and other *lyophilic* molecules in solution, and also with the theory of the stability of *lyophobic* colloidal dispersions in terms of the dispersive forces of attraction and the electrical double layer forces of repulsion between the particles.

15A Surface chemistry

1. ADSORPTION BY SOLIDS

Adsorption is the accumulation of a substance at an interface. It is met with at all types of interface, notably, gas–solid, solution–solid, solution–gas, solution α–solution β interfaces. Two principal types of adsorption are recognized; *physical adsorption* is non-specific, rapid and reversible, the *adsorbate* being held at the interface only by dispersion and polarization forces, while *chemisorption* is simply the formation of a two-dimensional compound, e.g. when gases are brought into contact with clean metallic surfaces. A subsidiary type of chemisorption is *ionic exchange adsorption*, as exemplified by the adsorption of cationic dyes by acidic oxides (e.g. silica) and of anionic dyes by basic oxides (e.g. ZnO); here an insoluble surface salt is formed. However, the unqualified term 'adsorption' will generally refer to a reversible, physical adsorption.

The amount of substance adsorbed at any surface decreases with rise of temperature, since all adsorption processes are exothermic. At constant temperature the amount adsorbed increases with the concentration of the adsorbate (in the gas phase or in solution), and the relationship between the amount adsorbed (x) and the concentration (c) is known as the *adsorption isotherm*. Only at very low concentration is x proportional to c. Generally the amount adsorbed increases less than proportionally to the concentration owing to the gradual saturation of the surface, and in many cases the isotherm can be fitted to an equation of the form $x=kc^n$, where $n<1$. This expression, generally known as the Freundlich adsorption isotherm, is empirical and lacks theoretical basis.

2. LANGMUIR ISOTHERM

The earliest and simplest theoretical model is that of Langmuir (1918) for gas adsorbed on solids. The surface is covered by a large number of sites on each of which an adsorbed molecule may sit. The sites are all equivalent and the adsorbed molecules are assumed not to interact with each other or to jump from site to site. The fraction of sites occupied is called θ. When each site is occupied $\theta=1$ and the surface is covered: no more adsorption occurs. The theory assumes that a gas molecule can stick on the surface only if it strikes an empty site; if it hits an occupied site it will be reflected back into the gas phase. The rate of condensation of gas molecules on the surface is proportional both to the rate of bombardment of the surface (i.e. proportional to the pressure p) and to the fraction of surface left uncovered (i.e. to $(1-\theta)$): rate of condensation $=k_1 p(1-\theta)$. The rate of evaporation from the surface depends on the number of sites occupied, so: rate of evaporation $=k_2\theta$. At equilibrium these two rates are equal so $k_1 p(1-\theta)=k_2\theta$ or

$$\theta = \frac{k_1 p}{k_2+k_1 p}$$

This is *Langmuir's adsorption isotherm*. It can also be written in the form $x=ap/(1+bp)$, where a, b are constants for a given gas, amount of absorbent and temperature. At low pressures when $bp \ll 1$, the amount adsorbed is proportional to the pressure.

3. B.E.T. ISOTHERM

Neither the Freundlich nor the Langmuir equation is capable of representing the S-shaped ('sigmoid') isotherms which are obtained when the adsorption of vapours on a porous solid is studied at relatively high vapour pressures (approaching saturation). A more complicated equation must then be used, the best-known being the isotherm of Brunaer, Emmett and Teller (1938).[1] This equation was arrived at on the assumption (confirmed by Harkins and Jura[2]) that *multimolecular adsorption* can take place. The equation is

$$\frac{p/p_0}{x(1-p/p_0)} = \frac{1}{x_m c} + \frac{(c-1)}{x_m c} \cdot \frac{p}{p_0}$$

where x is the amount of vapour adsorbed at a partial pressure p, p_0 is the saturation pressure, x_m is the amount of vapour which would be required to form a monomolecular layer over the surface and c is a constant. This equation (the 'B.E.T. isotherm') represents fairly satisfactorily the adsorption of water by textiles, silica gel, etc., and of many vapours by finely divided powders. It is much used for determining the surface areas of powders.

4. HEAT OF ADSORPTION

The 'reaction' involved in the adsorption of a molecule M is simply M (adsorbed) $=$ M (gas). ΔH, the molar enthalpy change for the reaction, is known as the *isosteric heat of adsorption*. The equilibrium constant for the reaction is the ratio of the activities of M in the two phases, i.e. $K = p/f(x/A)$ where A is the area of adsorbent and f is some function of x/A. Applying the van't Hoff equation

$$\left(\frac{\partial \ln p}{\partial(1/T)}\right)_{x/A} = \frac{-\Delta H}{R}$$

The equation is analogous to the Clausius–Clapeyron. Values of ΔH found for chemisorption range between 40 and 400 kJ mol^{-1}; for physical adsorption ΔH is generally lower than 40 kJ mol^{-1}.

EXPERIMENT

Determine the adsorption isotherm of oxalic acid from aqueous solution by charcoal.

Procedure. Prepare 1 dm^3 of 500 mol m^{-3} oxalic acid solution and titrate 10 cm^3 portions of it with approx. 100 mol m^{-3} $KMnO_4$ solution in the usual way (adding sulphuric acid and warming to 80°C). Hence calculate the true concentration of $KMnO_4$.

Clean and dry seven stoppered reagent bottles and weigh accurately into each about 2 g of finely ground charcoal: the kind sold for decolorizing solutions is particularly suitable. To the first bottle add 50 cm^3 of 500 mol m^{-3} oxalic acid and 50 cm^3 of water. Place the bottle in a shaking-machine and shake for successive $\frac{1}{4}$-hour periods. Take out a 10 cm^3 sample after each shaking and titrate it with the permanganate solution. These tests will show how rapidly the adsorption proceeds. Note the time required for a constant concentration to be reached. In the main set of experiments shake the bottles for at least 50% longer than the minimum time indicated by these preliminary tests.

Meanwhile, into the other six bottles pipette 100, 80, 60, 40, 20 and 10 cm^3 of the oxalic acid, and add 0, 20, 40, 60, 80 and 90 cm^3 of distilled water respectively, thus starting the experiments with 2 g samples immersed in 100 cm^3 of 500, 400, 300, 200, 100 and 50 mol m^{-3} oxalic acid. Shake the bottles for the time mentioned above, and then filter the contents through *small* filter-papers; reject the first 10–20 cm^3 portions of filtrate from each bottle in case there should be any adsorption of oxalic acid or of water by the filter paper. If time permits, this point should be investigated experimentally. It should be possible to collect at least 80 cm^3 of each filtrate; this quantity will provide two 25 cm^3 portions for duplicate titrations, and, if these do not agree, a third portion for a check.

Calculation. Calculate (a) the *final* concentration of the solution, C_s, (b) the weight of oxalic acid removed by adsorption, x, (c) the amount of adsorption per g of charcoal, x/m. Plot x/m against C_s. To test the applicability of the Freundlich isotherm plot $\log_e (x/m)$ against $\log_e C_s$. If a straight line is obtained, calculate the constants in the equation $x/m = kc^n$.

The applicability of the Langmuir isotherm can be tested similarly by plotting m/x against $1/C_s$.

EXPERIMENT
Investigate the adsorption of nitrogen on charcoal.

Apparatus and preliminary procedure. The apparatus is shown in Fig. 15A.1. S is a bulb containing about 2 g of the activated charcoal adsorbent. M is a mercury manometer. The space over limb b is kept at high vacuum by closing tap 7 after pumping out. The space bounded by taps 6, 7, 8 and 9 and the mercury in limb a is the *fixed dose volume* V_1. This is defined accurately when the mercury is at a fixed mark about 20 cm from the bottom of the manometer. V_1 should be about 150 cm^3, and can be determined by sharing gas between V_1 and a larger bulb of known volume. The dose volume is separated from S by tap 5.

R is a large gas (N_2) reservoir bulb; He a 'break-seal' bulb of helium (B.O.C.Co) used for the initial calibration and T_1 and T_2 liquid nitrogen cooled cold traps. The system is first evacuated by a mechanical rotary pump, and then by the mercury diffusion pump, keeping taps 2 and 13 closed and the remainder open (11 should remain closed after the calibration, i.e. after the helium seal is broken). A suitable vacuum is attained when no discharge is obtained from a high frequency leak tester. Slowly open tap 14 and allow air to enter the mercury reservoir: the mercury rises to record atmospheric pressure. Close tap 7, and then taps 4 to 13. The system should always be left evacuated, and air must never be allowed to enter the absorbent volume S.

Nitrogen is taken from a cylinder. The cylinder regulator is set to ≈ 5 psi, and the flow of gas adjusted by a needle valve so that a steady flow of nitrogen escapes through the bubbler. Tap 13 is cracked open so that the reservoir fills *slowly* without sucking mercury from the bubbler up the tube. When nitrogen is required, i.e. after the helium calibration, it is admitted to the dose volume through the trap T_1 which freezes out any moisture. The trap is cooled with liquid nitrogen, taps 8, 9, 10 and 11 are closed, and tap 12 is opened to fill

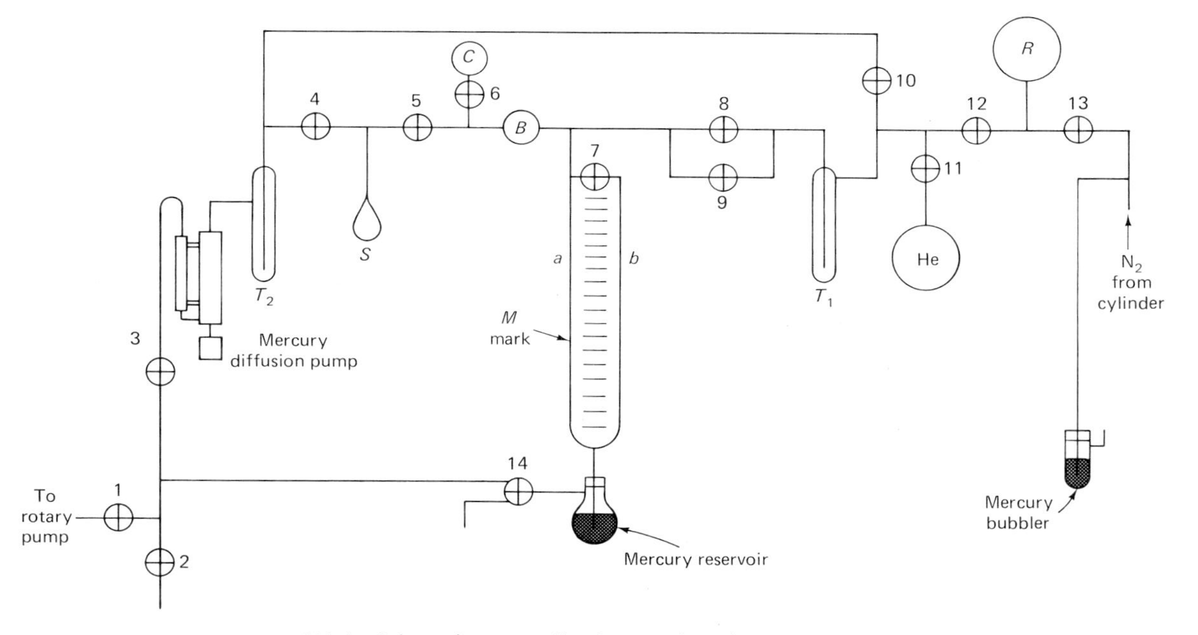

FIG. 15A.1 Schematic vacuum line for gas adsorption measurements.

the trap and then closed again. Nitrogen is admitted slowly to the dose volume through the small-bore tap 9, as described below.

Principle of the measurements. In this experiment the amount of adsorbed gas at varying equilibrium pressures has to be determined. This is done by expanding a certain quantity of nitrogen in the dose volume V_1, at some known pressure p_0, into the charcoal sample bulb which is kept at a fixed temperature. The equilibrium pressure p in the system is then measured after adsorption has occurred. If the pressure p is determined at the same volume V_1 as was p_0 than, since the room temperature is also constant, the amount of nitrogen adsorbed is directly proportional to the pressure change $p_0 - p$. A small dead-space correction is described below.

The pressure in the system (i.e. the difference in height between limbs b and a) is always measured after the mercury in column a has been adjusted to the fixed mark so that the dose volume is always the same. The adjustment is carried out by varying the pressure above the mercury in the mercury reservoir by means of the capillary bleed in tap 14. While this is being done, tap 3 on the backing side of the diffusion pump must be closed.

Theory. The amounts of gas adsorbed by the solid at the various equilibrium pressures may be calculated from the recorded values of p_o and p. The ideal gas equation can be employed with little error since the pressures are not high.

Let V_2 be the *dead space volume*, i.e. the volume between taps 4 and 5 less the actual volume of the charcoal itself. Then when a given mass of nitrogen at pressure p_0 in the dose volume V_1 is allowed to expand (by opening tap 5) into a total volume $V_1 + V_2$, we have

$$p_0 V_1 = p_t(V_1 + V_2) \tag{1}$$

However, because adsorption takes place when tap 5 is opened, we never measure p_t but only the final equilibrium pressure p.

Let us for convenience represent the amount of nitrogen adsorbed by U, where

$$U = n_{\text{ads}} \boldsymbol{RT} \tag{2}$$

Then, if we add superscripts 1 to indicate that only one dose of gas has been supplied

$$\begin{aligned} {}^1U &= {}^1p_t(V_1 + V_2) - {}^1p(V_1 + V_2) \\ &= {}^1p_0 V_1 - {}^1p(V_1 + V_2) \\ &= ({}^1p_0 - {}^1p)V_1 - {}^1pV_2 \end{aligned} \tag{3}$$

When the second dose of nitrogen is added, the equation corresponding to (1) becomes

$$^{2}p_0 V_1 + {}^{1}p V_2 = {}^{2}p_1(V_1 + V_2)$$

because the dead space now contains gas at the equilibrium pressure of the previous dose. The extra gas added in the second dose is

$$\begin{aligned} ^{2}U &= {}^{2}p_t(V_1+V_2) - {}^{2}p(V_1+V_2) \\ &= ({}^{2}p_0 - {}^{2}p)V_1 + ({}^{1}p - {}^{2}p)V_2 \end{aligned}$$

Thus the total amount of nitrogen added in the first two doses is

$$\begin{aligned} U &= {}^{1}U + {}^{2}U \\ &= V_1([{}^{1}p_0 - {}^{1}p] + [{}^{2}p_0 - {}^{2}p]) - {}^{2}pV_2 \end{aligned} \tag{4}$$

After the kth dose, therefore, the total quantity of gas adsorbed is

$$^{k}U = V_1 \sum_1^k ({}^{k}p_0 - {}^{k}p) - {}^{k}pV_2 \tag{5}$$

It is simpler at the beginning to express the adsorption uptake in terms of pressure only. Equation (5) may be written as

$$^{k}Q = \frac{^{k}U}{V_1} = \sum_1^k ({}^{k}p_0 - {}^{k}p) - \frac{^{k}pV_2}{V_1} \tag{6}$$

where ^{k}Q is the pressure the adsorbed gas would exert at temperature T and volume V_1 if it were not adsorbed. It follows that

$$^{k}n_{\text{ads}} = {}^{k}QV_1/RT \tag{7}$$

In Eq. (6) the term $\sum_1^k ({}^{k}p_0 - {}^{k}p)$ is found from the experiment. The correction term pV_2/V_1 is determined as a function of p by the same procedure but using helium as the gas. Helium is not adsorbed by charcoal at the temperatures involved: ^{k}Q in Eq. (6) is zero, so

$$\sum_1^k ({}^{k}p_0 - {}^{k}p)_{\text{He}} = {}^{k}pV_2/V_1 \tag{8}$$

The chief advantage of carrying out this calibration directly, instead of simply multiplying p by the ratio V_2/V_1, is that it allows for the fact that part of the dead space is at room temperature and part at the temperature of the adsorbent thermostat.

It is clear that T in Eqs (2) and (7) is the room temperature; this should therefore be measured.

Adsorption measurements. These are made for helium and nitrogen at $-45°C$ (charcoal bulb S cooled with a sludge of liquid and solid chlorobenzene prepared by stirring the liquid in a Dewar and adding dry ice *cautiously*) and at $-78°C$ (dry ice–acetone sludge). The temperatures are checked with an alcohol thermometer.

The procedure for determining the adsorption isotherm is as follows. Tap 9 is adjusted so as to draw sufficient gas from T_1 into the dose volume V_1 to exert a pressure of about 20 mm Hg. Tap 9 is shut and the mercury level in manometer arm a adjusted to the reference mark. The difference in the levels, i.e. the pressure p_0, is measured as accurately as possible. Tap 5 is opened with tap 4 closed to admit the gas to the charcoal sample; 2 min is allowed for the system to equilibrate. The mercury is adjusted to the reference mark and the equilibrium pressure p measured. Tap 5 is shut and more gas admitted to give a p_0 of about 40 mm Hg and the rest of the measurement repeated. Values of p should be determined for a total of about ten more suitably spaced values of p_0 up to at least 300 mm Hg.

Calculation. For each measurement calculate $\Delta p = p_o - p$ and the sum of this and the Δp's for the previous measurements in the series, i.e. $\sum_1^k {}^k\Delta p = \sum_1^k ({}^kp_o - {}^kp)$. For the helium runs this is equal to pV_2/V_1 (Eq. (8)); plot this against p at each temperature. For the nitrogen runs the value of pV_2/V_1 is obtained from this plot and subtracted from $\sum \Delta p$ to give Q (Eq. (6)) and hence ${}^kn_{\mathrm{ads}}$ (Eq. (7)). The *adsorption isotherm* is obtained by plotting n_{ads} against p. The *isosteric heat of adsorption* can be obtained by taking pairs of values of p at the same value of n_{ads} at the two temperatures and using the integrated form of Eq. (1) of section 15A.4, i.e. $\log_e (p_1/p_2) = -\Delta H/\boldsymbol{R}\,[(1/T_1)-(1/T_2)]$. The data should be used to test the applicability of the Freundlich and Langmuir equations (sections 15A(3) and 15A(4)). For the latter $\theta = k_1p/(k_2+k_1p)$, where θ is the fraction of sites occupied. If filling all the sites is equivalent to the adsorption of a monolayer n_m of gas, $\theta = n_g/n_m$ so $1/n_g = 1/n_m + (k_2/k_1n_mp)$. A plot of $1/n_g$ against $1/p$ tests the Langmuir isotherm: the intercept is n_m. In drawing the best straight line it should be noted that points at low values of p may show effects due to inhomogeneity of the adsorbent. From the value of n_m the surface area of the adsorbent can be calculated: the cross-sectional area of a molecule of nitrogen is $1{\cdot}6 \times 10^{-19}\ \mathrm{m}^2$.

OTHER EXPERIMENTS

1. The adsorption of an anionic detergent can be used to determine the relative surface areas of ZnO pigments.[3]

2. Reactions of salt solutions with ion-exchange resins in some ways simulate adsorption processes.[4]

3. The adsorption of simple anionic ('acid') dyes such as Orange II by the amphoteric fibre, wool, can be shown to increase with addition of HCl, demonstrating that the adsorption is an exchange reaction of the form $\text{Wool}^+\text{Cl}^- + \text{Na}^+\text{Dye}^- \rightarrow \text{Wool}-\text{Dye} + \text{NaCl}$. On the other hand, adsorption of 'direct cotton dyes' by cellulose is a case of direct adsorption of the whole dye salt and is therefore increased by addition of sodium chloride by virtue of the common ion effect.

4. Qualitative experiments may be made on the separation of dyes, natural pigments, etc., by *chromatography* on a column of alumina. Chromatography depends on slight differences in adsorption affinity.[5]

A solution of the dyes in a suitable solvent is allowed to pass through the vertical column. Bands of dye will form as the different dyes are continuously adsorbed and desorbed at different rates down the column. Paper chromatograms (using sheets of paper rather than alumina for the separation) are used to identify large organic molecules, especially those of biological importance.

5. GAS CHROMATOGRAPHY

Gas chromatography is a technique for separating a gas mixture into its constituents. The method depends upon the repeated distribution of the substances to be separated between a *moving* gas phase and a *stationary* solid or liquid phase packed into a column. The stationary phase may be an adsorbent (gas–solid chromatography) but is usually an absorbent liquid supported on an inert solid (gas–liquid partition chromatography), although the experimental method is essentially the same whichever technique is used.

As the constituents of the gas mixture emerge from the column, they may be detected in several ways. One often employed is the katharometer (thermal conductivity cell) which consists of an electrically heated thermistor (section 3A(5)) forming one arm of a Wheatstone bridge. As the constituents pass through the cell, the thermal conductivity of the gas surrounding the wire changes, producing variations in the temperature and hence the resistance of the thermistor. These variations unbalance the bridge and give an output signal which is measured on a potentiometric recorder (section 4C(4)).

A plot of detector response against time normally gives a Gaussian-shaped peak (Fig. 15A.2). The time elapsing between the injecting of a sample into the inert carrier gas (e.g. nitrogen) [point O] and

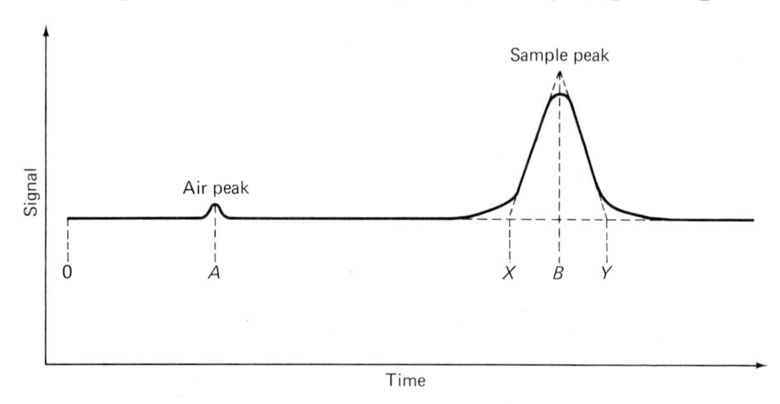

FIG. 15A.2 Gas chromatograph signal.

the peak maximum [point B] is the *retention time* t_R, and this is constant for a given sample substance, carrier gas, column, temperature and flow rate. *The retention volume* V_R, given by

$$V_R = \text{retention time } (t_R) \times \text{flow rate } (F)$$

is the volume of carrier gas required to sweep the sample substance from inlet to detector or, in other words, to elute it from the column. For a given substance, carrier gas, column and temperature, V_R is constant and independent of flow rate and the amount of substance injected. Thus identification of a constituent can be made by measuring its retention volume whilst quantitative analysis is also possible since (at constant flow rate) the area under the peak is generally proportional to the mass of substance present.

The small 'air peak' in Fig. 15A.2 at point A records the time taken for a relatively non-adsorbed sample (such as air or the carrier gas itself) to pass through the column, and includes contributions due to the interstitial volume of the column and the effective volumes of the sample injector and the detector. The time OA is called the *gas holdup time* and corresponds to the *gas holdup* V_M, the retention volume of a non-adsorbed gas. Since V_M is a contribution of the apparatus itself and is irrelevant to the gas–liquid partition phenomenon occurring in the column, we can now define an *adjusted retention volume* V'_R by

$$V'_R = V_R - V_M$$
$$= \text{time } AB \times \text{flow rate } F$$

No account has so far been taken of the fact that the carrier gas is compressible, and that its volume and therefore its velocity increase as it passes along the column. This is allowed for by introducing the *net retention volume* V_N:

$$V_N = j\,V'_R$$

where j, the *pressure-gradient correction factor*, is given by

$$j = \frac{3}{2}\left[\frac{(p_i/p_0)^2 - 1}{(p_i/p_0)^3 - 1}\right]$$

Here p_i and p_0 are, respectively, the inlet and outlet pressures of the carrier gas. Normally $p_0 = 1$ atm.

We are now in a position to calculate the *partition coefficient* of the sample between the stationary liquid phase and the gas phase. The definition, for the sample acting as solute, is

$$K = \frac{\text{weight of solute per cm}^3\text{ of stationary (liquid) phase}}{\text{weight of solute per cm}^3\text{ of mobile (gas) phase}}$$

$$= \frac{\text{volume of mobile (gas) phase per unit weight of solute}}{\text{volume of stationary (liquid) phase per unit weight of solute}}$$

$$= \frac{\text{volume of mobile (gas) phase per weight of solute (sample) used}}{\text{volume of stationary (liquid) phase per weight of solute (sample) used}}$$

$$= \frac{\text{retention volume corrected for gas hold up and pressure gradient}}{\text{volume of stationary (liquid) phase in the column employed}}$$

$$= \frac{V_N \rho}{w}$$

where ρ is the density and w the mass of the stationary (liquid) phase. Thus the retention time depends essentially upon the partition coefficient of the sample between carrier gas and liquid. Further theoretical reasoning allows the activity coefficient of the sample in the stationary phase to be derived, showing that gas chromatography can provide useful thermodynamic data.

The performance of the column in separating gases can be expressed in terms of the equivalent *number of theoretical plates*, n. (The same term is used to assess the efficiency of a distillation column in separating liquids.) For an ideal Gaussian peak, it can be shown that

$$n = 16\left(\frac{\text{retention time}}{\text{peak width}}\right)^2 = 16\left(\frac{OB}{XY}\right)^2$$

using the symbols from Fig. 15A.2.

EXPERIMENT
Gas chromatography.

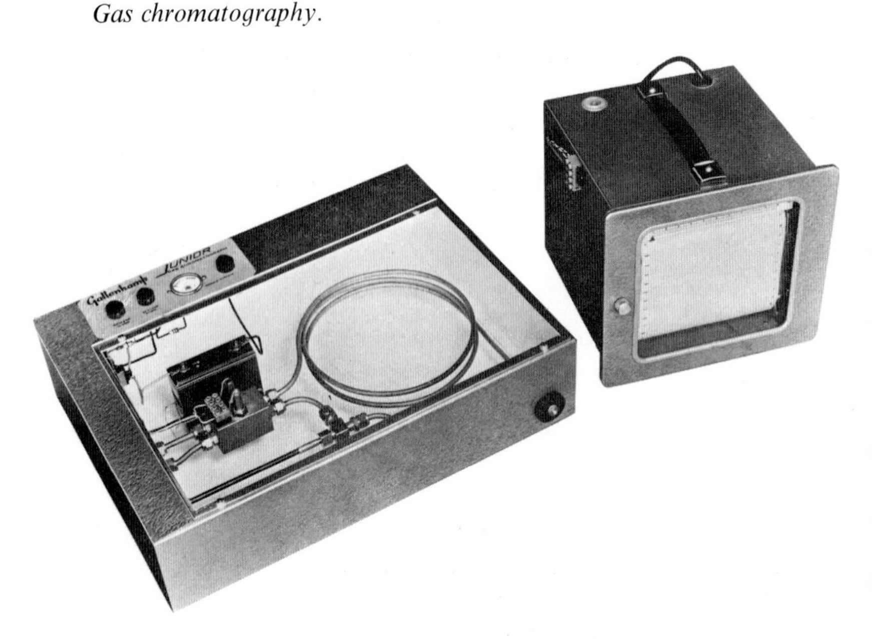

FIG. 15A.3 Student gas chromatograph. (Courtesy Gallenkamp, London.)

Apparatus. A suitable apparatus is shown in Fig. 15A.3, and can be purchased from Messrs Gallenkamp (London).

The carrier gas (nitrogen) is taken from a cylinder with regulator set to 5 psi, and the rate of flow controlled by a needle valve. Nitrogen enters the chromatograph and is then divided into two streams by a further needle valve (reference control) on the chromatograph: one stream flows through the reference channel of the detector whilst the other enters the column via the injection port. Each stream flows through a separate channel of the detector. Each detector channel contains an open bead thermistor; the thermistors form adjacent arms of a Wheatstone bridge, current for which is supplied by a 4·5 V dry battery. With only carrier gas flowing through the system, adjustment of the reference needle valve allows the gas flows through both channels to be equalized so that the wire temperatures and hence resistances are equal. On leaving the chromatograph, the gas from each channel passes through its own flowmeter. 'Meterate' flowmeter tubes (Glass Precision Eng. Ltd., available from Messrs Jencon's, London) are suitable: a float in a vertical tube is supported by gas flowing upwards. The position of the float indicates the flow-rate and calibration curves are supplied with the tubes.

The column consists of 4 ft of $\frac{1}{4}$-in. i.d. copper tubing packed with a mixture of 2 g of the absorbent liquid, silicone oil MS-550 (density 1 070 kg m^{-3}), supported on 10 g of Celite. Samples are introduced into the column by inserting the needle of a Hamilton (or other) hypodermic syringe through the silicone rubber septum in the injection port. A fresh septum is required at the start of the experiment.

Procedure. Switch on the electrical supplies to the bridge and to a potentiometric recorder. Set the bridge current to 5 mA. If necessary adjust the REFERENCE CONTROL so that the reference and sample streams have equal flow rates. This is necessary every time the flow rate from the gas cylinder is varied. When the 'reference control' is adjusted, the pressure will be slightly altered and must be readjusted. This is important for accurate results. Allow 5 min for conditions to stabilize.

Set the recorder pen to a suitable baseline using its own control; connect it to the chromatograph, set the sensitivity to 50 mV; re-position the pen on the baseline using the Baseline Control on the chromatograph. The THERMISTOR BALANCE control is preset to minimize the effect of temperature fluctuation and should not be used.

Samples of *n*-pentane, diethyl ether and a 1 : 1 mixture of the two are contained in tubes sealed with a septum. These are held vertically with the septum downwards, and the Hamilton syringe is filled by inserting the needle through septum. The syringe should be filled and emptied several times to exclude air. Take at least 10 μl (i.e. 10 μ dm^3 or 10^{-8} m^3) more liquid than is required, withdraw the syringe from the tube, expel a small amount to eliminate air locks in the needle, and wipe it. Suitable quantities are given below.

Push the needle through the silicone rubber septum in the injection port as far as it will go and inject the sample by rapidly, but smoothly, pushing the plunger through the required volume (*do not dribble it in*); start the recorder at the same time using a paper speed of about 3 cm min^{-1}. Stop the recorder when, after a few minutes, the peak has passed and the recorder pen has almost returned to zero. The retention times may be determined by measuring the distance of the peak from the injection point on the chart paper. The areas may be measured by counting squares or by cutting out and weighing.

The whole procedure is then repeated for other liquid samples. The following runs are suggested:

Run number	*1*	*2*	*3*	*4*	*5*	*6*
Flow rate, $cm^3 \, min^{-1}$	40	40	40	40	70	40
$10^{-9} \, m^3$ of *n*-pentane	5	10	—	—	—	—
$10^{-9} \, m^3$ of diethyl ether	—	—	5	10	10	—
$10^{-9} \, m^3$ of mixture	—	—	—	—	—	10

Also measure the gas holdup time by injecting a little air into the chromatograph.

Calculations. The following should be measured:

(a) the retention times
(b) the retention volumes
(c) the areas under the curves in arbitrary units
(d) these areas divided by the area found for run 1.

Do the results show that for the column used, with nitrogen as a carrier gas at room temperature:

(a) The retention time is independent of the amount of substance for a given flow rate?
(b) The retention volume is independent of flow rate?
(c) The retention time is the same for substances in a mixture?
(d) The peak area is proportional to the amount of substance added whether singly or in a mixture, using the same flow rate?
(e) The peak area for the same amount of one substance varies with flow rate?

The partition coefficient and the number of theoretical plates of the column can be calculated for both *n*-pentane and diethyl ether. The pressure-gradient correction factor j is 0·855 for a flow rate of 40 cm^3 min^{-1}.

6. ADSORPTION AT LIQUID SURFACES

Adsorption occurs at the gas–liquid interface and the liquid–liquid interface, although it cannot be measured as readily as adsorption by solids. Further, the adsorption is not strong unless the substance possesses a certain type of molecular structure. Substances which are strongly adsorbed at the air–water or oil–water interfaces are called *surface-active*, and are found to have an 'amphipathic' structure—that is, within the same molecule there are water-soluble (*hydrophilic*) and water-insoluble (*hydrophobic*) molecular groups. The *paraffin chain salts* (e.g. natural and synthetic soaps) are the best examples of highly surface-active substances. Simple inorganic salts are *negatively adsorbed*; in other words, water is more surface-active than salt.

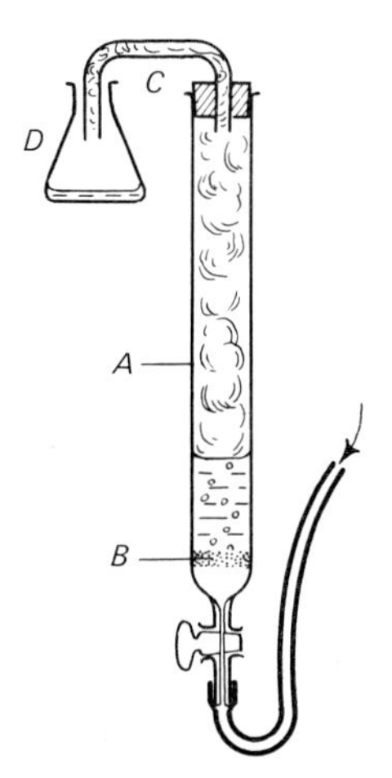

FIG. 15A.4 Demonstration of the adsorption of 'Carbolan Crimson' dye at the air–solution interface.

EXPERIMENT
Verify the adsorption of a surface-active dye at the air–water interface.

Procedure. Set up a large 'chromatography tube' *A*, Fig. 15A.4, having a coarse sintered disc *B* at the bottom. Fit it with a wide delivery tube C at the top. Prepare a dilute aqueous solution of 'Carbolan Crimson' (Imperial Chemical Industries Ltd), and pour enough into the column to half fill it. Pass a stream of air upwards through the column, and regulate the amount so that a column of froth rises slowly and passes over into the receiving flask D.

When sufficient liquid has been collected in D, pour a sample into a test-tube and place it beside similar test-tubes containing (i) the original dye solution, and (ii) the residue left in tube A.

Provided the solution used is not too concentrated, it will be observed that the liquid produced from the froth contains *more* dye than the original solution, and the residue contains less.

7. ADSORPTION AND SURFACE TENSION

It is a thermodynamic consequence of positive adsorption of a solute that the surface (or interfacial) tension is lowered; conversely, negatively adsorbed solutes (e.g. salts) cause a rise of surface tension. The Gibbs adsorption equation gives the quantitative relationship between concentration, adsorption and change of surface tension. For the special case of a very dilute solution of a non-ionic surface-active substance, this equation takes the simple form

$$\Gamma = -\frac{C}{RT} \cdot \frac{\mathrm{d}\gamma}{\mathrm{d}C}$$

where Γ is the mass of solute adsorbed *per unit area of the surface* from a solution of concentration C and surface tension γ. (In the exact form of the equation the *activity* of the solute takes the place of concentration.) This equation has been checked experimentally in several ways, the measurement of Γ being the chief difficulty. Conversely, a study of γ–C curves can be used to deduce the extent of adsorption.

EXPERIMENT

(a) Investigate the influence of chain-length on the surface activity of normal aliphatic alcohols. (b) Determine the surface tension–concentration curve for n-amyl alcohol, and hence calculate the adsorption.

Procedure. (a) For this part of the experiment any of the ordinary methods of measuring surface tension (chapter 6) may be used, since high accuracy is not needed; the apparatus described below for part (b) is equally suitable for (a).

Prepare 100 mol m^{-3} solutions of methyl, ethyl, *n*-propyl, *n*-butyl and *n*-amyl alcohols by delivering 0·40, 0·58, 0·75, 0·92 and 1·08 cm^3 of the alcohols respectively with a graduated 1-cm^3 pipette into clean 100-cm^3 standard flasks, and make the solutions up to the mark with distilled water. (If the amyl alcohol is found to leave a small residue of oily drops the solution should be discarded and the amyl alcohol purified by careful distillation with an efficient fractionating column.) Determine the surface tensions of the solutions at 20°C, and represent the results by a graph showing surface tension as a function of number of carbon atoms. The lowering of surface tension by a dilute solution of a substance can be taken as a measure of its 'surface activity'.

(b) For this part of the experiment it is necessary to measure small differences of surface tension, and, further, extreme care is needed to avoid minute amounts of contamination which would vitiate the results. Measurements can be made using the maximum bubble pressure method (section 6D(4)): for aqueous solutions the mercury run into the flask to increase the pressure can be replaced by water. The apparatus is calibrated with benzene, cleaned, and the cleanliness checked by measuring the surface tension of pure water (see appendix, Table A15). The surface tension for solutions with between 10 and 100 mol m^{-3} *n*-amyl alcohol is measured, working from the more dilute to the more concentrated. The solutions require vigorous shaking to dissolve the alcohol.

The all-glass differential U-tube apparatus shown in Fig. 15A.5 is also suitable, as small differences of liquid level can be measured by means of a travelling microscope; the whole apparatus can be immersed in a cleaning mixture before use. The two capillary tubes, *a* and *b*, are about 0·25 and 0·05 cm in radius, and should be selected for uniformity of bore before the apparatus is made, the length of a pellet of mercury being compared in different parts.

From the length of the thread and the weight of the pellet the radii r_1 and r_2 ($r_2 > r_1$) of the tubes are determined. Then $\gamma = a\{\Delta h - \frac{1}{3}(r_2 - r_1)\}$ where Δh is the difference in the heights of the liquid in the two tubes (section 6D(3)). The constant a is determined by calibration with a liquid of known surface tension.

After thorough cleaning in chromic–sulphuric acid mixture, the apparatus is washed with a jet of distilled water. From then on, care must be taken not to touch the ends *c* and *d* with the fingers, as invisible films of grease may travel on glass surfaces.

The tube is filled with distilled water through the funnel *d*, and the excess poured out through the side tube *c* by tilting the tube. Small quantities can be taken off with the aid of a clean strip of filter paper. The tube is then set up accurately vertical, preferably in a glass-sided tank of water at 20°C, and the difference between the heights of liquid in the two tubes is measured accurately with a travelling

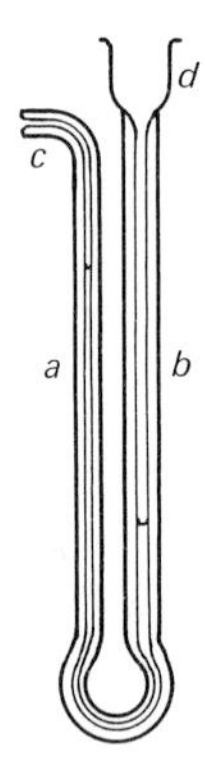

FIG. 15A.5 Differential U-tube for determining the influence of solutes on surface tension.

microscope. The tube is then rinsed out again with water, and another reading is taken. If the readings cannot be repeated to within 0·1 mm the apparatus is not clean.

Now proceed to determine the difference of level using the solutions of amyl alcohol, suggested above. Finally the apparatus is washed and dried, and then calibrated using benzene (Table A15).

Results. Plot γ against c and draw a smooth curve. The approximate forms of the Gibbs equation can be integrated to give

$$\Gamma = -\frac{1}{2{\cdot}303\ \boldsymbol{RT}}\frac{\mathrm{d}\gamma}{\partial \log_{10} c}$$

Plot γ against $\log_{10} c$ and hence obtain values of Γ from the slope. $\Gamma = 1/\boldsymbol{N}a$ where $\boldsymbol{N}$ is Avogadro's number and a is the area available to one adsorbed molecule in the surface. Γ tends to a constant value at high concentration. Estimate this limiting value from a plot of Γ against c, and hence the corresponding value of a. If the limiting value corresponds to a monolayer, the value of a will approximate to the cross-sectional area of a paraffin chain in crystalline paraffin, i.e. $1{\cdot}95 \times 10^{-19}$ m^2.

The analogue of the gas equation $PV = n\boldsymbol{R}T$ for adsorbed molecules in the surface is the two-dimensional equation of state $\pi A = kT$ (see section 15A(7)). π, the surface pressure, is $(\gamma_{solvent} - \gamma_{solution})$. A is the area through which the molecule can move, i.e. $a - b$ where $a = 1/(\Gamma N)$ is the available area and b is the effective cross-sectional area of one molecule. Then $(\gamma_{solvent} - \gamma_{solution})(1/\Gamma N - b) = kT$; rearranging

$$\frac{1}{\Gamma} = \frac{\boldsymbol{R}T}{(\gamma_{solvent} - \gamma_{solution})} - Nb$$

Plot $1/\Gamma$ against $1/(\gamma_{solvent} - \gamma_{solution})$. The slope should be $\boldsymbol{R}T$. b can be obtained from the intercept and compared with the values above.

8. THE SPREADING OF OILS

When a small drop of an insoluble oil is placed on a *clean* water surface it may behave in one of three ways: (a) remain as a *lens*, if it is a 'non-spreading' oil, (b) spread as a thin film which may show interference colours, until it is uniformly distributed over the whole surface of the water, forming a so-called 'duplex' film of appreciable thickness, (c) spread as an invisible monolayer, leaving excess oil as a lens in equilibrium.

The spreading behaviour is determined by the *cohesion* of the liquid (i.e. mutual attraction between its molecules) and its *adhesion* to water. Cohesion is measured by the surface tension of the liquid against air and adhesion to water by the interfacial tension, oil–water. By considering the forces acting at the circumference of a floating oil drop, the criterion for spreading can be seen to be that the *total* surface free energy must decrease.

The work done in extending a surface by an area dA is γdA. When the interface between two liquids, say oil and water, is increased by dA by the oil spreading, the area of the oil–air and oil–water interfaces increases by dA whilst that of the water–air decreases by dA. The total work done per unit area is the change in Gibbs free energy ΔG, i.e. $\Delta G = \gamma_o + \gamma_{ow} - \gamma_w$. For oil to spread on water ΔG must be negative. $W = -\Delta G = \gamma_w - \gamma_o - \gamma_{ow}$ is known as the spreading coefficient, and must be positive for spreading to occur. One must bear in mind that γ_o and γ_w will be to some extent different for the pure liquids and for the mutually saturated liquids. According to Harkins, even where the initial spreading coefficient is positive for the pure liquids, the *final* spreading coefficient is negative, i.e. initial spreading is followed by retraction into small lenses.

An example of a non-spreading oil is medicinal paraffin (for which the spreading coefficient is $-0{\cdot}0135\ \mathrm{J\,m^{-2}}$). The long hydrocarbon chains are hydrophobic ('water-hating') and consequently the interfacial tension γ_{ow} for paraffin–water is high. A heavy paraffin may be induced to spread on water by adding to it a small proportion of fatty material such as oleic acid or triolein (glycerol esterified with three oleic acid groups), the ester groups of which are sufficiently hydrophilic to reduce the interfacial tension.

EXPERIMENT

Investigate the spreading behaviour of benzene, 'Nujol' (medicinal paraffin) and 'Nujol' + 0·1% oleic acid on a clean water surface and on water contaminated by oleic acid.

Procedure. Thoroughly clean a large photographic dish with warm soap solution and wash it thoroughly in a sink under running water. Set the dish level in the sink and run in water until it overflows. This procedure secures, at least temporarily, a water surface which is substantially clean. To test for the presence or absence of contamination on a water surface one can sprinkle on a trace of pure talc powder and then slowly 'sweep' the surface with a long strip of 'Perspex' which has been cleaned with soap and freshly rinsed under a running tap. The 'Perspex' must not be touched with the fingers except at the very ends (which project well beyond the dish). If the water surface is contaminated one observes the talc move before the barrier actually reaches it. The ease with which a surface is contaminated can be readily demonstrated by touching a clean water surface bearing talc with a human hair or a tiny crystal of cetyl alcohol.

Having prepared a clean water surface, make the observations of spreading behaviour by placing one very small drop of the oil on the water. The dish must be washed thoroughly between each experiment. In the case of 'Nujol' + oleic acid, a duplex film showing interference colours can be obtained, and the colours can be changed by 'compressing' the film by means of the 'Perspex' strip.

Interpret the observations by reference to the following surface and interfacial tension data.

Liquid	*Surface tension*		*Interfacial tension against water*	*Surface tension of water satd. with liquid*
	dry	*satd. with* H_2O		
Benzene	29	29	35	62
'Nujol'	34	—	53	73
Oleic acid	33	—	16	—

$\gamma = entry \times 10^{-3}$, Nm^{-1}, *at* 20°C.

9. INSOLUBLE MONOLAYERS ON WATER

The invisible films demonstrated in the last experiment to spread spontaneously with oleic acid or cetyl alcohol are known to be one molecule thick. Stable monolayers are obtained with substances possessing both hydrophobic and hydrophilic portions in the same molecule. The latter may be $-COOH$, $-OH$, $-NH_2$, $-CN$, etc. Such head groups are attracted to water because of their electric dipole, and they serve to anchor the molecules on the water surface, while the hydrophobic portion of the molecule—generally a hydrocarbon chain—is highly insoluble in water. Not all such substances, however, spread spontaneously over water to any appreciable extent. Many long chain aliphatic acids and alcohols, e.g. cetyl alcohol, spread spontaneously as monolayers, even from the crystalline solid. If the chain is too long, however, the cohesion of the hydrocarbon groups out-balances the spreading tendency of the hydrophilic groups, and the substance can only be obtained as a monolayer (if at all) by the use of a dilute solution in some spreading volatile solvent such as benzene. If the chain contains fewer than ten carbon atoms, the molecule becomes soluble in water.

Monolayers on water exhibit the following properties. (a) They can be 'handled' as though they were an elastic skin on the water. They can be retained intact on the water if the sides of the vessel are made hydrophobic, and the monolayer can then be compressed or expanded by hydrophobic barriers moved on the surface. Waxed paper strips, chromium-plated rods or strips of 'Perspex' can be used as barriers. (b) Monolayers exert a two-dimensional pressure which can move light floating particles such as talc powder or a fine vaselined thread. (c) Monolayers lower the surface tension of the water on which they float. The reduction of surface tension is called the '*surface pressure*' ($\pi = \gamma_w - \gamma$) of the monolayer, since it is equal to the two-dimensional pressure mentioned in (b). The surface pressure varies with the area (A), and π–A curves for different substances show a variety of different forms corresponding roughly to solid, liquid and gaseous states (see section 15A(7)).

On compressing a dilute monolayer, π at first increases rather slowly with decreasing A but on further decreasing A, π starts to rise more rapidly as the film approaches a close-packed state. Beyond this point, the surface pressure increases very rapidly for a slight decrease in A; if the film is compressed strongly beyond the close-packed state, it crumbles, visible pellicles being formed.

The surface pressure has most often been measured by the horizontal *film balance* (Langmuir 1917, Adam 1926), in which the actual thrust on a barrier is determined by means of a torsion balance. In recent years the *vertical* film balance, developed by Dervichian (1935) and Harkins and Anderson (1937)[6] has come into general use. This is based on Wilhelmy's surface-tension method, and is shown diagrammatically in Fig. 15A.6. A is a thin glass cover-slip, suspended

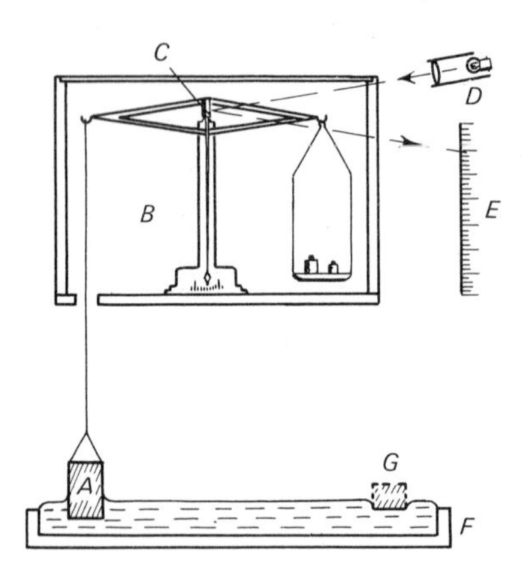

FIG. 15A.6 Principle of the vertical surface balance.

from the balance B, the beam of which carries a small galvanometer mirror C. A dips into a shallow trough of water F, the edges of which are hydrophobic. The balance is brought to its normal zero by means of weights, and then the monolayer is applied to the water surface. There is a reduction of downward pull on the cover-slip of $\pi \times p$, where p is the perimeter of the plate. The balance therefore rises on the left by an amount determined chiefly by the thickness of the plate (reduced buoyancy). The deflection of the beam, previously calibrated in terms of grams, is read accurately by a lamp D and scale E. One then proceeds to move the barrier G to a new position and read the corresponding deflection of the spot of light on the scale E, thus obtaining values of π at different film areas. An apparatus for automatically recording π–A graphs has been developed on this principle. An electronic microbalance is a convenient way of determining the force on a Wilhelmy plate directly.

10. A SIMPLE APPARATUS FOR DETECTING SURFACE FILMS

For many purposes a very simple method can be used for measuring surface pressures approximately, use being made of the capillary rise of a liquid between two parallel glass plates separated by a small gap. The meniscus between the plates is, of course, in connection with the surface of the trough *via* the narrow strips of water at the edges of the plates, and consequently any change in the surface tension of the water is registered by a corresponding change in the level of the meniscus.*

A convenient form of the 'film indicator' is shown in Fig. 15A.7. It is made from two strips of best quality plate glass 10 × 1 × 0·6 cm. Plate glass is usually nearly optically plane; it is easily tested by

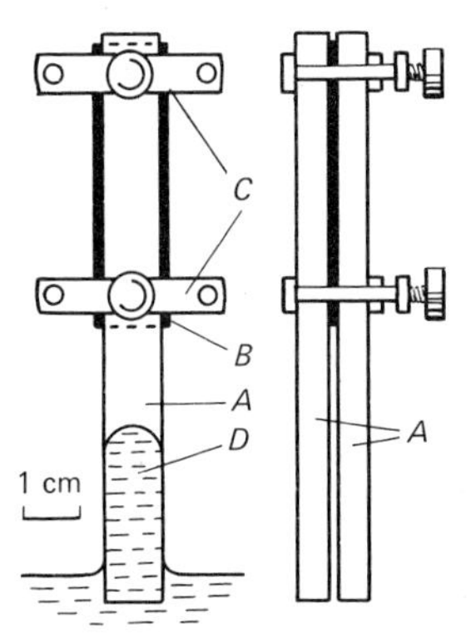

FIG. 15A.7 Simple apparatus for detecting surface films.

* It is interesting to record that this idea was used as long ago as 1890 by Lord Rayleigh, who employed it to demonstrate to an audience at the Royal Institution the reduction in the surface tension of water by traces of oil, an image of the meniscus being projected on a screen. The parallel plate apparatus may therefore be said to be the earliest 'film balance'. Rayleigh later abandoned it because he suspected, on the authority of Quincke, that the angle of contact might not be zero—a matter which is no longer in doubt, at least for a receding meniscus.

observing the interference bands produced when two clean pieces are 'wrung' together. Thinner glass is usually inferior in flatness.

The plates are separated by a gap of about 0·6 mm obtained by inserting over the upper half a spacer made of three microscope cover-glasses $5 \times 1{\cdot}2 \times 0{\cdot}02$ cm. Each of these is selected for uniformity of thickness by measurement with a micrometer, and it is not difficult to find three which together form a spacer which does not vary by more than 0·005 mm from end to end. The plates are clamped together by two ordinary screw clips (C) with rubber washers. This arrangement ensures that the gap between the lower half of the plates is uniform. The dimensions of the apparatus are not critical.

Since it is difficult to cut plate glass exactly to size, the inner edges of the strips may be bevelled off at 45° by grinding with carborundum powder until the two pieces are parallel and of the same width. Once made, the device need not be dismantled; it is scrupulously cleaned before use by immersing the lower half in cleaning mixture.

Clean water rises about 2·5 cm in an apparatus of the size recommended; the meniscus is of the form shown in Fig. 15A.7, D, and makes an excellent mark on which to focus the cross-wires of a travelling microscope. On the other hand, it is very difficult to focus a microscope on the wide water level in the trough, but this difficulty is avoided by arranging a thin glass fibre to dip into the water at an angle between the glass device and the microscope, when the fibre and its reflection form an arrow pointing to the precise water surface. If the capillary rise between the plates is read to the nearest 0·005 cm the sensitivity of the apparatus is about 10^{-4} N m^{-1}. If the device is required merely to indicate the effect of a film on the surface tension it is only necessary to observe the position of the meniscus, as the absolute capillary rise may not be needed.

When the film is applied and compressed in the usual way the meniscus falls progressively, and its level is measured at a series of diminishing film areas. Like the vertical balance, the capillary apparatus is not suitable for measurements with increasing film areas, as a monolayer may be deposited on the glass by the meniscus as it falls, making the glass hydrophobic.

In using the device for the approximate or comparative measurements for which it is intended, it is sufficient to assume that the rise is proportional to the surface tension, as for infinite plates,

$$h = \frac{2\gamma}{bg\rho}$$

where b is the distance between the plates and ρ the density. Thus the initial reading with clean water is taken as a calibration of the apparatus by assuming the surface tension of water at the temperature of the experiment. This is not accurately true, partly because the plates are narrow and partly because the spacing is too wide for the meniscus to be semi-circular in cross-section.

EXPERIMENT

Determine the surface area occupied by a close-packed, condensed monolayer of cholesterol.

Procedure. Thoroughly clean a shallow rectangular 'Perspex' trough with hot soap solution, followed by washing under the tap. Thereafter keep the trough covered with a glass plate when not in use and avoid touching any part of its working area. Level the trough and fill it with tap water until the water stands slightly above the level of the flat sides. Connect the overflow pipe to the sink. Meanwhile, prepare about 100 cm^3 of freshly distilled benzene, collecting direct into a clean stoppered 100 cm^3 flask, and taking some to rinse all parts of the apparatus beforehand to remove possible contamination present on the glass.

Weigh out as accurately as possible about 25 mg of cholesterol into a scrupulously cleaned 25 cm^3 stoppered standard flask, and make up to the mark with the freshly prepared benzene.

Before applying the film, the surface of the water in the trough must be cleaned by repeatedly 'wiping' it with freshly rinsed chromium-plated or 'Perspex' rods. Any contaminating film may be collected and confined behind a barrier at one end of the trough. To test the cleanliness of the surface, sprinkle on a little talc powder and observe it carefully while a barrier is moved along the trough; the presence of a film is indicated by movement of the talc particles before the barrier actually reaches them. Scrupulous care is needed to get a properly clean surface. (A simple test with talc will show how a film spreads immediately from any ordinary object which is allowed to touch the water.) Some tap water contains traces of surface active impurities (detergents). If repeated 'wiping' fails to remove surface contamination, the distilled water should be shaken with activated charcoal to remove impurities, and then redistilled before use.

As soon as the surface is proved clean, set up the glass plate film indicator (freshly cleaned and rinsed under the tap) at one end of the trough, and focus a travelling microscope on the meniscus without delay. Then drop on to the water about 0·05 cm^3 of the cholesterol solution, preferably measured by a calibrated micrometer syringe (section 6D(6)), but otherwise by a micropipette or small weight pipette.

Advance a barrier slowly in stages along the trough, thus slowly compressing the film. Measure the area occupied by the film and the position of the meniscus in the indicator at each stage. When the film becomes nearly close-packed the surface tension, and hence the meniscus, will fall rather rapidly with decrease of area. Obtain enough points to plot a curve of 'capillarity' against area. Hence calculate the area occupied by a molecule of cholesterol. Repeat the whole experiment with slightly different quantities of cholesterol.

Calculate the approximate thickness of the film, assuming that the cholesterol monolayer has the same density as cholesterol crystals.

11. THE ELECTRICAL DOUBLE LAYER

Electrical double layers are formed at interfaces by preferential adsorption of one of the ions of an electrolyte. Two extreme cases can be distinguished; chemically *inert* surfaces (e.g. paraffin, oil, quartz) receive a 'charge' only by adsorption of a foreign ion, whereas intrinsically *ionogenic surfaces* (e.g. sparingly soluble salts, proteins, clays, etc.) release ions by simple interaction with water.

Almost invariably, substances in contact with dilute aqueous solutions are found to carry a 'charge' as shown by the fact that small particles of the material will migrate through water in the presence of an electric field—the phenomenon of *electrophoresis*. However, the 'charge' is not a net electrostatic charge detectable by an electroscope, but a double layer of ions. An excess of ions of one charge is firmly attached to the surface, while an equivalent quantity of ions of the opposite charge exists free in the solution. The free ions are held in the vicinity of the surface by electrostatic attraction. Electrophoresis and the three other *electro-kinetic* effects (namely, *electro-osmosis*, *streaming potentials* and *sedimentation potentials*) arise when attempts are made to shear off the mobile part of the double layer from the surface. The four electro-kinetic effects are physically equivalent, and any one of them may be used to study the nature of the electrical double layer. The basic theory of these effects was given by Helmholtz, but the quantitative interpretation of them is not entirely satisfactory, and results of electro-kinetic measurements should be reported directly in terms of the measured quantities, e.g. the rate of migration of particles in electrophoresis (see section 15B(2)).

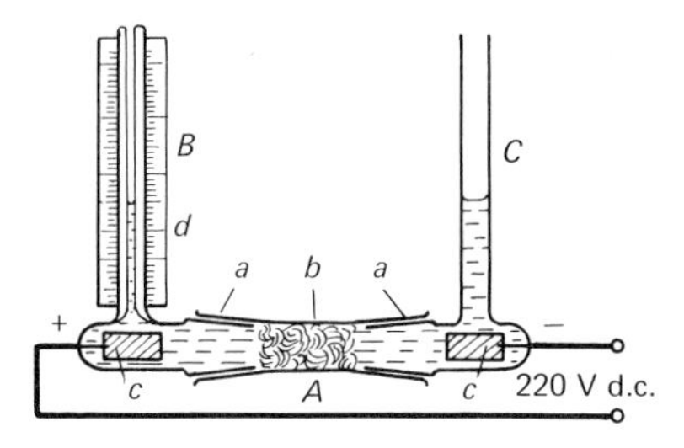

FIG. 15A.8 Apparatus for study of electro-osmosis.

EXPERIMENT

Determine by means of electro-osmosis the sign of the charge on cellulose in contact with water and with various dilute salt solutions.

Procedure. The simple apparatus shown in Fig. 15A.8 may be used for semi-quantitative measurements of electro-osmosis. It consists of a U-tube, constructed in three pieces, *A*, *B* and *C*, which can be fitted together by means of the ground glass joints *a*,*a*. Platinum electrodes *c*,*c* are sealed into each limb, and the side *B* is of semi-capillary tubing to magnify the rate of movement of the liquid. A scale *d* is placed against this tube.

Macerate some filter paper with a little distilled water, and pack a plug of the paste fairly tightly into the middle of tube *A*. Then fit the tubes together and fill the whole apparatus with distilled water, taking care to exclude air bubbles. Arrange the height of water so that the meniscus in *B* stands half-way up the scale, and allow the apparatus to stand until the levels are steady. Read the meniscus on the scale.

Now switch a 220 V d.c. source (or dry batteries), noting the polarity. Read the position of the meniscus at suitable intervals of time and then reverse the connections and measure the reversed movement. Verify that the rate of electro-osmosis is approximately proportional to the applied potential.

Disconnect the leads and dismantle the apparatus, but without disturbing the filter-paper plug. Re-connect *A* and *C* and fill *C* with a 0·01 mol m^{-3} solution of thorium nitrate, allowing the solution to percolate through the plug. Then repeat the measurements of electro-osmosis. Similar measurements may be made with very dilute (0·01 to 1 mol m^{-3}) solutions of KCl, KOH, $K_4Fe(CN)_6$, a basic dye such as methylene blue and a direct cotton dye such as Chlorazol Sky Blue. A fresh plug of filter paper must be used each time, and therefore the flow with water should be measured first so that the rates can be compared.

Results. Express the results as the rate of electro-osmotic flow per volt of applied potential, giving direction. Cellulose itself is feebly negatively charged—it carries some carboxyl groups. This charge can be increased by adsorption of OH^- ions or polyvalent anions, or decreased and even reversed by polyvalent cations. 'Indifferent' electrolytes such as KCl reduce the electro-kinetic effects without appreciably changing the charge density.

Similar experiments may be made with other materials; for example, the iso-electric point of wool can be determined in such an apparatus, using dilute buffer solutions of different pH. At low pH values wool is cationic ($-NH_3^+Cl^-$) while at high pH values it is anionic ($-COO^-Na^+$). At the iso-electric point the number of $-NH_3^+$ and $-COO^-$ groups is equal.

The same type of apparatus can be used for measuring streaming potentials. Liquid from an external reservoir is flowed through the plug, and the p.d. set up between the two electrodes is measured on an electronic voltmeter or pH meter.

BIBLIOGRAPHY 15A: Surface chemistry

Weissberger, Vol. I, Pt 1, chapter XIII.
Adam, *Physics and Chemistry of Surfaces*, 3rd edn, 1941 (Oxford University Press, London).
Gregg and King, *Adsorption, Surface area and Porosity*, 1967 (Academic Press, London).
Thomson and Webb, *Heterogeneous Catalysis*, 1968 (Oliver and Boyd, Edinburgh).
Ambrose and Ambrose, *Gas Chromatography*, 1961 (Newnes, London).
Weissberger, Vol. XIII.
Harkins, *The Physical Chemistry of Surface Films*, 1952 (Reinhold, New York).
Burdon, *Surface Tension and the Spreading of Liquids*, 1949 (Cambridge University Press, London).
Butler (ed.), *Electrical Phenomena at Interfaces*, 1951 (Methuen, London).
Schwartz *et al.*, *Surface Active Agents and Detergents*, **1**, 1949; **2**, 1958 (Interscience, New York).
Shaw, *Electrophoresis*, 1969 (Academic Press, London).
Bier, *Electrophoresis*, 1959 (Academic Press, New York).
Weissberger, Vol. I, Pt 4, chapter XLVII.

[1] Brunaer, Emmett and Teller, *J. Amer. Chem. Soc.*, 1938, **60**, 309.
[2] Harkins and Jura, *ibid.*, 1944, **66**, 1366.
[3] Ewing and Rhoda, *Analyt. Chem.*, 1950, **22**, 1453.

[4] Kunin and Myers, *J. Amer. Chem. Soc.*, 1947, **69**, 2874; Bishop, *J. Phys. Chem.*, 1950, **54**, 697.
[5] Cassidy, in *Weissberger, op. cit.*, Vols V and X; Lederer and Lederer, *Chromatography: A Review of Principles and Applications*, 2nd edn, 1957 (Elsevier, Amsterdam).
[6] Harkins and Anderson, *J. Amer. Chem. Soc.*, 1937, **59**, 2189.

15B Colloids

1. For the present purpose the term 'colloids' will be restricted to *dispersions* of one phase in another. At one extreme one finds suspensions of solids in liquids, or of liquids in liquids—emulsions. Particles in the 'sub-sieve range' i.e. below about 50 μm ($\mu = 10^{-6}$) are sometimes called colloidal. With solid particals of sub-microscopic size ($<0{\cdot}1$ μm) the dispersion is called a *sol*. The dividing line between a colloidal solution and a crystalloidal solution is equally vague and arbitrary; Graham's original criterion was the ability of the latter to diffuse through a parchment membrane, but one can readily prepare membranes (e.g. of collodion) with various pore sizes. True solutions of substances of high molecular weight (e.g. proteins) exhibit many of the characteristic properties of colloids. Others of low molecular weight owe their colloidal properties to the formation of *aggregates* of colloidal size (e.g. dyes, soaps).

A useful broad classification of colloids is into *lyophobic* and *lyophilic* types. Although many actual colloids are of intermediate character, these concepts are clear-cut and assist in understanding the behaviour of colloids. A lyophobic colloid is essentially a dispersion of one phase in another of *unlike chemical type*. Such colloids therefore possess large interfacial free energy and are intrinsically unstable, or, at least, metastable. They owe their temporary stability to stabilizing agents, and, if these agents are removed, the colloids *coagulate*. The stabilizing agent may be an electrical double layer of adsorbed ions, or a film of adsorbed lyophilic colloid.

Lyophilic colloids are those which disperse readily in the solvent on account of solvation. They are, in fact, thermodynamically stable, molecular solutions, but exhibit colloidal properties on account of their high molecular weights (e.g. rubber in benzene, gelatine in water).

2. LYOPHOBIC COLLOIDS

The chief characteristics of lyophobic colloids can best be illustrated by the following semi-quantitative experiments with emulsions (oil–water dispersions) and with two typical inorganic sols.

EXPERIMENT

Investigate the properties of emulsions by means of the following experiments, and interpret the observations.

(a) *Dispersion of liquid in liquid: emulsifying agents.* Attempt to disperse 5 cm^3 of light paraffin oil (kerosene) in 100 cm^3 of each of the following solutions, first by vigorous shaking and then by 'homogenizing' the suspensions in a domestic blender: (i) distilled water, (ii) 1% sodium oleate solution, (iii) 1% gelatine solution, (iv) 1% suspension of bentonite (a finely divided clay) in distilled water. Also, pour a layer of olive oil as carefully as possible on to a solution which is approximately 1 mol m^{-3} with respect to sodium oleate and 1 mol m^{-3} with respect to sodium chloride.

Note the stability of the suspensions formed; distinguish between 'creaming' and 'breaking'. Note the scattering of light by a very dilute emulsion.

(b) *Particle size.* Dilute 10 cm^3 of each of the above emulsions (well mixed) with an equal volume of water in a test-tube. By means of a pipette remove 1 cm^3 of the dilute emulsion from the middle of the tube and add to it 1 cm^3 of warm 10% gelatine solution. Place a small drop of this final mixture on a microscope slide, put on a clean cover-glass, and allow the gel to set.

Examine the slides under a microscope with magnification of about 500 ×. Describe the state of dispersion by noting the size of the largest and smallest oil drops observed and the chief range of size. The size can be found by reference to a scale placed in the ocular of the microscope, and the magnification determined by use of a stage micrometer slide.

Careful examination of a very finely dispersed, dilute emulsion under high magnification will reveal the existence of *Brownian motion*—the continual random displacement of very small particles, caused by molecular bombardment.

(c) *Emulsion types.* Homogenize 50 cm^3 of paraffin in 50 cm^3 of 1% sodium oleate solution. Divide the emulsion into two equal parts and add to one a quantity of magnesium sulphate. Determine the emulsion type (that is, whether oil-in-water or water-in-oil) by the following tests: (i) add a few drops of the emulsion to water and to oil, and see with which it mixes readily. (ii) Add to one portion of emulsion an oil-soluble dye (e.g. 'Oil Red'), and to another portion a water-soluble dye (e.g. methylene blue). The electrical conductivity can also be used to indicate whether oil or water is the *continuous phase*. Hydrophilic emulsifiers (sodium oleate, bentonite) favour oil-in-water emulsions; hydrophobic emulsifiers (magnesium oleate, carbon black) favour water-in-oil.

(d) *Viscosity of emulsions.* Prepare several emulsions of kerosene in 1% sodium oleate solution, using different proportions of kerosene, and measure their viscosities in an Ostwald viscometer (section 6C(3)). The densities of the emulsions can be calculated from the volume composition of the emulsions, on the assumption that there is no volume change when oil is dispersed in water.

Why is an emulsion more viscous than either of its constituent liquids?

The viscosity of emulsions is not a very reproducible quantity since it depends on the degree of dispersion and *age* of the emulsion. Many colloids show *ageing*.

Furthermore, emulsions exhibit 'anomalous' or 'non-Newtonian' viscosity—that is, the apparent coefficient of viscosity depends on the *rate of shear* at which the measurement is made. With emulsions, the higher the rate of shear, the lower the apparent viscosity owing to the breakdown of the *structural viscosity*. (See section 15C for the study of anomalous viscosity.)

(e) *Interfacial tension.* The effect of emulsifying agents (sodium oleate, gelatine) on the oil–water interfacial tension can be determined *approximately* by the drop-pipette (Fig. 15B.1(a)). This has a bulb

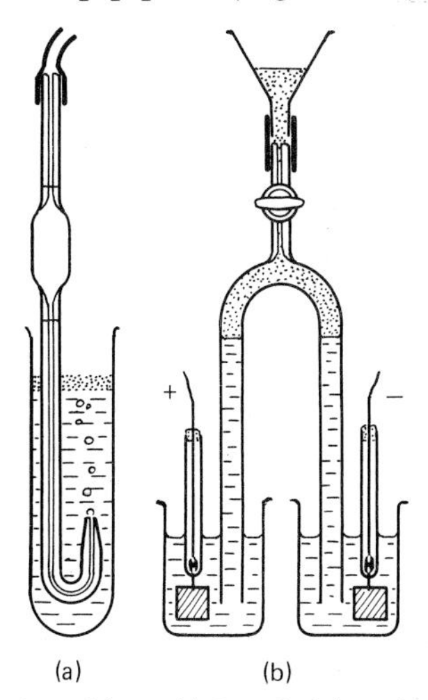

FIG. 15B.1 Properties of emulsions. (a) Interfacial tension. (b) Electrophoresis.

of capacity of about 1 cm^3 and an up-turned jet which is ground smooth and sharp. With a given pair of liquids (oil–water) the *drop number* for a fixed volume is approximately inversely proportional to the interfacial tension. (For methods of measuring interfacial tensions accurately, see section 6D.)

Most emulsifying agents lower the interfacial tension greatly and thus facilitate dispersion. Extremely low interfacial tensions (10^{-4} N m^{-1}) lead to 'spontaneous emulsification'. A stable adsorbed film is, however, the chief factor in obtaining high *stability* in an emulsion.

(f) *Electrophoresis*. The stability of emulsions containing sodium oleate is closely associated with the presence of an electrical double layer which causes neighbouring droplets to repel each other, thus preventing coalescence. The presence of the double layer can be demonstrated most easily by electrophoresis, using an inverted U-tube as shown in Fig. 15B.1(b). The tube is first filled with the dispersion medium, dilute sodium oleate solution. The emulsion is then slowly introduced through the capillary stopcock from a small funnel. When sharp boundaries have been formed in both limbs, the tap is closed and the position of the boundaries is marked. The electrodes are then connected to the 220 V d.c. supply, and the direction and rate of migration of the boundaries is noted. Calculate the speed of migration approximately in m s^{-1}/(V m^{-1}).

It is evident that the oleate ions are strongly adsorbed as the oil–water interface, while the sodium ions provide the free 'counter-ions' (*gegenions*) in the aqueous phase.

INORGANIC SOLS

Classical colloid chemists devoted much attention to the methods of preparation and the properties of sols of metals, hydrated oxides, and sulphides. Only two examples, however, will be given here.

EXPERIMENT
Prepare and study ferric hydroxide and arsenious sulphide sols.

(a) *Preparation of colloidal ferric hydroxide*. To a half-saturated, freshly prepared solution of ferric chloride add, in drops and with stirring, an approximately 2 000 mol m^{-3} solution of ammonium carbonate, until the precipitate formed just ceases to be dissolved. Filter if necessary and dialyse the liquid in a cellophane tube against distilled water until only a trace of chloride can be detected in the water.

Or, add gradually to 200 cm^3 of boiling water, 20 cm^3 of a 2% solution of ferric chloride. Red-brown ferric hydroxide sol is produced by hydrolysis. When cold, free it from hydrochloric acid by dialysis.

(b) *Preparation of arsenious sulphide sol.* Boil about 1 g of arsenious oxide with 500 cm^3 of water until all is dissolved, and allow to cool. (Crystalline arsenious oxide dissolves only very slowly; the amorphous form dissolves more rapidly.) Pass hydrogen sulphide, washed by bubbling through water, into the solution until the latter is saturated. Free the solution from excess of H_2S by means of a stream of hydrogen, and filter.

(c) *Electrophoresis.* Determine the sign of the charge on the sols by a U-tube moving-boundary apparatus (i.e. the inverse of the one used for emulsions—Fig. 15B.1(b)). By attaching a millimetre scale to the limb of the U-tube the rate at which the boundary moves can be determined, and knowing the voltage applied to the electrodes and their distance apart, the rate of movement for a potential gradient of 1 $V\ m^{-1}$ can be calculated.

Most hydrous oxide sols are positive, while metal sulphide sols are generally negative. However, it is possible to stabilize some sols (e.g. AgI) with either sign, by using a very slight excess of a salt of the cation or anion.

(d) *Mutual precipitation of colloids.* The mutual precipitation of colloids of opposite charge can be studied by mixing the sols together in various proportions. Mutual precipitation takes place only when the colloids are mixed together in certain proportions within rather narrow limits. When excess of one or other colloid is present precipitation does not occur or does so only incompletely.

Prepare sols of As_2S_3 and of dialysed $Fe(OH)_3$, of 0·5% concentration (referred to As_2S_3 and Fe_2O_3 respectively). By means of graduated 10 cm^3 pipettes mix in a series of test-tubes 1 cm^3 As_2S_3 and 9 cm^3 $Fe(OH)_3$; 3 cm^3 As_2S_3 and 7 cm^3 $Fe(OH)_3$, etc., and determine the relative concentrations at which complete precipitation occurs, and the supernatant liquid remains colourless after the mixtures have been allowed to stand for one or two hours.

In the above case the optimum amounts will be found to be about 9 cm^3 As_2S_3 and 1 cm^3 $Fe(OH)_3$. Vary the proportions by fractions of a millilitre on either side of the optimum point.

(e) *Precipitation of lyophobic colloids by electrolytes.* Lyophobic colloids are generally very sensitive to electrolytes; in some cases even quite small quantities of an electrolyte produce precipitation. The 'precipitation value' or the concentration of an electrolyte necessary to cause precipitation depends on the nature of the colloid, on the method of its preparation, concentration, etc., as well as on the nature and valency of the electrolyte ions. In the case of positively charged colloids the precipitation value depends on the valency of the electrolyte anion and is largely, although not entirely, independent of the nature of the cation; in the case of negatively charged colloids it is the valency of the cation that is of primary importance. The higher the valency, the lower is the precipitation value.

EXPERIMENT

Determine the precipitation values of NaCl, $CaCl_2$ *and* $LaCl_3$ *for arsenious sulphide sol.*

Prepare the sol of arsenious sulphide as described above using a 0·5% solution of arsenious oxide. An approximate value of the precipitation concentration may be obtained by running from a burette, into a given volume (say, 50 cm^3 of the sol, solutions of known concentration of the different salts.

Having obtained approximate values for the precipitation concentrations, prepare solutions of the salts of appropriate concentration and proceed as indicated below.

(a) NaCl. Prepare a 200 mol m^{-3} solution of sodium chloride (11.7 kg m^{-3}), and from this prepare a series of 10 cm^3 portions of diluted solution by mixing, in small test-tubes, 1, 3, 5, 7, 9 cm^3 of NaCl solution with 9, 7, 5, 3, 1 cm^3 of distilled water. In five test-tubes of as nearly as possible the same size, and well cleaned by steaming, place 10 cm^3 of the arsenious sulphide sol, and add to each of these in turn one of the five portions of diluted NaCl solution. After each addition mix well by inverting the tube two or three times, and then place in a rack for a definite time, say, one or two hours. At the end of this time ascertain in which cases precipitation has occurred, and so determine the concentration of NaCl necessary to produce precitation.

In the above case, 1 cm^3 of the original NaCl solution contains 0·0002 mole (or 0·2 millimole), and since this amount is then finally diluted to 20 cm^3 the concentration of the NaCl in the mixed solution will be 10 mol m^{-3}; and in the other cases the concentration will be 30, 50, 70, 90 mol m^{-3}. It will be found that in the case of the last two solutions, precipitation has taken place, but not in the case of the first two. In the case of the concentration 50 mol m^{-3}, a turbidity will probably be observed. If necessary, further experiments with a concentration of 40 and 60 mol m^{-3} can be carried out in order to fix the value more closely.

(b) $CaCl_2$. Proceed in the same way as for NaCl, starting with a 5 mol m^{-3} solution of $CaCl_2$ (0·555 kg m^{-3}). Dilute this in portions as above.

(c) $LaCl_3$. Prepare a 5 mol m^{-3} solution of $LaCl_3$ (1·22 kg m^{-3}). For stock solution dilute this ten times, and then dilute in portions as before.

Similar experiments can be carried out with ferric hydroxide sol. In this case use potassium chloride, sulphate, ferrocyanide and ferricyanide as precipitating electrolytes.

(f) *Protective action of lyophilic colloids.* The precipitating action of electrolytes on lyophobic sols is diminished by the presence of lyophilic colloids (gelatin, albumin, starch, etc.); the latter are therefore said to exercise a *protective action.* To illustrate this, one may study the concentration of gelatin which will just prevent the precipitation of As_2S_3 by a certain definite concentration of electrolyte (e.g. NaCl). Solutions of gelatin are made up and by systematic dilution (as in the previous experiment) and mixing with a definite volume of As_2S_3 sol, one determines the concentration which just prevents precipitation when 1 cm^3 of a 10% NaCl solution is added.

3. LYOPHILIC COLLOIDS

Lyophilic colloids comprise soluble macromolecules and aggregated molecules of smaller size. For obvious reasons they show the following characteristics of colloids: scattering of light (Faraday-Tyndall beam), low osmotic pressure, slow diffusion, electrophoresis (in the case of colloidal electrolytes), dialysis, high viscosity, reversibility (to drying), and, in some cases, moderate sensitivity to electrolytes.

3.1. (a) *Macromolecular colloids.* As an example of a wholly lyophilic colloid one may consider polystyrene in toluene. The polymer, like the monomer, is truly soluble in the 'like-natured' solvent, but, because of its flexible, chain-like structure it causes a great increase of viscosity. The longer the chains, the greater the viscosity; this provides a method of determining the molecular weight of different polystyrenes. In a solvent of less favourable nature such as toluene–methanol mixtures, the chains become coiled instead of randomly extended, and the polymer has less influence on the viscosity. If a small proportion of the bifunctional monomer, divinyl benzene, is copolymerized with styrene one obtains a three-dimensional polymer network. This cannot dissolve because of the *cross-links* which prevent unlimited extension, and the material behaves as a *gel* of definite *limited swelling* when placed in contact with the solvent.

Swelling gels can exert very large *swelling pressures*, but there is generally a slight *net contraction* when a gel imbibes solvent, especially in the case of the imbibition of water by materials containing hydrophilic groups. There is then an evolution of heat—*the heat of imbibition*.

Some supposedly typical lyophilic colloids become partially lyophobic if their electrical double layers are neutralized. For example, most proteins form stable solutions only in certain pH ranges; at their iso-electric point, or in the presence of large concentrations of simple salts or smaller concentrations of specific salts, they are precipitated just like lyophobic sols. It is evident that proteins consist of a framework which is substantially hydrophobic, together with ionizable groups which, when dissociated, bestow a net hydrophilic nature. Partial removal of charge often produces *gels* in such materials; the structure is probably an open sponge-like, aggregate, 'cross-linked' by weak secondary forces such as hydrogen bonds. Unlike the cross-linked polystyrene gels mentioned above, such gels are unstable, and show the phenomenon of *syneresis*, that is, a shrinkage with exudation of dilute colloidal solution.

Apart from osmometry[1] and electrophoresis,[2] the study of lyophilic colloids does not involve any special experimental techniques not already mentioned elsewhere. The qualitative phenomena can be studied with gelatin. This protein can be used to demonstrate (a) limited swelling in water at low temperatures, (b) swelling pressure, (c) heat of imbibition, (d) volume contraction on imbibition (by means of a dilatometer), (e) the influence of salt on gelation (Hofmeister series of ions), (f) syneresis of weak gels on standing, (g) non-Newtonian viscosity of solutions and ageing, (h) influence of salts on viscosity ('electro-viscous effect'), (i) effect of pH on viscosity and gelation, and (j) electrophoretic mobility (measured by microscopic observation of very small particles of quartz on which the protein is adsorbed).

3.2. (b) *Aggregation colloids.* Solutions of soaps and synthetic detergents (the so-called 'paraffin chain salts'), and also of many dyes, show colloidal properties such as anomalous osmotic pressure and diffusion rate and scattering of light. These solutions contain aggregates of colloidal size, known as *micelles*. At extremely low concentrations true crystalloidal solutions are formed, but, owing to the sharp onset of aggregation for a micelle containing many molecules, there is a fairly distinct concentration above which the solutions are almost entirely colloidal. This point is known as the *critical micelle concentration* (c.m.c.). At this point many properties of the solution, e.g. conductivity, f.p., surface tension, etc., show a change. Micelle formation is strongly favoured by addition of salts, since electrolytes reduce repulsion between ionized particles.

Solutions containing soap micelles exhibit the phenomenon of *solubilization*; this is the ability to dissolve appreciable quantities of substances which are practically insoluble in pure water, e.g. hydrocarbons, azobenzene. It is apparent that the substances are incorporated in the hydrophobic interior of the micelles. Solubilization shows a sharp increase above the c.m.c. By using certain dyes which exhibit a different colour when in solution in organic solvents from that shown in water, it is possible to determine the critical micelle concentration by a simple titration and to study the influence of salts and solubilized substances on this concentration.[3]

EXPERIMENT

Determine the influence of salts on the critical micelle concentration of an anionic soap.

Method. Prepare an approximately 0·01 mol m^{-3} solution of pinacyanol chloride (or of Rhodamine 6G) in distilled water. To a measured volume of it add an accurately known quantity of a pure anionic soap, sufficient to change the colour to the shade typical of the solubilized dye. Then 'titrate' this solution with the above dye solution until the change of shade is observed. Calculate the concentration of the soap at this point; this is the critical micelle concentration in the absence of salts. Suitable soaps are potassium laurate, potassium myristate, sodium cetyl sulphate, and their critical concentrations in water are about 20, 6 and 0·01 mol m^{-3} respectively, showing that aggregation is favoured by increased chain length. It is necessary to have very pure samples of the compounds, otherwise an indistinct end-point is obtained.

Repeat the measurement, but with the previous addition of varied measured quantities of KCl, K_2SO_4 and $BaCl_2$.

Results. As with other colloid phenomena dependent on the electrical double layer, it is found that the number of charges on the ion having the same sign as the micelle has little influence, whereas micelle formation is more strongly favoured, with consequent lowering of the critical concentration, by polyvalent ions of sign *opposite* to that carried by the colloidal particles. This can be explained by a 'closing-up' of the diffuse double layer by polyvalent gegenions.[4] In order to compare the effects of salts on different soaps it is found convenient to plot log (c.m.c.) against log (total gegenion concentration), a straight line being generally obtained.

BIBLIOGRAPHY 15B: Colloids

Kruyt (ed.), *Colloid Science*, 2 vols, 1949, 1952 (Elsevier, Amsterdam).
Alexander and Johnson, *Colloid Science*, 2 vols, 1949 (Oxford University Press, London).
McBain, *Colloid Science*, 1950 (D. C. Heath, Boston).
Jirgensons and Straumanis, *Short Textbook of Colloid Chemistry*, 1954 (Pergamon, London).
Mysels, *Introduction to Colloid Chemistry*, 1959 (Interscience, New York).

[1] Weissberger, Vol. I, Pt 1, chapter XV.
[2] See Bibliography 15A.
[3] Corrin, Klevens and Harkins, *J. Chem. Phys.*, 1946, **14**, 480.
[4] Verwey and Overbeek, *Theory of the Stability of Lyophobic Colloids*, 1948 (Elsevier, Amsterdam).

15C Rheology of colloids

1. One of the most common characteristics of many dispersions and materials of high molecular weight is their anomalous mechanical behaviour when subjected to deformation or flow. Rheology is the study of such phenomena. At present it is largely an empirical science, lacking quantitative explanation in structural terms, but the rheological properties of colloids are often of great practical importance, as, for example, in the case of paints, creams, plastics, printing inks, adhesives, and also in the food industry.

An *ideal solid*, when subjected to a *stress*, immediately assumes a deformation, such that the *strain* (e.g. extension) is proportional to the stress (Hooke's Law). When the stress is removed the original form is exactly recovered. Metals behave as ideal solids up to fairly high stresses, whereas most organic solids obey Hooke's law only at very small stresses.

An *ideal liquid* ('Newtonian fluid') cannot withstand the smallest stress, but flows at a rate which is proportional to the applied stress (e.g. pressure): in other words, it shows a constant coefficient of viscosity (section 6C(1)). All pure, homogeneous, liquids are

Newtonian, but colloidal solutions are generally non-Newtonian, especially when concentrated. Suspensions, pastes, gels and emulsions frequently behave in a manner which is intermediate between a solid and a liquid.

2. Liquids can be divided into classes according to their rheological behaviour:

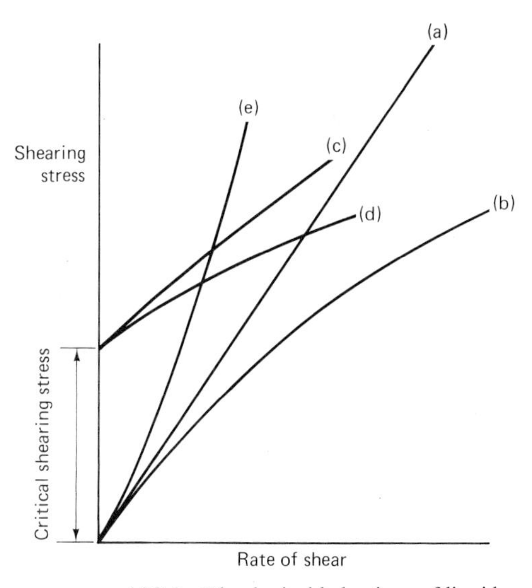

FIG. 15C.1 Rheological behaviour of liquids.

2.1. *Newtonian liquids*. These show a simple proportionality between the stress and the rate of shear, the proportionality constant being the viscosity of the liquid (Fig. 15C.1(a)).

2.2. *Liquids possessing 'structural viscosity'*. These show a stress vs rate of shear relationship where the apparent viscosity decreases with increasing rate of shear (Fig. 15C.1(b)). This arises frequently with solutions of polymers, in which the molecules are orientated parallel to the lines of flow at high rates of shear thereby reducing the viscosity.

2.3. *Liquids possessing 'plasticity'*. These are not permanently deformed by the shearing stress unless it exceeds some constant value, the 'critical shearing stress' (Fig. 15C.1(c)). Their behaviour can also be recognized experimentally by the failure of the shear stress to return to zero when the shear rate is reduced to zero.

If increments of stress beyond the 'critical shearing stress' produce proportionate increments in the rate of shear, i.e. the graph is linear, then the liquid is known as a Bingham body (Fig. 15C.1(c)). If the plastic material does not give a linear graph then it is said to be non-Binghamian (Fig. 15C.1(d)).

2.4. '*Thixotropic*' *liquids.* (e.g. bentonite paste.) These are fluid under shearing, but set to pastes on removal of the shear. 'Thixotropy' differs from 'structural viscosity' in that the apparent viscosity is time-dependent. Thus, if the rapidly sheared, and hence limpid, fluid is suddenly subjected to a lower rate of shear the apparent viscosity does not immediately rise. The substance takes time before it recovers its 'structure' and sets to a gel.

2.5. '*Dilatant*' *materials.* These become stiffer on being sheared (Fig. 15C.1(e)). 'Dilatancy' arises in pastes in which the liquid just fills the interstices of the close-packed arrangement of the solid particles. In order for the paste to move the solid particles adopt a more 'open' arrangement, and the amount of liquid is now insufficient to fill the voids. The paste now behaves rather more as a dry powder. The most frequently encountered example of this is wet sand, where movement makes the sand appear to be dry.

2.6. '*Viscoelastic*' *materials.* These lie between the Newtonian fluid, for which the stress is proportional to the *rate of shear*, and the ideal elastic solid, for which the stress is proportional to the *shear*. Such materials when subjected to a constant stress show an immediate elastic deformation followed by a further slow deformation ('creep'). Long-chain polymers frequently give rise to *high elasticity*, as in rubber. This is due to the tendency of the chains to assume a random-coiled distribution. Solutions of chain-like molecules frequently show the property of being readily drawn out into threads ('Spinnbarkeit').

3. RHEOLOGICAL MEASUREMENTS

The phenomena mentioned above are a few of many rheological phenomena which may be encountered with colloids. Because of their technical importance, instruments[1] have been devised to measure many of them, although generally in relative rather than absolute terms. For example, there is not yet any accepted absolute definition of thixotropy, but this property can readily be detected and studied by a rotating concentric cylinder viscometer. This instrument is probably the best for most kinds of 'pseudo-fluids', since the rate of shear is virtually constant in the narrow annulus between the cylinders. End effects in this apparatus can be eliminated by using a guard-ring fixed to the inner cylinder. Unfortunately the apparatus is complicated to construct and set up accurately, and requires a large quantity of material.

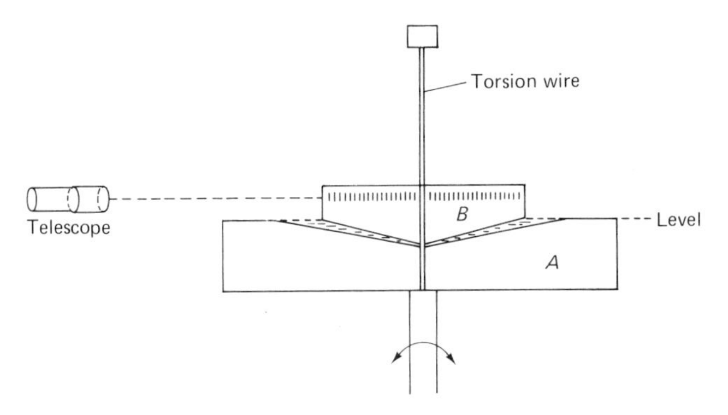

FIG. 15C.2 Rotating cone viscometer.

A simpler yet satisfactory instrument is based on concentric cones (Fig. 15C.2), the outer of which (A) can be rotated at any one of a number of constant speeds ($\approx$1–60 rev min^{-1}) by a geared-down constant-speed induction motor. The inner cone (B) is suspended by a constantan torsion wire, the torque being proportional to the angular deflection within the range of this experiment. In the narrow gap between the two cones, the relative velocity gradient is constant, for the gap width and the relative velocity of the surfaces of the two cones are proportional to the distance from the axis. The angular deflection of the torsion wire is determined by observing a circular scale on the edge of cone B either through a fixed telescope or moving relative to a pin mounted independently close to the scale.

Thus the apparatus provides a definite, constant rate of shear throughout the material. It is also simple to fill and to clean, and needs only small quantities of material. The range of the apparatus can be extended by using inner cones of different angles and also by using suspension wires of different torsional constants.

EXPERIMENT

Investigate the rheological properties of liquids using the rotating cone viscometer.

The viscometer is described in the previous section. In use, care should be taken not to kink the torsional wire by swinging cone *B* out of the vertical. The instrument is levelled with the aid of a spirit level placed on the lower cone, the legs of the stand being adjusted until the lower cone is level in two directions at right angles. The suspended cone (*B*) is then adjusted to be above the centre of the lower cone (*A*) by loosening the screws holding the suspension head and displacing the head until the point of *B* lies exactly in the centre of *A*. The screws are then tightened.

Cone *B* is now lowered until it touches *A*, and then very carefully raised until contact is just broken—as indicated by connecting an electric lamp across the upper and lower cones. A quantity of the material to be studied is added. The level of liquid should not rise on to the cylindrical part of *B*. Readings of the deflection are now taken at various speeds. For each speed the deflection is taken to be the mean of the deflections taken with the cone *A* rotating clockwise and anti-clockwise, although these two values should be in good agreement. The motor should be stopped to change gear.

Plot the shearing stress (i.e. torsional deflection) against rate of shear (i.e. rate of rotation) for the following materials, and hence classify their behaviour:

(a) Glycerol.
(b) A solution of 3·5 g Perspex in 100 cm^3 pyridine.
(c) 50 cm^3 ethylene glycol + 40 g bentonite. *Do not leave materials (b) and (c) in the cone longer than necessary.*
(d) 13 g potato starch plus 10 cm^3 cold water. If, in the examination of the starch–water system, the deflection appears to be getting very large do *not* carry on with the experiment.

The use of capillary methods is open to the criticism that the *rate of shear* is not uniform across the tube, and only a mean rate of shear can be stated; however, this disadvantage is largely counter-balanced for routine work by the simplicity of the technique; even with rotating cylinder viscometers the results cannot, at present, be given quantitative interpretation.

In order to vary the rate of shear in a capillary viscometer, one uses an applied air pressure to drive the liquid instead of relying on gravity (as in the ordinary Ostwald viscometer, section 6C(3)). The air pressure is measured on a manometer, and the rate of flow of the fluid is measured either by movement of a meniscus or by means of a sensitive gas flow-meter applied to the air displaced as the fluid moves.

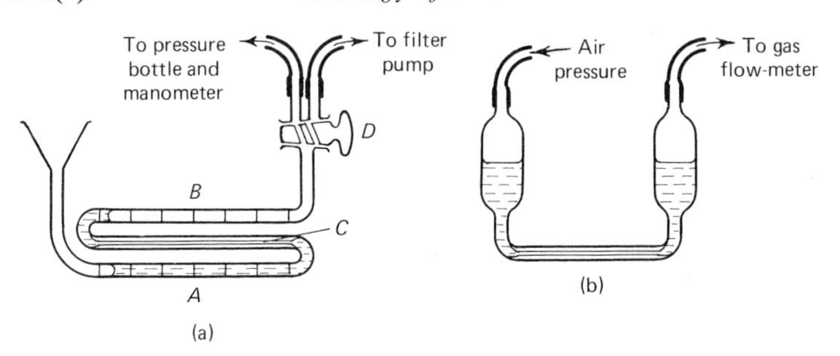

FIG. 15C.3 Simple apparatus for the study of non-Newtonian fluids. (a) Tsuda's viscometer. (b) Capillary plastometer.

Figure 15C.3(a) shows the horizontal capillary viscometer used by Tsuda[2] for studying the structural viscosity of hydrophilic sols. Two lengths of tube, *A* and *B*, of equal, uniform bore, are separated by a length of capillary tube *C*. Liquid is forced from *A* to *B* by an applied air pressure from a reservoir. The rate of flow is obtained by timing the movement of the liquid meniscus as it passes marks on tubes *A* or *B*. Tubes *A*, *B* and *C* lie in the same horizontal plane.

Figure 15C.3(b) shows a capillary tube 'plastometer' suitable for studying more viscous suspensions and pastes;[3] the rate of flow can be measured by means of a 1 cm^3 soap-bubble flow-meter (section 6B(1)).

EXPERIMENT
Determine the influence of applied pressure on the flow-rate of a 1% starch sol.

Procedure. Prepare a 1% starch solution by heating a starch paste on a water-bath for 1 h; filter, if necessary. A viscometer of the form shown in Fig. 15C.3(a) is required. Its dimensions should be such that flow-times of about 10–100 s are obtained with water, using applied pressures of about 5–50 cm of water gauge; e.g. the capillary should be about 0·3 mm radius and 10 cm long, while *A* and *B* are of about 2·5 mm radius. A pipette to deliver a suitable quantity of liquid should be made.

Set the viscometer with its tubes horizontal in a water-bath or thermostat. Connect one arm of the three-way tap *D* to a Winchester bottle (preferably immersed in a tank of water to eliminate temperature fluctuations) and to a long U-tube manometer. A side connection with stopcock must be provided for the purpose of pumping air into the bottle by means of a bicycle-pump with valve, and another stopcock is convenient for letting out air. The other arm of tap D is connected to a filter-pump. The pump is kept running. By turning D to the first position, the liquid in the viscometer can be pushed from B to A under a measured pressure; by reversing the tap connections, it can be drawn back again by the reduced pressure in readiness for a repetition of the measurement at the same or another pressure.

First make a series of measurements of the rate of flow with distilled water at a range of applied air pressures. The product of pressure × time of flow should remain constant so long as streamline flow is occurring. Avoid excessive flow-rates where this relation is found not to hold, as turbulence is then occurring.

Repeat the measurements with the same volume of starch solution.

Results. The *average* rate of shear, β, of a liquid flowing through a tube is given by

$$\beta = \frac{8}{3} \cdot \frac{V}{\pi r^3 t}$$

where V is the volume of liquid flowing in time t through a tube of radius r. The 'specific viscosity' of a solute is defined by

$$\frac{\eta_{\text{solution}} - \eta_{\text{solvent}}}{\eta_{\text{solvent}}} = \eta_{\text{sp}}$$

Plot η_{sp} against β. Wolfgang Ostwald found empirically that for many colloidal sols log (rate of flow) is a linear function of log (applied pressure).

Other experiments can be made with the same apparatus using solutions of methyl cellulose, gelatin, etc.; 0·5% agar solutions show a change of viscosity on standing for a few hours. Solutions of long-chain polymers in non-ionizing solvents, e.g. polystyrene in toluene, will be found to behave as Newtonian liquids in spite of their high η_{sp}. The apparatus shown in Fig. 15C.3(b) may be used to study plasticity in bentonite paste or a 'solid' emulsion (e.g. 50% medicinal paraffin–water, emulsified while heated with 'self-emulsifying' glyceryl monostearate), or dilatancy with ground quartz suspensions.

BIBLIOGRAPHY 15C: Rheology of colloids

Fischer, *Colloidal Dispersions*, 1950 (John Wiley, New York).

[1] Scott Blair, *A Survey of General and Applied Rheology*, 1944 (Pitman, London); Ward and Oldroyd, *Textbook on Rheology*, 1960 (Butterworths, London).

[2] Tsuda, *Koll. Zeit.*, 1928, **45**, 325; Alexander and Hitch, *Biochem. Biophys. Acta*, 1952, **9**, 229.

Appendix 1
A brief bibliography of physico-chemical techniques not referred to in the text

References to Weissberger refer to the 3rd Edn (1959, 1960).

Crystallography
Weissberger, Vol. I, Pt 1, chapters XIX, XX.
Evans, *Crystal Chemistry*, 2nd edn, 1964 (Cambridge University Press, London).
Phillips, *Introduction to Crystallography*, 3rd edn, 1964 (Longmans, London).
Hartshorne and Stuart, *Crystals and the Polarizing Microscope*, 3rd edn, 1950 (Arnold, London).

Dielectric constants, dielectric loss, etc.
Weissberger, Vol. I, Pt 3, chapter XXXVIII.

Diffusion
Weissberger, Vol. I, Pt 2, chapter XVI.
Shewmon, *Diffusion in Solids*, 1963 (McGraw-Hill, New York).
Yost, *Diffusion in Solids, Liquids and Gases*, 1952 (Academic Press, New York).

Dipole moments
Weissberger, Vol. I, Pt 3, chapter XXXIX.
Lefevre, *Dipole Moments*, 3rd edn, 1953 (Methuen, London).
Smith, *Electric Dipole Moments*, 1955 (Butterworths, London).

Electron diffraction
Weissberger, Vol. I, Pt 2, chapter XXVI.
Pinsker, *Electron Diffraction*, 1953 (Butterworths, London).

Electron microscopy
Weissberger, Vol. I, Pt 2, chapter XXIV.
Haine and Cosslett, *The Electron Microscope*, 1960 (Spon, London).
Hirsch, Howie, Nicholson, Pashley and Whelan, *Electron Microscopy of Thin Crystals*, 1965 (Butterworths, London).

Electron spin resonance, electron paramagnetic resonance
Weissberger, Vol. I, Pt 4, chapter XLII.
Ayscough, *Electron Spin Resonance in Chemistry*, 1967 (Methuen, London).
Alger, *Electron Paramagnetic Resonance*, 1968 (Wiley-Interscience, New York).

Experimental methods for gases

Melville and Gowenlock, *Experimental Methods for Gas Reactions*, 1964 (Macmillan, London).

Glass-blowing and laboratory arts

Lang (ed.), *Laboratory and Workshop Notes*, 4 vols, 1949–1960 (Arnold, London).

Robertson, Fabian, Crocker and Dewing, *Laboratory Glass-Working for Scientists*, 1957 (Butterworths, London).

High temperatures

Lebeau, *Les Hautes Temperatures et leurs utilisations*, 2 vols, 1950 (Masson, Paris).

Bockris *et al.*, *Physico-Chemical Measurements at High Temperatures*, 1959 (Butterworths, London).

Infra-red spectroscopy

Weissberger, Vol. I, Pt 3, chapter XXIX.

Cross, *Practical Infra-red spectroscopy*, 1960 (Butterworths, London).

Herzberg, *Infra-red and Raman Spectra of Polyatomic Molecules*, Pt 2 of *Molecular Spectra and Molecular Structure*, 1945 (Van Nostrand, New York).

Magnetochemistry

Weissberger, Vol. I, Pt 4, chapters XLII, XLIII.

Selwood, *Magnetochemistry*, 2nd edn, 1956 (Interscience, New York).

Mass spectrometry

Weissberger, Vol. I, Pt 4, chapter LI.

Duckwork, *Mass Spectrometry*, 1958 (Cambridge University Press, London).

Robertson, *Mass Spectrometry*, 1954 (Methuen, London).

Biemann, *Mass Spectrometry* (*Organic Chemical Applications*), 1962 (McGraw-Hill, New York).

Microscopy (optical)

Weissberger, Vol. I, Pt 2, chapter XXI.

Allen, *Photomicrography*, 1958 (van Nostrand, New York).

Martin and Johnson, *Practical Microscopy*, 3rd edn, 1958 (Blackie, London).

Light-scattering

Weissberger, Vol. I, Pt 3, chapter XXXII.

Stacey, *Light-scattering in Physical Chemistry*, 1956 (Butterworths, London).

Van de Hulst, *Light-scattering by Small Particles*, 1957 (John Wiley, New York).

Micro-wave spectroscopy

Weissberger, Vol. I, Pt 4, chapter XL.

Gordy and Cook, *Microwave Molecular Spectra*, 1970 (Wiley-Interscience, New York). (Pt. II, Vol. IX, *Chemical Applications of Spectroscopy*, *Techniques of Organic Chemistry*, Ed. Weissberger.)

Ingram, *Spectroscopy at Radio and Microwave Frequencies*, 2nd edn, 1967 (Butterworths, London).

Townes and Schawlow, *Microwave Spectroscopy*, 1955 (McGraw-Hill, New York).

Neutron diffraction

Weissberger, Vol. I, Pt 2, chapter XXVII.

Bacon, *Applications of Neutron Diffraction*, 1963 (Pergamon and Macmillan, Oxford and New York).

Nuclear magnetic resonance

Weissberger, Vol. I, Pt 4, chapter XLI.

Pople, Schneider and Bernstein, *High-resolution Nuclear Magnetic Resonance*, 1959 (McGraw-Hill, New York).

Radiochemistry

Weissberger, Vol. I, Pt 4, chapter XLII.

Cook and Duncan, *Modern Radiochemical Practice*, 1952 (Oxford University Press, London).

Friedlander and Kennedy, *Nuclear Radiochemistry*, 2nd edn, 1964 (John Wiley, New York).

Overman and Clark, *Radio-isotope Techniques*, 1960 (McGraw-Hill, New York).

Ultracentrifuge

Weissberger, Vol. I, Pt 2, chapter XVII.

Svedberg and Pederson, *The Ultracentrifuge*, 1940 (Oxford University Press, London).

Ultra-violet spectroscopy

Jaffé and Orchin, *Theory and Applications of Ultraviolet Spectroscopy*, 1962 (John Wiley, New York).

Murrell, *Theory of Electronic Spectra of Organic Molecules*, 1963 (Methuen, London).

Vacuum technique

Dushman, *Scientific Foundations of Vacuum Technique*, 1949 (John Wiley, New York).

Sanderson, *Vacuum Manipulation of Volatile Compounds*, 1948 (John Wiley, New York).

Yarwood, *High Vacuum Technique*, 3rd edn, 1955 (Chapman and Hall, London).

X-ray diffraction crystallography

Weissberger, Vol. I, Pt 2, chapter XXV.

Klug and Alexander, *X-ray Diffraction Procedures*, 1954 (John Wiley, New York).

Peiser, Rooksby and Wilson, *X-ray Diffraction by Polycrystalline Materials*, 1960 (Chapman and Hall, London).

Lipson and Steeple, *Interpretation of X-ray Powder Diffraction Patterns*, 1970 (Macmillan, London).

Appendix 2
Tables of some physico-chemical constants

TABLE A1 Melting points, boiling points, densities, refractive indices, viscosities and surface tensions of some common organic compounds (see also A14 and A15)

	m.p.	*b.p.*	ρ_{15}	ρ_{20}	ρ_{25}	n^{D}_{15}	n^{D}_{20}	n^{D}_{25}	η_{25}	γ_{15}	γ_{20}	γ_{25}
acetic acid	16·56	111·72	—	1 049·23	1 043·65	—	1·37160	1·36995	10·40 (30°)	2·795	2·742	2·634
acetone	−95·35	56·20	795·97	790·79	785·08	1·36157	1·35880	1·35604	3·07	2·392	2·332	2·201
iso-amyl acetate	—	142·0	—	874·6	—	—	1·4014 (18°)	—	—	—	—	—
n-amyl alcohol	−78·5	138·00	818·37	—	807·6 (30°)	1·41173	—	1·40815	29·87 (30°)	2·603	2·560	2·472
aniline	−6·10	184·7	1 026·13	1 021·83	1 017·4	1·5887	1·5862	1·5940	36·40	—	4·34 (19°)	4·79
benzene	5·53	80·099	884·2	879·03	873·63	1·50439	1·5010	1·4980	5·99	2·955	2·887	2·749 (30°)
benzoic acid	122·37	250	—	—	1 316 (28°)	—	—	—	—	—	—	—
bromobenzene	−30·82	156·06	1 501·7	1 495·00	1 488·24	1·56252	1·55972	1·55709	9·85 (30°)	3·693	3·634	3·509
n-butyl alcohol	−89·8	117·7	813·37	809·7	805·7	1·40118	—	1·3972	22·71 (30°)	2·490	2·452	2·398
carbon disulphide	−111·93	46·24	1 270·55	1 263·2	1 255·9	1·63189	1·6281	—	3·65 (20°)	3·307	3·225	3·079
carbon tetrachloride	−22·80	76·7	1 603·70	1 593·97	1 584·29	1·4631	1·46005	1·45704	9·03	2·736	2·675	2·615
chlorobenzene	−45·2	131·72	1 111·72	1 106·54	1 101·18	1·52748	1·5246	1·5219	7·56	3·386	3·325	3·265
chloroform	−63·49	61·2	1 498·47	1 489·2	1 479·8	1·44858	1·4455	—	5·14 (38°)	2·86	2·72	2·655
cyclo-hexane	6·55	80·74	783·0	778·4	773·8	1·42886	1·4263	1·4233	8·9	2·564	2·51	2·382 (30°)
p-dichlorobenzene	53·2	173·8	—	—	1 249·5 (55°)	—	—	—	—	—	—	3·07 (68°)
di-ethyl ether	−116·3	34·6	719·25	713·5	707·8	1·35555	—	—	2·24	1·762	1·705	1·650
ethyl acetate	−39·50	77·15	906·9	900·7	894·6	—	1·37243	1·3728	4·25	2·436	2·375	2·255 (30°)
ethyl alcohol	−114·49	78·30	793·6	789·3	785·1	1·3634	1·3613	1·3593	10·68	2·276	2·230	2·185

	m.p.	*b.p.*	ρ_{15}	ρ_{20}	ρ_{25}	η^{D}_{15}	η^{D}_{20}	η^{D}_{25}	η_{25}	γ_{15}	γ_{20}	γ_{25}
ethyl bromide	−118·8	38·4	1 470·8	1 461·2	1 451·5	1·42756	1·4242	—	3·79	2·483	2·41	2·345
ethylene glycol	−13·0	197·9	1 117·0	1 113·69	1 109·9	1·43312	1·4313	1·43063	133·5 (30°)	—	—	—
glycerol	18·2	290·5	1 264·3	1 261·3	—	1·47547	—	—	5 870 (30°)	—	—	—
n-hexane	−95·325	68·67	663·9	659·5	654·80	1·37744	1·3750	1·3722	2·937	1·942 (10°)	1·840	1·738 (30°)
methyl acetate	−98·1	56·9	940·0	934·2	927·9	—	1·3614	—	3·64	—	2·48	2·41
methyl alcohol	−97·8	64·65	796·09	791·14	786·53	1·33057	1·3287	—	5·44	2·299	2·255	2·228
methyl bromide	−93·72	3·46	—	—	—	—	—	—	—	—	—	—
naphthalene	80·28	217·95	—	—	975·2 (85°)	—	—	—	—	—	—	—
nitrobenzene	5·72	210·66	1 208·3	1 203·3	1 198·33	—	1·5523	1·5499	16·34 (30°)	4·395	4·335	4·217
n-pentane	−129·75	−11·82	631·1	626·3	621·3	1·3604	1·3572	1·3548	2·296	1·663	1·605	1·494 (30°)
phenol	40·8	181·84	—	—	1 057·1 (41·5°)	—	—	1·5404	3·20 (41·5°)	—	—	3·577 (50°)
n-propyl alcohol	−126·1	97·2	807·5	803·5	796·0	—	—	1·3840 (23·6°)	22·37 (20°)	2·409	2·375	2·289 (30°)
pyridine	−41·7	115·3	987·8	983·15	978·24	1·51246	1·51018	1·50745	8·84	3·686	3·625	3·570
tetrachloroethylene	−22·35	121·1	1 613·09	—	1 606·40 (30°)	—	1·5057	—	7·98 (30°)	3·286	3·232	3·127
toluene	−95·03	110·626	871·6	866·94	862·30	1·49985	1·49693	1·49413	5·24 (30°)	—	2·853	2·732 (30°)
trichloroethylene	−86·4	86·9	—	1 464·2	—	—	1·4775	—	5·32	—	2·95	2·85
o-xylene	−25·2	144·411	884·3	880·1	876·0	—	1·50545	1·50295	6·93 (30°)	3·075	3·003	2·962 (26°)
p-xylene	13·26	138·351	865·35	861·0	856·7	—	1·49582	1·49325	5·68 (30°)	2·891	2·831	2·722 (30°)

Most entries from: Timmermans, *Physico-Chemical Constants Of Pure Organic Compounds*, 1950, 1965, Vols I and II (Elsevier, Amsterdam).
Units: All temperatures, °C; density, ρ in kg m^{-3}; viscosity, entry $\times 10^{-4} = \eta$ in kg m^{-1} s^{-1}; surface tension, entry $\times 10^{-2} = \gamma$ in N m^{-1}.

TABLE A2 Transition temperatures of some salt hydrates

Salts in equilibrium		*°C**
$Na_2CrO_4.10H_2O$	$+Na_2CrO_4.6H_2O$	19·53
$Na_2CrO_4.10H_2O$	$+Na_2CrO_4.4H_2O$	19·99
$Na_2CrO_4.6H_2O$	$+Na_2CrO_4.4H_2O$	25·90
$Na_2CO_3.10H_2O$	$+Na_2CO_3.7H_2O$	32·02
$Na_2SO_4.10H_2O$	$+Na_2SO_4$	32·38
$NaBr.2H_2O$	$+NaBr$	50·67
$MnCl_2.4H_2O$	$+MnCl_2.2H_2O$	58·09
$SrCl_2.6H_2O$	$+SrCl_2.2H_2O$	61·34

* Temperatures given are on the hydrogen scale. Mercury-in-glass thermometers give higher readings; the differences vary with the glass, and are about 0·05° at 10°C and 80°C, and about 0·1° between 30 and 60°C.

TABLE A3 Eutectic ('cryohydric') temperatures of some salts in equilibrium with ice*

Salt	*Eutectic temp. (°C)*	*Salt*	*Eutectic temp. (°C)*
$Na_2SO_4.10H_2O$	− 1·10	$NaNO_2.\frac{1}{2}H_2O$	−19·5
$Na_2CO_3.10H_2O$	− 2·05	$NaCl.2H_2O$	−21·1
KNO_3	− 2·85	$NaBr.2H_2O$	−28
KCl	−10·65	$CaCl_2.6H_2O$	(−51)
KBr	−12·5		

* See Seidell, *Solubilities of Inorganic and Metal Organic Compounds*, 3rd edn, 1940 (van Nostrand, New York).

TABLE A4 Density of water at various temperatures

Temperature *t* (°C)	*Density*	*Difference in density for* +0·1°C, *in units of the third decimal place, for the range t to* (*t*+1)°C
0	999·846	—
4	999·972	− 0·8
5	999·964	− 2·3
6	999·941	− 3·8
7	999·903	− 5·3
8	999·850	− 6·6
9	999·784	− 8·1
10	999·703	− 9·4
11	999·609	−10·7
12	999·502	−12·0
13	999·382	−13·3
14	999·249	−14·5
15	999·104	−15·6
16	998·948	−16·8
17	998·780	−18·0
18	998·600	−19·1
19	998·409	−20·2
20	998·207	−21·2
21	997·995	−22·3
22	997·772	−23·2
23	997·540	−24·2
24	997·298	−25·3
25	997·045	−26·2
26	996·783	−27·1
27	996·512	−28·0
28	996·232	−28·9
29	995·943	−29·7
30	995·646	—

TABLE A5 Vapour pressure of water at various temperatures (in k Nm^{-2} and (mm Hg))

°C	0	2	4	5	6	8
0	0·61 (4·6)	0·70 (5·3)	0·81 (6·1)	0·86 (6·5)	0·93 (7·0)	1·06 (8·0)
10	1·22 (9·2)	1·39 (10·5)	1·59 (12·0)	1·70 (12·8)	1·81 (13·6)	2·06 (15·5)
20	2·33 (17·5)	2·63 (19·8)	2·97 (22·3)	3·15 (23·7)	3·34 (25·1)	3·77 (28·3)
30	4·22 (31·7)	4·73 (35·5)	5·30 (39·8)	5·59 (42·0)	5·91 (44·4)	6·59 (49·5)
40	7·35 (55·2)	—	—	—	—	—
45	9·56 (71·8)	—	—	—	—	—
50	12·31 (92·4)	—	—	—	—	—
55	15·71 (117·9)	—	—	—	—	—
60	19·89 (149·3)	—	—	—	—	—
65	24·97 (187·4)	—	—	—	—	—
70	31·13 (233·6)	—	—	—	—	—
75	38·51 (289·0)	—	—	—	—	—
80	47·32 (355·1)	—	—	—	—	—
85	57·79 (433·6)	—	—	—	—	—
90	70·07 (525·8)	—	—	—	—	—
95	84·48 (633·9)	—	—	—	—	—

TABLE A6 Temperature–e.m.f. tables for thermocouples

Temperature (°*C*)	*Copper–constantan*	*Chromel–alumel*	*Platinum/13% rhodium–platinum**
−200	−5·54	−5·75	—
−100	−3·35	−3·49	—
0	0	0	0
10	0·39	0·40	0·054
20	0·79	0·80	0·111
30	1·19	1·20	0·170
40	1·61	1·61	0·231
50	2·03	2·02	0·295
100	4·28	4·10	0·644
200	9·29	8·13	1·463
300	14·86	12·21	2·392
400	20·87	16·39	3·397
500	—	20·64	4·460
600	—	24·90	5·571
700	—	29·14	6·735
800	—	33·31	7·952
900	—	37·36	9·209
1 000	—	41·31	10·510
1 200	—	—	13·222
1 400	—	—	16·039
1 600	—	—	18·855

* After Barber, *Proc. Phys. Soc.*, 1950, **63**, 492.

The tables give the e.m.f. (in millivolts) generated by a thermocouple with one junction at 0°C and the other at the temperature shown. Interpolation is normally linear. For accurate work reliance should not be placed on the tables (except with platinum thermocouples), but individual couples should be calibrated at fixed temperatures (see Table 3A.1).

TABLE A7 Salts for controlling relative humidity at 25°C

The table gives the relative humidity (i.e. (v.p. of solution/v.p. of water) × 100) for *saturated* solutions of the salts.

Salt	*r.h.*	*Salt*	*r.h.*
$NaOH.H_2O$	7·0	$SrCl_2.6H_2O$	70·8
$LiCl.H_2O$	11·1	$NaNO_3$	73·7
$K(C_2H_3O_2).1{\cdot}5\,H_2O$	22·5	$NaCl$	75·3
$MgCl_2.6H_2O$	33·0	KBr	80·7
$K_2CO_3.2H_2O$	42·8	KCl	84·3
$LiNO_3.3H_2O$	47·1	$BaCl_2.2H_2O$	90·2
$Mg(NO_3)_2.6H_2O$	52·9	KNO_3	92·5
$NaBr.2H_2O$	57·7	$K_2Cr_2O_7$	98·0

Stokes and Robinson, *Ind. Eng. Chem.*, 1949, **41**, 2013.

TABLE A8 Short list of prominent spectral lines in the visible region. Wavelength in nm

Flame spectra

Barium	*Boric acid*	*Calcium*	*Caesium*	*Lithium*
513·7	519·3	422·7	455·5p	610·4
534·7	544·0	554·4	459·3p	670·8
553·5p	548·1	618·2p	621·3	
		620·3p	672·3	
			697·3	

Potassium	*Rubidium*	*Sodium*	*Strontium*	*Thallium*
404·4	420·2p	589·0p	407·0p	535·0
404·7	421·6p	589·6p	460·8	
766·8p	629·9		606·0	
769·9p	780·0		638·7	
	794·8		662·8	

Spectra of electric arcs or discharges

Argon	*Cadmium*	*Copper*	*Helium*	*Hydrogen*	*Mercury*	*Neon*
415·9	467·8	402·3	388·9	410·2	404·7	614·3
419·2	480·0	406·3	447·1	434·0	407·8	626·7
419·8	508·6	458·7	587·6	486·1	435·8	638·3
420·1	634·8	510·6		656·3	546·1	640·2
425·9		515·3			577·0	650·7
470·3		521·8			579·1	
603·1		570·0			690·8	
		578·2				

p marks particularly 'persistent' line. See Bibliography 9B for sources of more detailed tables.

TABLE A9 Solubilities of some sparingly-soluble salts at 25°C. Solubilities are given in $kg\ m^{-3}$

Salt	*Solubility*	*Salt*	*Solubility*
lead chloride	10·8	silver chromate	0·029
lead iodide	0·76	silver bromide	0·00014
lead sulphate	0·045	silver thiocyanate	0·00017
silver benzoate	2·61	thallous chloride	3·85
silver chloride	0·00195		

Seidell, *Solubility of Inorganic and Metal Organic Compounds*, 3rd edn, 1940 (van Nostrand, New York).

TABLE A10 Mean ion activity coefficient of some common electrolytes at 25°C

Valence type	*Electrolyte*	*Molality* [*mol* (*kg* $H_2O)^{-1}$]			
		0·001	0·01	0·1	1
1:1	HCl	0·965	0·904	0·796	0·809
	HNO_3	0·96	0·90	0·79	0·72
	LiCl	0·96	0·90	0·79	0·77
	KCl	0·965	0·902	0·771	0·611
	KOH	0·96	0·90	0·80	0·76
	KNO_3	0·965	0·896	0·723	0·44
	$AgNO_3$	0·96	0·892	0·733	0·43
	*	0·966	0·901	0·758	0·55
1:2	$BaCl_2$	0·88	0·72	0·50	0·39
	H_2SO_4	0·84	0·54	0·38	0·19
	**	0·89	0·72	0·44	0·21
1:3	$LaCl_3$	0·85	0·64	0·31	0·34
2:2	$CuSO_4$	0·74	0·41	0·15	0·04
2:3	$La_2(SO_4)_3$	—	—	0·04	0·02

*, ** Average behaviour of 1:1 and 1:2 electrolytes; see Güntelberg formula, sections 13A(3, 4).

For further data see Landolt–Bornstein table; and Stokes and Robinson, *Trans. Faraday Soc.*, 1949, **45**, 612.

TABLE A11 Reduction of barometric readings to 0°C

When the barometer has a brass scale, the corrected barometric height B_0 is given by the expression,

$$B_0 = B\left[1 - \frac{(\beta - \alpha)t}{(1 + \beta t)}\right]$$

where B and t are the observed height and temperature in °C; $\alpha = 0{\cdot}0000184$, the coefficient of linear expansion of brass; $\beta = 0{\cdot}0001818$, the coefficient of cubical expansion of mercury.

In the following table are given the corrections in millimetres to be *subtracted* from the observed barometric height.

Temp. (°C)	*Barometric readings*			
	720	740	760	780
10	1·17	1·21	1·24	1·27
12	1·41	1·45	1·49	1·53
14	1·64	1·69	1·73	1·78
16	1·88	1·93	1·98	2·03
18	2·11	2·17	2·23	2·29
20	2·34	2·41	2·47	2·54
22	2·58	2·65	2·72	2·79
24	2·81	2·89	2·97	3·05
26	3·04	3·13	3·21	3·30
28	3·28	3·37	3·46	3·55
30	3·51	3·61	3·71	3·80

TABLE A12 Reduction of barometric readings to latitude 45° and sea-level

In the following table are given the corrections (D) in millimetres to reduce the barometric readings, corrected for temperature, to the values at latitude 45°. In latitudes between 0° and 45°, the corrections must be subtracted, and in latitudes between 45° and 90°, they must be added.

Latitude	0° 90°	5° 85°	10° 80°	15° 75°	20° 70°	25° 65°	30° 60°	35° 55°	40° 50°	45° 45°
D	1·97	1·94	1·85	1·70	1·51	1·27	0·98	0·67	0·34	0·00

For each 1 000 m above sea level, 0·24 mm must be *subtracted* from the observed reading.

TABLE A13 Molar ionic conductances at infinite dilution

(Λ = entry $\times 10^{-3}$, $m^2\ \Omega^{-1}\ mol^{-1}$)

Cation	*18°C*	*25°C*	*Anion*	*18°C*	*25°C*
Ag^+	5·43	6·35	Br^-	6·72	7·74
H^+	31·4	3·50	Cl^-	6·55	7·55
K^+	6·46	7·45	I^-	6·64	7·62
Na^+	4·35	5·09	NO_3^-	6·17	7·06
NH_4^+	6·46	7·39	OH^-	17·2	19·2
Ba^{++}	11·0	13·0	Ac^-	3·5	4·08
Ca^{++}	10·2	12·0	$C_2O_4^{--}$	12·6	14·6
Pb^{++}	12·3	14·4	SO_4^{--}	13·6	15·8
Mg^{++}	9·0	10·62	CrO_4^{--}	14·4	16·72

TABLE A14 Viscosities of liquids

(Viscosity = entry $\times 10^{-4}$, $kg\ m^{-1}\ s^1$ = entry, millipoise)

Substance	*0°C*	*10°C*	*15°C*	*20°C*	*25°C*	*30°C*	*40°C*	*50°C*
aniline	122·0	64·5	53·0	42·7	36·4	31·1	23·6	18·6
alcohol, methyl	8·08	6·90	6·23	5·92	5·44	5·15	4·49	3·95
alcohol, ethyl	17·72	14·66	13·0	12·00	10·78	10·03	8·34	7·02
benzene	9·00	7·57	6·96	6·47	5·96	5·61	4·92	4·36
carbon tetra-chloride	13·51	11·38	10·4	9·75	9·03	8·48	7·46	6·62
chloroform	6·99	6·25	5·96	5·63	—	5·10	4·64	4·24
toluene	7·72	6·71	6·23	5·90	—	5·25	4·71	4·26
water	17·865	13·037	11·369	10·019	8·909	7·982	6·540	5·477

TABLE A15 Surface tension of liquids

(γ = entry $\times 10^{-2}$, Nm^{-1})

Substance	*0°C*	*10°C*	*15°C*	*20°C*	*25°C*	*30°C*	*40°C*	*50°C*
water	7·564	7·422	7·349	7·275	7·197	7·118	6·956	6·791
ethyl alcohol	2·405	2·314	2·276	2·227	2·185	2·143	2·060	1·980
ethyl acetate	2·65	—	2·436	2·39	—	2·255	—	2·02
carbon tetra-chloride	2·938	2·800	2·736	2·677	2·614	2·553	2·441	2·314
ethyl ether	—	—	1·762	1·701	1·650	1·595	—	1·347
benzene	3·158	3·022	2·955	2·888	—	2·756	2·626	2·498

INTERNATIONAL ATOMIC WEIGHTS, 1969. [Reprinted (with omission of actinides) from 25th Compt. Rend. IUPAC, Jan. 1970.] These atomic weights are based on $^{12}C = 12$.

Name	*Symbol*	*Atomic weight*	*Name*	*Symbol*	*Atomic weight*
Aluminium	Al	26·982	Molybdenum	Mo	95·94
Antimony	Sb	121·76	Neodymium	Nd	144·24
Argon	Ar	39·948	Neon	Ne	20·179
Arsenic	As	74·921	Nickel	Ni	58·71
Barium	Ba	137·34	Niobium	Nb	92·906
Beryllium	Be	9·012	Nitrogen	N	14·007
Bismuth	Bi	208·98	Osmium	Os	190·2
Boron	B	10·81	Oxygen	O	15·999
Bromide	Br	79·90	Palladium	Pd	106·4
Cadmium	Cd	112·40	Phosphorus	P	30·974
Caesium	Cs	132·91	Platinum	Pt	195·09
Calcium	Ca	40·08	Potassium	K	39·102
Carbon	C	12·011	Praseodymium	Pr	140·91
Cerium	Ce	140·12	Rhenium	Re	186·2
Chlorine	Cl	35·453	Rhodium	Rh	102·91
Chromium	Cr	51·996	Rubidium	Rb	85·47
Cobalt	Co	58·933	Ruthenium	Ru	101·1
Copper	Cu	63·55	Samarium	Sm	150·4
Dysprosium	Dy	162·50	Scandium	Sc	44·96
Erbium	Er	167·26	Selenium	Se	78·96
Europium	Eu	151·96	Silicon	Si	28·09
Fluorine	F	18·998	Silver	Ag	107·868
Gadolinium	Gd	157·25	Sodium	Na	22·990
Gallium	Ga	69·72	Strontium	Sr	87·62
Germanium	Ge	72·59	Sulphur	S	32·06
Gold	Au	196·97	Tantalum	Ta	180·95
Hafnium	Hf	178·49	Tellurium	Te	127·60
Helium	He	4·0026	Terbium	Tb	158·93
Holmium	Ho	164·93	Thallium	Tl	204·37
Hydrogen	H	1·0080	Thorium	Th	232·04
Indium	In	114·82	Thulium	Tm	168·93
Iodine	I	126·90	Tin	Sn	118·69
Iridium	Ir	192·22	Titanium	Ti	47·90
Iron	Fe	55·85	Tungsten	W	183·85
Krypton	Kr	83·80	Uranium	U	238·03
Lanthanum	La	138·91	Vanadium	V	50·94
Lead	Pb	207·2	Xenon	Xe	131·30
Lithium	Li	6·941	Ytterbium	Yb	173·04
Lutetium	Lu	174·97	Yttrium	Y	88·91
Magnesium	Mg	24·31	Zinc	Zn	65·37
Manganese	Mn	54·94	Zirconium	Zr	91·22
Mercury	Hg	200·59			

	0	1	2	3	4	5	6	7	8	9	1	2	3	4	5	6	7	8	9
10	0000	0043	0086	0128	0170						4	9	13	17	21	26	30	34	38
						0212	0253	0294	0334	0374	4	8	12	16	20	24	28	32	37
11	0414	0453	0492	0531	0569						4	8	12	15	19	23	27	31	35
						0607	0645	0682	0719	0755	4	7	11	15	19	22	26	30	33
12	0792	0828	0864	0899	0934	0969					3	7	11	14	18	21	25	28	32
							1004	1038	1072	1106	3	7	10	14	17	20	24	27	31
13	1139	1173	1206	1239	1271						3	7	10	13	16	20	23	26	30
						1303	1335	1367	1399	1430	3	7	10	12	16	19	22	25	29
14	1461	1492	1523	1553							3	6	9	12	15	18	21	24	28
					1584	1614	1644	1673	1703	1732	3	6	9	12	15	17	20	23	26
15	1761	1790	1818	1847	1875	1903					3	6	9	11	14	17	20	23	26
							1931	1959	1987	2014	3	5	8	11	14	16	19	22	25
16	2041	2068	2095	2122	2148						3	5	8	11	14	16	19	22	24
						2175	2201	2227	2253	2279	3	5	8	10	13	15	18	21	23
17	2304	2330	2355	2380	2405	2430					3	5	8	10	13	15	18	20	23
							2455	2480	2504	2529	2	5	7	10	12	15	17	19	22
18	2553	2577	2601	2625	2648						2	5	7	9	12	14	16	19	21
						2672	2695	2718	2742	2765	2	5	7	9	11	14	16	18	21
19	2788	2810	2833	2856	2878						2	4	7	9	11	13	16	18	20
						2900	2923	2945	2967	2989	2	4	6	8	11	13	15	17	19
20	3010	3032	3054	3075	3096	3118	3139	3160	3181	3201	2	4	6	8	11	13	15	17	19
21	3222	3243	3263	3284	3304	3324	3345	3365	3385	3404	2	4	6	8	10	12	14	16	18
22	3224	3444	3464	3483	3502	3522	3541	3560	3579	3598	2	4	6	8	10	12	14	15	17
23	3617	3636	3655	3674	3692	3711	3729	3747	3766	3784	2	4	6	7	9	11	13	15	17
24	3802	3820	3838	3856	3874	3892	3909	3927	3945	3962	2	4	5	7	9	11	12	14	16
25	3979	3997	4014	4031	4048	4065	4082	4099	4116	4123	2	3	5	7	9	10	12	14	15
26	4150	4166	4183	4200	4216	4232	4249	4265	4281	4298	2	3	5	7	8	10	11	13	15
27	4314	4330	4346	4362	4378	4393	4409	4425	4440	4456	2	3	5	6	8	9	11	13	14
28	4472	4487	4502	4518	4533	4548	4564	4579	4594	4609	2	3	5	6	8	9	11	12	14
29	4624	4639	4654	4669	4683	4698	4713	4728	4742	4757	1	3	4	6	7	9	10	12	13
30	4771	4786	4800	4814	4829	4843	4857	4871	4886	4900	1	3	4	6	7	9	10	11	13
31	4914	4928	4942	4955	4969	4983	4997	5011	5024	5038	1	3	4	6	7	8	10	11	12
32	5051	5065	5079	5092	5105	5119	5132	5145	5159	5172	1	3	4	5	7	8	9	11	12
33	5185	5198	5211	5224	5237	5250	5263	5276	5289	5302	1	3	4	5	6	8	9	10	12
34	5315	5328	5340	5353	5366	5378	5391	5403	5416	5428	1	3	4	5	6	8	9	10	11
35	5441	5453	5465	5478	5490	5502	5514	5527	5539	5551	1	2	4	5	6	7	9	10	11
36	5563	5575	5587	5599	5611	5623	5635	5647	5658	5670	1	2	4	5	6	7	8	10	11
37	5682	5694	5705	5717	5729	5740	5752	5763	5775	5786	1	2	3	5	6	7	8	9	10
38	5798	5809	5821	5832	5843	5855	5866	5877	5888	5899	1	2	3	5	6	7	8	9	10
39	5911	5922	5933	5944	5955	5966	5977	5988	5999	6010	1	2	3	4	5	7	8	9	10
40	6021	6031	6042	6053	6064	6075	6085	6096	6107	6117	1	2	3	4	5	6	8	9	10
41	6128	6138	6149	6160	6170	6180	6191	6201	6212	6222	1	2	3	4	5	6	7	8	9
42	6232	6243	6253	6263	6274	6284	6294	6304	6314	6325	1	2	3	4	5	6	7	8	9
43	6335	6345	6355	6365	6375	6385	6395	6405	6415	6425	1	2	3	4	5	6	7	8	9
44	6435	6444	6454	6464	6474	6484	6493	6503	6513	6522	1	2	3	4	5	6	7	8	9
45	6532	6542	6551	6561	6571	6580	6590	6599	6609	6618	1	2	3	4	5	6	7	8	9
46	6628	6637	6646	6656	6665	6675	6684	6693	6702	6712	1	2	3	4	5	6	7	7	8
47	6721	6730	6739	6749	6758	6767	6776	6785	6794	6803	1	2	3	4	5	5	6	7	8
48	6812	6821	6830	6839	6848	6857	6866	6875	6884	6893	1	2	3	4	4	5	6	7	8
49	6902	6911	6920	6928	6937	6946	6955	6964	6972	6981	1	2	3	4	4	5	6	7	8
50	6990	6998	7007	7016	7024	7033	7042	7050	7059	7067	1	2	3	3	4	5	6	7	8

	0	1	2	3	4	5	6	7	8	9	1	2	3	4	5	6	7	8	9
51	7076	7084	7093	7101	7110	7118	7126	7135	7143	7152	1	2	3	3	4	5	6	7	8
52	7160	7168	7177	7185	7193	7202	7210	7218	7226	7235	1	2	2	3	4	5	6	7	7
53	7243	7251	7259	7267	7275	7284	7292	7300	7308	7316	1	2	2	3	4	5	6	6	7
54	7324	7332	7340	7348	7356	7364	7372	7380	7388	7396	1	2	2	3	4	5	6	6	7
55	7404	7412	7419	7427	7435	7443	7451	7459	7466	7474	1	2	2	3	4	5	5	6	7
56	7482	7490	7497	7505	7513	7520	7528	7536	7543	7551	1	2	2	3	4	5	5	6	7
57	7559	7566	7574	7582	7589	7597	7604	7612	7619	7627	1	2	2	3	4	5	5	6	7
58	7634	7642	7649	7657	7664	7672	7679	7686	7694	7701	1	1	2	3	4	4	5	6	7
59	7709	7716	7723	7731	7738	7745	7752	7760	7767	7774	1	1	2	3	4	4	5	6	7
60	7782	7789	7796	7803	7810	7818	7825	7832	7839	7846	1	1	2	3	4	4	5	6	6
61	7853	7860	7868	7875	7882	7889	7896	7903	7910	7917	1	1	2	3	4	4	5	6	6
62	7924	7931	7938	7945	7952	7959	7966	7973	7980	7987	1	1	2	3	3	4	5	6	6
63	7993	8000	8007	8014	8021	8028	8035	8041	8048	8055	1	1	2	3	3	4	5	5	6
64	8062	8069	8075	8082	8089	8096	8102	8109	8116	8122	1	1	2	3	3	4	5	5	6
65	8129	8136	8142	8149	8156	8162	8169	8176	8182	8189	1	1	2	3	3	4	5	5	6
66	8195	8202	8209	8215	8222	8228	8235	8241	8248	8254	1	1	2	3	3	4	5	5	6
67	8261	8267	8274	8280	8287	8293	8299	8306	8312	8319	1	1	2	3	3	4	5	5	6
68	8325	8331	8338	8344	8351	8357	8363	8370	8376	8382	1	1	2	3	3	4	4	5	6
69	8388	8395	8401	8407	8414	8420	8426	8432	8439	8445	1	1	2	2	3	4	4	5	6
70	8451	8457	8463	8470	8476	8482	8488	8494	8500	8506	1	1	2	2	3	4	4	5	6
71	8513	8519	8525	8531	8537	8543	8549	8555	8561	8567	1	1	2	2	3	4	4	5	5
72	8573	8579	8585	8591	8597	8603	8609	8615	8621	8627	1	1	2	2	3	4	4	5	5
73	8633	8639	8645	8651	8657	8663	8669	8675	8681	8686	1	1	2	2	3	4	4	5	5
74	8692	8698	8704	8710	8716	8722	8727	8733	8739	8745	1	1	2	2	3	4	4	5	5
75	8751	8756	8762	8768	8774	8779	8785	8791	8797	8802	1	1	2	2	3	3	4	5	5
76	8808	8814	8820	8825	8831	8837	8842	8848	8854	8859	1	1	2	2	3	3	4	5	5
77	8865	8871	8876	8882	8887	8893	8899	8904	8910	8915	1	1	2	2	3	3	4	4	5
78	8921	8927	8932	8938	8943	8949	8954	8960	8965	8971	1	1	2	2	3	3	4	4	5
79	8976	8982	8987	8993	8998	9004	9009	9015	9020	9025	1	1	2	2	3	3	4	4	5
80	9031	9036	9042	9047	9053	9058	9063	9069	9074	9079	1	1	2	2	3	3	4	4	5
81	9085	9090	9096	9101	9106	9112	9117	9122	9128	9133	1	1	2	2	3	3	4	4	5
82	9138	9143	9149	9154	9159	9165	9170	9175	9180	9186	1	1	2	2	3	3	4	4	5
83	9191	9196	9201	9206	9212	9217	9222	9227	9232	9238	1	1	2	2	3	3	4	4	5
84	9243	9248	9253	9258	9263	9269	9274	9279	9284	9289	1	1	2	2	3	3	4	4	5
85	9294	9299	9304	9309	9315	9320	9325	9330	9335	9340	1	1	2	2	3	3	4	4	5
86	9345	9350	9355	9360	9365	9370	9375	9380	9385	9390	1	1	2	2	3	3	4	4	5
87	9395	9400	9405	9410	9415	9420	9425	9430	9435	9440	0	1	1	2	2	3	3	4	4
88	9445	9450	9455	9460	9465	9569	9474	9479	9484	9489	0	1	1	2	2	3	3	4	4
89	9494	9499	9504	9509	9513	9518	9523	9528	9533	9538	0	1	1	2	2	3	3	4	4
90	9542	9547	9552	9557	9562	9566	9571	9576	9581	9586	0	1	1	2	2	3	3	4	4
91	9590	9595	9600	9605	9609	9614	9619	9624	9628	9633	0	1	1	2	2	3	3	4	4
92	9638	9643	9647	9652	9657	9661	9666	9671	9675	9680	0	1	1	2	2	3	3	4	4
93	9685	9689	9694	9699	9703	9708	9713	9717	9722	9727	0	1	1	2	2	3	3	4	4
94	9731	9736	9741	9745	9750	9754	9759	9763	9768	9773	0	1	1	2	2	3	3	4	4
95	9777	9782	9786	9791	9795	9800	9805	9809	9814	9818	0	1	1	2	2	3	3	4	4
96	9823	9827	9832	9836	9841	9845	9850	9854	9859	9863	0	1	1	2	2	3	3	4	4
97	9868	9872	9877	9881	9886	9890	9894	9899	9903	9908	0	1	1	2	2	3	3	4	4
98	9912	9917	9921	9926	9930	9934	9939	9943	9948	9952	0	1	1	2	2	3	3	4	4
99	9956	9961	9965	9969	9974	9978	9983	9987	9991	9996	0	1	1	2	2	3	3	3	4

Index

The entry in bold type indicates the chapter, section or sub-section; this is followed by the page on which the first reference occurs. A List of experiments follows the Contents at the beginning of the book.